AF335714

CONTENTS

AERIAL BANNER TOWING

by Alan C. Taylor

THE BARNSTORMERS

by Don Dwiggins

PILOT'S DIGEST OF FAA REGULATIONS

2nd Edition

by John L. Nelson

Modern Aviation Library
BLUE RIDGE SUMMIT, PA. 17214

Aerial Banner Towing

by Alan C. Taylor

Copyright © 1981 by TAB BOOKS Inc.

Library of Congress Cataloging in Publication Data

Taylor, Alan C
Aerial banner towing.

   Includes index.
   1. Aeronautics in advertising. 2. Airplanes—
Piloting. 3. Towing. I. Title.
TL722.5.T39   629.132'5216   81-325
ISBN 0-8306-2303-5   AACR1

Preface

As a first time author, I am somewhat amused by the observation that the "Preface" of a book is necessarily the last division to be written. Notwithstanding the fact that I have personally never been able to wade through such an introductory note in its entirety, the notion confronts me that there is an obligatory prelude for the reader, if for no other purpose than to set the tempo for what lies ahead. With the assumption that the title of this book is an adequate harbinger of the chapters to follow, I rather choose to utilize this space for miscellaneous comments that fit nowhere else.

This book is the product of some two years of experience with banner towing. In the beginning, it was distressing to discover that the subject has never been pursued (to my knowledge) to any great extent in aviation literature, save for a few miscellaneous magazine articles within the past decade. This is indeed curious, for it seems unthinkable that such a potentially lucrative flying avocation would remain undocumented for so long. Pilots always seem to be searching for ways to fly and/or own an airplane with someone else picking up the tab. "Lease-back" schemes seem to enjoy a front row position for this, but I have yet to meet a lease-back participant in recent times who has actually seen a black balance sheet with this arrangement. The potential to generate a true profit is too limited because of high overhead and a relatively low hourly return. The potential for profit-taking in the banner towing industry seems more realistic because of a substantially higher hourly net for aircraft flying time. There is no guarantee of profit from pulling banners, but the potential is clearly there.

To me, there is also an aesthetic flaw in renting an airplane to the public at large. On a daily basis, I could not maintain any semblance of mental health with the knowledge that some unidentified solo student was out there beating the hell out of the runway with my airplane. (Incidentally, some argue that towing banners is tough on the towing plane. Admittedly, the additional strain puts more miles on an aircraft than equivalent free air time. However, the argument usually terminates when the punishment of rental aircraft is brought up for comparison.)

Derry Air witnessed its genesis during the summer of 1978, orginally as a partnership between my very closest friend, Roger Baker, and myself. Ever since our college days in the 1960s at the University of Kentucky, Roger and I have spent countless hours dreaming up all kinds of implausible "get-rich" schemes. This one surfaced as a not-quite-so-crazy idea that seemed to have at least a reasonable basis to form a viable business.

The first six months, from a business standpoint, were absolutely terrible as we reached deeply into our pockets each month to pay the bills. At this point, we mutually agreed that the net result of this venture was doing nothing more than supporting my flying habit, as my partner was not a licensed pilot. The doomed business relationship concluded when I bought the whole mess, since I didn't want to give up the venerable old Stearman I had grown so fond of.

On my own, things didn't change a bit for another three months. As the spring of 1979 rolled around, a few jobs scored enough to pay the bills for the very first time. Business gradually accelerated to a saturation level as the summer began. Still, I had barely scratched the surface of the market potential, as my other commitments would not allow but a flimsy effort in the department of scrounging around for customers. Nonetheless, my flying fetish was being satisfied with others footing the bill, and that *alone* was a gratifying outcome.

Although this is basically a very easy aviation specialty, there are many things that can go wrong. I remember my first towing experience as a tremendous success. The careful planning and thinking through the pickup procedures, the reading and re-reading of the scant little pamphlet that accompanied the banner equip-ment, the mental dry runs of each operations detail—all paid off with a perfect aerial pickup on the very first shot! The mental "high" that followed was perhaps paralleled by only two experiences in my aviation career: (1) my first solo and (2) returning alive from an air combat tour in Viet Nam. The only difference was that this euphoric state was short lived as I rapidly plunged into one of the all time lows of my life. While reveling in the glory of my latest conquest, I was suddenly inundated by the awesome responsibility that this "hobby" beheld, above and beyond the matter of flying. The consequences of dropping the banner by accident became a perverse distortion in my mind. After a sleepless night, I told my wife that I would honor the contract I had begun and would never tow again thereafter. On top of this, another sobering event came to pass in this first week of operation as I was to experience my first landing with a banner in tow. Adverse weather forced me to abort fully one-half of the contract, which seemed a blessed relief at the time. Somehow, I managed to regain confidence in myself and the equipment, completing this first job with the urge to go on. (Aborting for weather was not a *total* disaster in this case, as we were being paid in tickets, rather than money for this advertising service.)

So why then, did I write this book? Perhaps I would like to save others some of the aggravations I endured through a lack of reference sources, though I sincerely doubt that I am *that* altruistic. At any rate, the text is now available for those who care to assimilate the information and reap whatever benefits it may offer. Other than reading and practicing, there is only one other way to learn about banner towing—write a book about it.

Alan C. Taylor

Contents

Acknowledgements

This book expresses the labor and cooperation of scores of individuals who must go unnamed for lack of space. There must be made a special mention of those who directly participated in its creation.

I would like to thank my father, Charles P. Taylor, C.P.A. for writing the excellent accounting summary in Appendix B.

The photographic excellence is the product of Mark and Doreen Ross, both of whom contributed many hours of their personal time and talents to this work.

Will Walden of Gasser Banners, Inc., Nashville, Tennessee, provided answers to many technical questions in addition to a hospitable tour of the manufacturing facilities. Some of the information in this book is to be credited to the contents of *Instruction Booklet for Gasser Banner Equipment*, a product of their company.

Finally, I would like to express my appreciation to my wife, Elisabeth Taylor (not the real one) for her editorial assistance as well as her devotion as a standby crew for the past two years.

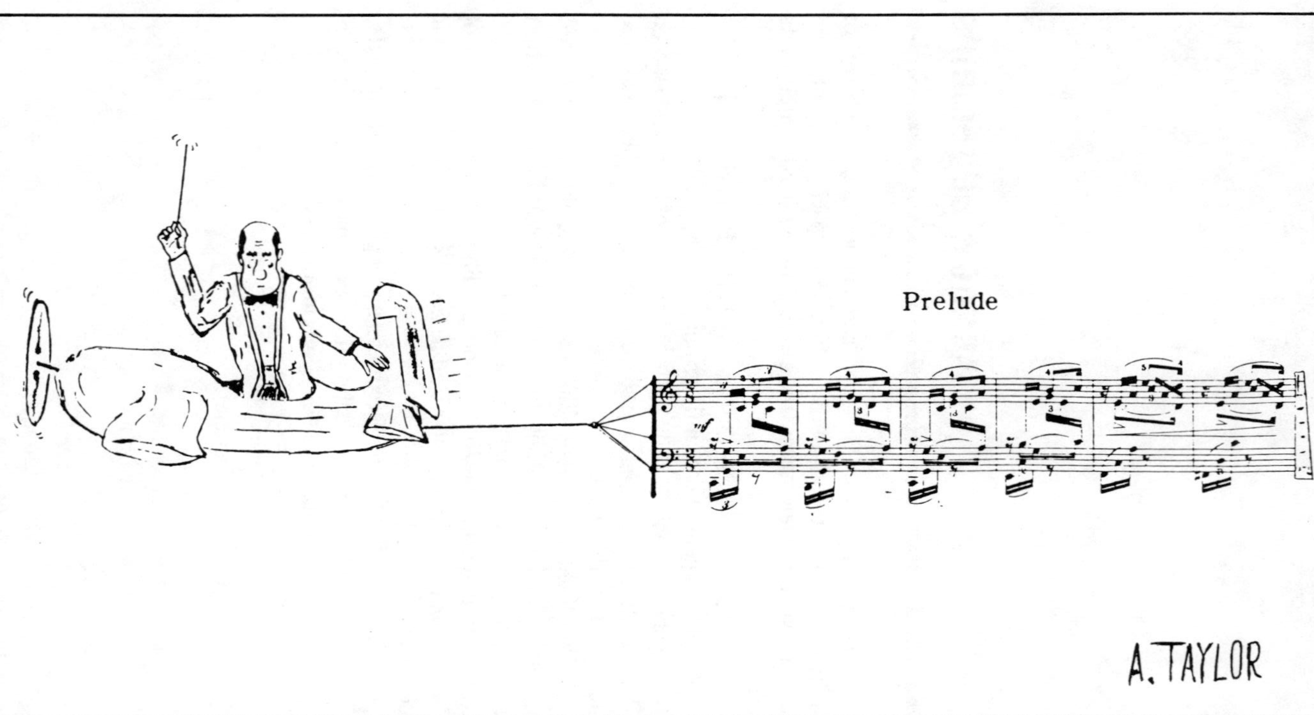

Prelude
A. TAYLOR

Chapter 1
Prelude

As a basic first step, a self-assessment of personal goals must be integrated with the potential market for banner services in order to arrive at realistic expectations for this endeavor.

THE MARKET

The closer you are to an urban center, the more likely you will be able to begin a banner towing service and still expect reasonable returns for a modest effort. This is not to suggest that rural areas automatically preclude the thought of opening up shop. It rather suggests that a careful market survey is in order before deciding to commit substantial funds to a business of this type in a rural setting. On the other hand, you would be ill-advised to make a similar commitment strictly on the basis of a location in an urban environment. Again, a market survey represents a clear advantage in planning for an active operation. Competition must be considered, and it may be very difficult to exist as a newcomer in a setting where your competitors are well established. It is, however, hard to conceive of an existing monopoly so tightly controlled that it would preclude an aggressive entrepreneur from entering the market with a minimum of salesmanship and a lot of hustling.

The local newspaper(s) should lend a wealth of good ideas about where to start. It doesn't naturally follow that the larger the ad, the more likely you are to catch some of those bucks.

Remember, the smaller advertisements may earmark the budget watchers, and *price* is one of the selling features of the banner industry.

Radio and television also hold a plethora of potential customers. Peruse the advertising rate sheets available from the audio-visual media for a breath-taking view of the advertising dollars spent in today's market. Look for the penny-pinchers during off hour broadcasting spots (non prime-time) on the less popular stations. The big prime-time spenders may also welcome a supplement to their advertising campaigns.

You will find that many businesses will refer you to various advertising agencies. Get the name of the specific account executive and *personally* visit him or her at the advertising agency for the information you seek regarding a particular client.

A brief shopping list of prospective aerial advertising customers is presented here for convenience, but is by no means exhaustive.

- ☐ Advertising agencies
- ☐ Athletic departments (schools, colleges)
- ☐ Amusement centers
- ☐ Automobile dealers
- ☐ Banks & savings associations
- ☐ Concert halls and promoters
- ☐ Churches
- ☐ Furniture outlets
- ☐ Land developers
- ☐ Political candidates (Fig. 1-1).
- ☐ Radio stations/television stations (especially during the "rating seasons")
- ☐ Real estate agencies
- ☐ Record shops & hi-fi centers
- ☐ Shopping centers
- ☐ Theaters

Although the business community is logically the bread and butter of an aerial towing service, the general public should not be dismissed from the potential pool of customers. For starters, there are a number of folks quite willing to spend a couple of hundred bucks for a practical joke or novelty message without batting an eyelash. To expand this list, add birthdays, anniversary greetings, retirement send-offs, company picnics—even wedding proposals! A few examples here may stimulate thought in other directions.

**HAPPY 25TH MOM & DAD
GET WELL, CAROLYN
CONGRATULATIONS JIM AND JANET
HAPPY BIRTHDAY JASON**

**WELCOME HOME BOB
CINDY, WILL YOU MARRY ME?
DINNER IS ON THE TABLE, HARRY** (*for the golf widow)*

For instance, wedding engagements are announced in the newspaper well in advance of the event, so a phone call or letter to the parents of the prospective bride and groom can offer a unique and memorable wedding gift at a price they can't afford to pass up.

With such a variety of potential clients to sample, any meaningful survey will be a formidable undertaking. It may consist of anything from a simple telephone sampling to a more statistically-oriented questionnaire aimed at the business community. Regardless of the method, the results must provide some basis upon which to risk the investment in time and equipment for the expected returns in a realistic setting. If you can exhaustively sample *all* of the suggested prospects mentioned here, you indeed live in a small town.

To summarize, the "market" is a variable entity dependent on the circumstances of your location and community situation. *Your* survey will be contingent upon the market and the degree of sophistication you wish to attain to satisfy yourself of an adequate potential to support an aerial banner towing business at the level you desire.

GOALS

Inasmuch as individual goals will, perhaps more than anything, dictate a decision to enter the banner towing industry, an attempt to introduce some advantages of the business at various levels of participation is hereby presented.

Banner Towing To Advertise Your Own Business

If an individual restricts his towing activity to the promotion of his own business, it isn't necessary to obtain a Commerical Pilot Certificate to do so. The FAR's permit a Private Pilot Certificate holder to excercise towing privileges in this capacity, provided that he adheres to the additional criteria (i.e. FAR 91.18, etc.) required of his commerical counterparts. Furthermore, the cost of

Fig. 1-1. Banner towing business is where you find it. Here the author's Stearman is used to solicit votes for a gentleman seeking the office of Constable.

operation is somewhat attenuated in this type of endeavor, presuming that this individual needs only a limited amount of equipment without having to stock the large inventory of letters to meet the demands of advertising for the general public. This is an excellent inroad to aircraft ownership with "write-off" potential.

Banner Towing To Defray Aircraft Ownership Costs

Aircraft ownership in today's inflationary economy is expensive by anyone's yardstick. Leaseback and rental agreements almost always turn out to be losing propositions in terms of cash flow, but admittedly better than nothing if you don't mind *your* plane being flown by every hamburger in the county. On a per hour basis, the revenue generated from banner towing operations is easily *five to ten times* the payback allowance net on the average leasing arrangement.

Banner Towing As A Second Business or Hobby

Distinct tax advantages are available to individuals through such mechanisms as the investment tax credit and depreciation, as well as others. Handsome financial rewards await the serious entrepreneur. See Chapter 11 for additional comments.

Banner Towing To Complement An Existing Aviation Business

For those already operating a flight school or other related flying service, banner towing should be a welcome addition to the business. Again, on a per hour basis, receivables from the banner towing services will more than likely outshine rentals and even charter revenues on comparable aircraft.

Banner Towing As A Livelihood

There is no doubt that a successful aerial towing operation holds a handsome living for the *serious aviator* in the *right market*. However, even the most optimistic business venture needs time to "get off the ground." Making business contacts and developing a reputation come only through hard work, adequate exposure, and patience with the calendar. One should consider an alternative source of income until it becomes apparent that a viable profession is, indeed, at hand.

Introductory comments out of the way, let's go on to look at the specifics of banner towing operations.

THE TOWING
ENVIRONMENT
A. TAYLOR

Chapter 2
The Towing Environment

If you are at all encouraged by the suggestions of the preceding chapter with its overtones related to you as a pilot, aircraft owner, and business tycoon, then a modest amount of homework still lies ahead. The purpose of this chapter is to acquaint the reader with the milieu in which he must work, with some emphasis on the pitfalls which may lurk in the weeds for the uninitiated. For the sake of clarity, the discussion will be categorized into five sections, representing a variation on the familiar "man-media-machine" theme so popular in aviation pedagogy.

THE PILOT

Banner towing is a time consuming business. It is this issue, as related to the pilot, that is the central subject to be addressed here.

Individual circumstances are far too variable to justify an exhaustive discussion to exploit the availability of time for each reader. Consider, however, the mechanics of the events that must be executed in order to launch a banner for, say, one hour. Although it seems incredibly obvious, the pilot must be available to fly during that hour. For the line instructor, this probably presents no problem at all. For someone whose principal business may be outside the aviation industry, this may pose an enormous problem to, say, get time off work to meet a 2 p.m. target time for an advertising customer. It will become increasingly clear in subsequent chapters that *generous* amounts of time are an ever-present ingredient in nearly all aspects of a safe operation. This includes, but is in no way limited to: banner assembly; travel to and from the airport; launch set-up procedures; banner recovery requirements; equipment repair, maintenance; storage, and—oh, by the way, business solicitation. Suffice it to say that the pilot must have excellent flexibility in scheduling if he is to meet the demands of his banner clients and the business.

Additional comments regarding the pilot and the FAR's are outlined in Chapter 3.

THE AIRCRAFT

A wide spectrum of aircraft are currently in use for banner towing operations throughout the country. Many single-engine aircraft are suitable for towing in "stock" condition, while others may require minor modifications to circumvent overheating problems. The adaptability of any aircraft to the towing task can be estimated by consideration of the following characteristics.

Speed. Any aircraft used for towing purposes must have excellent low speed flight characteristics (under 80 MPH). Furthermore, it must not present significant overheating problems at low speeds in hot weather with the load imposed by the additional drag of the banner. The latter problem can be overcome in some aircraft by the installation of an engine oil cooler and/or auxiliary cowl flaps. The engine cowling may be completely removed from some planes without affecting safety of flight. Planes with border-line low speed tolerance may become quite satisfactory with the inflight use of flaps and/or extended gear. The bottom line, however, is *safety* and any marginal considerations should be carefully researched with the help of the aircraft manufacturer if at all possible. The issue of exceeding 80 MPH lies solely with the wear and tear on the banner equipment.

Power. Sufficient power to haul a reasonable banner length is desirable for obvious reasons. The definition of "reasonable" is left up to the reader, but experience warns that anything less than a capacity to tow at least 25 letters is somewhat restrictive. For practical purposes, the drag of a banner depends on the number and size of the letters used as well as the style of the connector rods. Therefore, matching the aircraft's performance with the type of equipment is an important consideration. Although many aircraft logically demonstrate a direct ratio between power and speed, there is a variety of birds with *both* low speed comfort *and* a generous powerplant. These are most often seen in glider towing ops and agricultural dusting rigs.

The RPM limits of the engine should be scrutinized, especially for mid-range RPM restrictions. The Continental R-670 on the Stearman, for example, has a cruise restriction range (1500 - 1650 RPM) with a max cruise at around 1900 RPM. The latter basically sets the upper limit of the banner length. A very short banner, however, requires the pilot to throttle back to the edge of the cruise restriction range to avoid exceeding the prudent operating limits of the banner equipment (80 MPH). Thus, a built-in *lower limit* for the banner length can be inherent in the powerplant restrictions and should be watched for when choosing a suitable aircraft.

A climb propeller is superior to a cruise prop for towing operations. A flat pitch and wide diameter propeller yields the most efficient combination for cruise at towing speeds. Other uses of the aircraft (e.g. cross-country, flight instruction, aerobatics, etc.) may make trade-offs necessary in this department.

Aircraft Equipment. Two-way radio communication is not required if towing operations are exclusively outside controlled airspace. However, since that would indeed be the exception, a radio is mandatory equipment for all practical purposes, as well as an additional safety consideration. A transponder is likewise highly recommended for convenience, safety, and to minimize limitations and unnecessary detours in today's ever-congested skies.

Refer to Chapter 4 for a discussion of tow hitch equipment.

Pizzazz. Advertising clients will not overlook the type of aircraft used to display their messages. An "attention-getter" such as a Waco or a Stearman will more likely promote business than will a common, garden variety single-engine machine. By the way, helicopters can be used quite effectively in the banner business.

A guide for comparative purposes is presented in Table 2-1 courtesy of Gasser Banners, Inc. Keep in mind that actual capabilities will vary slightly, depending on the usual variables such as density altitude, temperature, gross weight, prop pitch, etc.

THE AIRFIELD

Not every airfield is suitable for banner towing operations. Although it is not necessary to use the airstrip property for launch and recovery operations, the hassle of remote site pick-ups clearly dictates on-site launching as the superior way. Appropriate

Aircraft Type	5 ft.	7 ft.
Maule 220	50	40
Piper J-3-85	24	17
Piper PA-12-115	32	23
Piper PA-18-150	50	35
Piper PA-22-150	32	23
Piper PA-28-140	26	18
Piper PA-25-235	60	40
Champion 7ECA-115	32	23
Champion 7KCAB-150	42	29
Champion 7GCBC-150	45	31
Stearman 220	46	32
Waco UPF-7-220	36	25
Cessna 150/150 HP	38	27
Cessna 170/172	32	23
Cessna 180/182	38	27
Cessna Ag Wagon 300	70	50
Bell 47-G-210	40	28
Hughes 300-180	36	25

Table 2-1. Sign Capacity: Number of Characters For Various Aircraft (courtesy Gasser Banners, Inc.).

planning and ample research into the airfield you intend to use will be clearly rewarding and may avert a double migraine headache.

The airfield must be suitable for the towing plane. This may seem insultingly obvious, but the subject here is not runway length or hangar space. There is a *business* involved that orchestrates the assembly or raw materials into a product. One of those raw materials is aviation fuel—otherwise stated *gas*. In today's capricious fuel market, we are becoming all too familiar with the regional and local problems with fuel supply. And the aviation industry (particularly general aviation) *isn't* exempt just because we pay a premium price for the stuff. It's one thing to be on the ground fantasizing about flight on a beautiful spring day. It's entirely another to be in a similar situation with four hundred bucks riding on empty tanks. Some of the more peripheral airstrips literally "get left out" when the petrol trucks make their rounds. All other factors being equal, the wise planner will find out where they make their *first* stop, and who has the *biggest* tanks.

An appropriate airfield ideally has a suitable space for the aerial pick-up technique for launching the banner. Consider the approach and departure obstacle clearances (see Chapter 5) along with any restrictions that may be specified in the waiver or Certificate of Authorization (see Chapter 3). Don't overlook the watershed and drainage situation, as well as the general condition of the field. The runway may be high and dry but the surrounding terrain may be a swamp eleven months out of the year. Picking up a sopping wet banner is hard on the equipment, and there are some aesthetic drawbacks to tromping through cow chips to retrieve your banner. High grass or weeds can damage the characters during either launch or recovery, so the general field maintenance (mowing, etc.) should be thoroughly investigated.

Other uses of the proposed airport may interfere with a towing operation. As previously alluded, what may appear to be the ideal location at one time of the year may be used for cattle grazing or cultivation at other times. Airshows, parachute clubs, helicopter operations, and so on, may interfere on an intermittant basis so that a complete itemization of airport activities and schedules will be worth the effort.

Not every airport manager is intrinsically ecstatic about banner pullers. Indeed, there are more than a handful out there who just flat won't allow it—*period*! In any case, a personal visit with the airport manager is recommended. Some gentle persuasion may be necessary, especially if he isn't familiar with the nuts and bolts of the mission. A brief didactic session may help the cause with particular emphasis on your commitment to safety as well as the fact that runway movements won't come to a screeching halt just because a banner is being launched.

As previously mentioned, there is no mandate to have a launch/recovery site within the physical boundaries of the airport. Verily, there will be situations that arise that will make it advantageous *not* to snag a banner on home soil. However, as will be highlighted in later chapters, there are reasons to abort a launch maneuver with the implications of landing again to set up for another try. Close proximity to the landing strip can make the difference between a mission completed and a big *zero* in the bank.

THE AIRSPACE

A survey of the proposed territory over which towing operations are contemplated should be conducted with specific points in mind. The location of the exceptional target areas, e.g. football stadiums, outdoor gathering places, etc., is best scrutinized with the airspace above, specifically looking for conflicts with other known traffic. Airport traffic areas and approach courses deserve the most attention in this respect. High density terminal areas may make a banner service impractical, if not impossible. Air Traffic Control is the best source for advice on this subject, and a personal visit will be enlightening. ATC is generally fan-n-n-n-tastic about cooperating with commercial aviation, but there *are* areas in which the banner tower is not welcome unless just passing through.

It bears repeating that two-way radio communications and a transponder (with code capability applicable to the nearby TCA group type) should be standard equipment on a towing plane. The investment will pay for itself many times over in due time.

THE WEATHER

Banner towing is permitted in VFR conditions (not special VFR) during the hours of sunrise to sunset. The serious prospective banner baron will assess the annual weather pattern and climate for the area of intended service to certify general alignment with the goals previously discussed in Chapter 1.

The regional climate should be taken into consideration if you have yet to select an airplane. A predominantly cold climate will make your socks roll up and down in an open cockpit classic. Likewise, a three hour mission during the summer months in a hot, humid closed cabin can take the edge off of *anyone's* personality.

See Chapter 8 for additional comments on weather conditions.

The preceding discussion serves primarily as an outline for a sensible preliminary investigation of the environment in which the prospective banner towing operator must serve. The individual circumstances of the reader must be integrated meaningfully into the scenario in order to effect a responsible and realistic evaluation *prior* to the commitment of resources.

FAA REGS
A. TAYLOR

Chapter 3
FAA Regs

Although the FAR's are quite non-specific in regulating banner towing operations, they give the FAA the freedom and authority to be *very* specific about the nuts and bolts of any given operation. Furthermore, they have a lot to say about the hardware involved—at least from the tow-hitch and its installation on back to the tip of the towline. This chapter shall deal with the input of aviation's regulatory authority on the subject of banner towing. For the sake of clarity, the narrative will be categorized by general subjects.

THE PILOT

The vast majority of towers will operate with an unrestricted commercial ticket, defined in Part 61 of the FAR's. A valid and current Class II Medical Certificate is, of course, required to be in the possession of the pilot engaging in commercial aviation.

It is possible to get a limited commercial certificate *without* holding an instrument rating. Thus, a private pilot who meets the other criteria of eligibility, flight proficiency, and aeronautical knowledge can side-step at least part of the aeronautical experience requirements (i.e., the instrument rating) and qualify for a commercial license with a restrictive endorsement. FAR 61.129 should be carefully referenced for this option.

FAR 61.118 (a) permits a private pilot to tow banners in conjunction with a business, provided he does *not* carry passengers or cargo for hire. Thus, it isn't necessary to secure a commercial ticket if towing activity is appropriately limited, say, to advertising one's own business. A Class III Medical Certificate is adequate for this purpose.

FAR 91.18 AND THE CERTIFICATE OF WAIVER

Banner towing is only briefly mentioned in Part 91 of the FAR's which, in fact, prohibit towing other than gliders except under specifically approved circumstances:

91.18 Towing: other than under 91.17.

(a) No pilot of a civil aircraft may tow anything with that aircraft (other than under 91.17) except in accordance with the terms of a certificate of waiver issued by the administrator.

(b) An application for a certificate of waiver under this section is made on a form and in a manner prescribed by the Administrator and must be submitted to the nearest Flight Standards District Office.

The general wording in (a) gives the Administrator: 1) the authority to specify the requirements for the issue of a certificate of waiver and 2) the right to modify towing activity by listing terms and conditions pertinent to the local flying environment.

The language in (b) is loose enough to allow the local FSDO to exercise an option, at the discretion of the Administrator, to require a demonstration of proficiency as a condition of waiver approval.

Figure 3-1 shows an example of an application for Certificate of Waiver of FAR Part 91.18. The reverse side, which essentially applies to airshows and special events, will be left blank on your application, except for the date and signature. This form is submitted in triplicate. Figures 3-2 and 3-3 show a copy of a Certificate of Waiver and the Special Provisions attachment. The waiver can be written to include more than one pilot and/or multiple aircraft.

The special provisions attachment (Fig. 3-3) raises an interesting point. A reasonable person would be justified in developing a "Catch 22" situation, for if the FSDO requires a demonstration of proficiency for a waiver approval, *and* one can't legally tow a banner without a waiver, *and* passenger carrying is prohibited—then how the hell do you get checked out to qualify for a waiver? Fortunately, the term "passenger" does *not* apply to a crewmember. You therefore can fly as a crewmember under someone else's waiver, ergo, take lessons. This may not be a viable option if there are no tow operators in the area or if the operator of a local monopoly doesn't care to have you trained so you can cut in on the market. The details of this type of problem can be easily worked out with the local FSDO.

The guidelines for the issuance of the special provisions section of the Certificate of Waiver come from the *General Aviation Operations Inspector's Handbook* (March, 1975) which states:

215. BANNER TOWING BY AIRPLANES. Processing a waiver of Section 91.18 will require close coordination with the Maintenance Unit of the District Office. While most banner-tow applicants utilize standard attaching hooks with either electrical or mechanical release devices, a maintenance inspector should determine that the attaching device and aircraft airworthiness meet the requirements of all applicable FARs. The following are a few guidelines that will assist the GAOI in processing a banner-tow waiver application:

a. Aircraft Checks. Since banner-towing operations require flights over or within close proximity to congested areas, it would be advisable to require inspection of the aircraft prior to each operation. The inspection should include a functional check of the tow hook and release mechanism.

b. Banners. The inspector must determine that the banner would not create a hazard to persons or property if deliberately or inadvertently dropped. Most banners are constructed so that they will float gently to the ground when released. However, if there is any doubt, a demonstration should be required prior to the issuance of a waiver. If it is found that a hazard exists, the waiver may still be issued with a provision that prohibits the towing over other than open land and water, including the takeoff, landing, and en route phase of the operation. If the operator cannot comply with this requirement, the waiver must be denied.

c. Airports. It should be determined that the airport to be used has a suitable pick-up and drop area for the banner that is at least 500 feet from any active runway and areas used by airport visitors or employees (other than those engaged in the operation).

d. Competency. The inspector should satisfy himself that all pilots listed on the application who will be engaged in the operation are competent to perform their assigned duties. This may be satisfied by previous knowledge or a demonstration with the inspector observing from the ground. In this way, he can also evaluate the competency of any necessary ground personnel.

DEPARTMENT OF TRANSPORTATION
FEDERAL AVIATION ADMINISTRATION

CERTIFICATE OF WAIVER
OR AUTHORIZATION
APPLICATION

To: Federal Aviation Administration

Form Approved: OMB No. 04 R0073

APPLICANTS—DO NOT USE THESE SPACES

REGION	DATE

ACTION
☐ APPROVED ☐ DISAPPROVED *(Explain under "Remarks")*

SIGNATURE OF AUTHORIZED FAA REPRESENTATIVE

INSTRUCTIONS

Submit this application in triplicate (3) to any FAA Office.

Applicants requesting a Certificate of Waiver or Authorization for an air meet will complete all items and certification on this form and will attach a properly marked map or diagram of the operations area. This map or diagram must be to scale, and distances must be shown. It must include race courses, obstructions, grandstands; congested areas, parking areas, dead lines, police stations; ambulance, fire-truck, crash-wagon, and control stations. Application for air meets should be submitted not less than thirty (30) days prior to the requested beginning date of the proposed operation.

Applicants requesting a Certificate of Waiver or Authorization for activities other than an air meet, will complete items 1 through 7 only and the certification on the reverse.

1. NAME *(First, Middle, and Last)*

2. PERMANENT MAILING ADDRESS	HOUSE NUMBER AND STREET, OR ROUTE NO.	POST OFFICE	STATE	TELEPHONE

3. TO AUTHORIZE NONOBSERVANCE OF FEDERAL AVIATION REGULATIONS, SECTION *(Indicate sections which prohibit proposed operation)*

4. IN PERFORMANCE OF *(Describe proposed operation and purpose thereof in detail)*

(If necessary, attach supplement to continue)

5. AREA OF OPERATION

6. FOR THE PERIOD OF—				BEGINNING *(Date)*	ENDING *(Date)*	BETWEEN THE HOURS OF—
HOURS	DAYS	WEEKS	MONTHS			

7. AIRCRAFT MAKE AND MODEL	IDENTIFICATION MARK	OWNER	ADDRESS (STREET, CITY, STATE)

WHILE BEING FLOWN BY THE FOLLOWING PILOTS:

NAME	ADDRESS (STREET, CITY, STATE)	CERTIFICATE NUMBER AND RATING

FAA Form 7711–2 (11/76)

Fig. 3-1. FAA form 7711-2: Certificate of Waiver or Authorization Application.

| 8. THE AIR MEET WILL BE SPONSORED BY |

9. PERMANENT MAILING ADDRESS	HOUSE NUMBER AND STREET OR ROUTE NO.	POST OFFICE	CITY	STATE

10. POLICING *(What provision will be made for policing the meet?)*

11. EMERGENCY FACILITIES

☐ PHYSICIAN   ☐ AMBULANCE   ☐ FIRE TRUCK   ☐ CRASH WAGON
☐ OTHER *(Specify)* ____

12. AIR TRAFFIC CONTROL *(Describe method of controlling traffic, including provision for arrival and departure of scheduled aircraft)*

13. SCHEDULE OF EVENTS *(Include arrival and departure of scheduled aircraft and other open port periods; unforeseen changes and revisions to be subject to approval of local inspector)*

HOUR	DATE		PILOT RATING

(If sufficient space is not available, the entire schedule of events may be submitted on separate sheets, in the order and manner indicated above.)

The undersigned applicant accepts full responsibility for the strict observance of the terms of the Certificate of Waiver, and understands that the authorization contained in such certificate will be strictly limited to the above-described operations.

I CERTIFY that the foregoing statements are true.

_______________________________________   _______________________________________
(DATE)   (SIGNATURE OF APPLICANT)

REMARKS:

CERTIFICATE OF WAIVER OR AUTHORIZATION

ISSUED TO

DERRY AIR, INCORPORATED

ADDRESS
11005 GREENWILLOW DRIVE
HOUSTON, TEXAS 77025

This certificate is issued for the operations specifically described hereinafter. No person shall conduct any operation pursuant to the authority of this certificate except in accordance with the standard and special provisions contained in this certificate, and such other requirements of the Federal Aviation Regulations not specifically waived by this certificate.

OPERATIONS AUTHORIZED

TOWING: ADVERTISING BANNER

AREA OF AUTHORIZATION: STATE OF TEXAS

LIST OF WAIVED REGULATIONS BY SECTION AND TITLE

FAR 91.18 Towing: Other than 91.17

STANDARD PROVISIONS

1. A copy of the application made for this certificate shall be attached to and become a part hereof.
2. This certificate shall be presented for inspection upon the request of any authorized representative of the Administrator of the Federal Aviation Administration, or of any State or municipal official charged with the duty of enforcing local laws or regulations.
3. The holder of this certificate shall be responsible for the strict observance of the terms and provisions contained herein.
4. This certificate is nontransferable.

NOTE.—This certificate constitutes a waiver of those Federal rules or regulations specifically referred to above. It does not constitute a waiver of any State law or local ordinance

SPECIAL PROVISIONS

Special Provisions Nos. __1__ to __15__, inclusive, are set forth on the attached sheets.

This certificate is effective from __11-15-80__ to __11-15-81__, inclusive, and is subject to cancellation at any time upon notice by the Administrator or his authorized representative.

BY DIRECTION OF THE ADMINISTRATOR

SOUTHWEST
(Region)

WILLIAM V. GRAY
(Signature)

November 15, 1980
(Date)

General Aviation Inspector
SW-FSDO-62 (Title) Houston, Texas

FAA Form 7711-1 (7-74)

Fig. 3-2. FAA form 7711-1: Certificate of Waiver or Authorization.

e. Altitudes. Minimum safe altitudes of Section 91.79 should not be waived for banner-tow operations. Low-level operations generally reduce the advertising effectiveness of banner towing and most operators would not request a waiver of

GENERAL OPERATING AND FLIGHT RULES

A handful of Part 91 FAR's deserve to be underscored for the in-depth discussion of aerial banner towing. This regulatory potpourri is by no means an exhaustive index, but rather a selection of subjects intended to stimulate further reading.

91.17 Towing: Gliders. This section delineates the operation and conduct of glider towing procedures. Applicable portions of Part 61 regarding pilot qualifications should be carefully referenced if both types of towing activities are anticipated.

91.24: ATC Transponder Equipment. With the ever increasing radar vigil demanded by today's high density air traffic areas, one can hardly operate without undue inconvenience without a transponder. If you are looking to purchase equipment, you would be wisely advised to consider future trends in your local area of intended operations. TCA's are gobbling up airspace at an unprecedented rate, requiring progressively sophisticated equipment. Be certain that your transponder is appropriate to any TCA Group you may have occasion to fly in, with an eyebrow raised for the future. It should be noted that the operation of a transponder requires periodic equipment testing (FAR 91.177) each 24 months.

91.79 Minimum Safe Altitudes. Towing activity holds no exemption from the guidelines for altitude restrictions set forth in this section. Although 1,000 feet above the highest obstacle and 2,000 feet horizontal clearance is required in a "congested" area, the definition of the latter isn't objectively specified. This should be of little consequence to the banner operator, however, since most of his activity will be around "congested" areas anyway and his choice of banner equipment should be made with the notion in mind that the banner copy should be easily read at the higher altitudes without problem.

Keep in mind that the minimum safe altitude anywhere is at a height so as to be able to execute a safe emergency landing in the event of a powerplant failure. The beach, therefore, is not an appropriate place to relax altitude standards for at least two reasons: 1) A beach in the summertime is a congested area and an open assembly of persons. Even though your towing groundtrack is somewhat displaced from the shoreline and out over the water, this situation should be treated as a "congested area". 2) A safe, emergency landing in the ocean is a contradiction of terms.

SPECIAL PROVISIONS FAR 91.18 AND/OR FAR 91.39 (d) FOR BANNER TOWING AND NIGHT AERIAL ADVERTISING. (AIRPLANES AND/OR HELICOPTERS)

1. This Certificate of Waiver and attached special provisions or copies thereof, shall be carried in the aircraft during all aerial advertising operations.

2. Passenger carrying is prohibited.

3. Prior to each flight, the aircraft shall receive a preflight inspection by the tow pilot. This inspection shall be an operational check of the engine and all flight control systems as recommended by the manufacturer. The tow hook and release mechanism shall receive a functional check, if appropriate.

4. Towing operations are restricted to the hours between official sunrise and sunset.

5. Operations will not be conducted unless the ceiling is at least 1500 feet and the visibility is at least five miles. (Operations under special VFR weather minimums are prohibited).

6. The aircraft shall receive an annual or 100-hour inspection within the previous 100-hour of time-in-service.

7. The banner shall not be picked up or dropped closer than 500 feet to any active runway or other areas being actively used by aircraft, airport tenants, visitors or employees, or other persons or vehicles not engaged with the operation.

8. Prior to and during aerial advertising operations within controlled airspace (control zone, terminal control area, airport traffic area, or any controlled airspace) each person planning and conducting such operations shall establish and maintain two-way radio communications with the appropriate Air Traffic Control facility for instructions or clearances as required by Air Traffic. (No flights will be conducted in a terminal control area unless prior approval is obtained from the appropriate Flight Standards field office chief and Air Traffic Control facility chief or their representatives).

9. The holder of the waiver shall insure that the pilot is properly certificated and is qualified and competent to perform aerial advertising operations. The holder of the waiver will also insure that the pilot is briefed on all provisions of the waiver. The pilot will then certify in writing to the holder of this waiver that he has read and understands all of the special provisions.

10. The holder of this Certificate of Waiver shall establish and maintain at his home base a current list of aircraft and pilots to be used under the terms of this waiver and the certification statements of each pilot required under Special Provision No. 9.

11.   Each pilot shall hold at least a commercial pilot certicate with an appropriate category and class rating.

12.   The operation shall be executed in accordance with a planned course of action with emphasis on selection and availability of emergency landing areas. Due to the added weight and steeper power-off glide ratio of the aircraft as a result of the special equipment installation, special precautions will be exercised by the pilot to insure compliance with FAR 91.79(a), (b), and (c). FAR 91.79(d) is not applicable for helicopters under the terms of this waiver. Operations over congested areas or open air assemblies of persons must be no lower than 1000 feet above the highest obstacle within a horizontal radius of 2000 feet and operations elsewhere shall be in compliance with FAR Section 91.79(c).

13.   If operations are authorized outside the boundaries of the issuing Flight Standards field office, those operations may be conducted only after coordinating in advance with the appropriate Flight Standards field office chief or his representative and complying with any/all additional special provisions imposed by that office.

14.   When engaged in aerial advertising operations over cities, towns, or settlements listed on current sectional aeronautical charts, appropriate community officials will be notified in advance of flight over their community.

15.   Flight directly over large assemblies of persons, such as sports arenas or stadiums located near large metropolitan areas, is prohibited while engaged in aerial advertising operations. The flight path will consist of a left-hand pattern, with no turns in excess of a 30-degree bank, and with no abrupt change in altitude or direction of flight. Crossing over, under, or within 500 feet of other aircraft engaged in aerial advertising is prohibted.

NAME DERRY AIR, INC.
EFFECTIVE FROM 11-15-80 TO 11-17-81
HOUSTON, FLIGHT STANDARDS DISTRICT OFFICE #62

Fig. 3-3. Sample of an actual Special Provisions supplement to a Certificate of Waiver. Regional variation is to be expected.

DEPARTMENT OF TRANSPORTATION
FEDERAL AVIATION ADMINISTRATION

MAJOR REPAIR AND ALTERATION
(Airframe, Powerplant, Propeller, or Appliance)

Form Approved
Budget Bureau No. 04-R060.1

FOR FAA USE ONLY

OFFICE IDENTIFICATION

INSTRUCTIONS: Print or type all entries. See FAR 43.9, FAR 43 Appendix B, and AC 43.9-1 (or subsequent revision thereof) for instructions and disposition of this form.

1. AIRCRAFT	MAKE	MODEL
	SERIAL NO.	NATIONALITY AND REGISTRATION MARK
2. OWNER	NAME *(As shown on registration certificate)*	ADDRESS *(As shown on registration certificate)*

3. FOR FAA USE ONLY

4. UNIT IDENTIFICATION				5. TYPE	
UNIT	MAKE	MODEL	SERIAL NO.	REPAIR	ALTERATION
AIRFRAME	••••••••••••••••••••• *(As described in item 1 above)* •••••••••••••••••••••				
POWERPLANT					
PROPELLER					
APPLIANCE	TYPE MANUFACTURER				

6. CONFORMITY STATEMENT

A. AGENCY'S NAME AND ADDRESS	B. KIND OF AGENCY	C. CERTIFICATE NO.
	☐ U.S. CERTIFICATED MECHANIC ☐ FOREIGN CERTIFICATED MECHANIC ☐ CERTIFICATED REPAIR STATION ☐ MANUFACTURER	

D. I certify that the repair and/or alteration made to the unit(s) identified in item 4 above and described on the reverse or attachments hereto have been made in accordance with the requirements of Part 43 of the U.S. Federal Aviation Regulations and that the information furnished herein is true and correct to the best of my knowledge.

DATE	SIGNATURE OF AUTHORIZED INDIVIDUAL

7. APPROVAL FOR RETURN TO SERVICE

Pursuant to the authority given persons specified below, the unit identified in item 4 was inspected in the manner prescribed by the Administrator of the Federal Aviation Administration and is ☐ APPROVED ☐ REJECTED

BY	☐ FAA FLT. STANDARDS INSPECTOR	☐ MANUFACTURER	☐ INSPECTION AUTHORIZATION	OTHER *(Specify)*
	☐ FAA DESIGNEE	☐ REPAIR STATION	☐ CANADIAN DEPARTMENT OF TRANSPORT INSPECTOR OF AIRCRAFT	

DATE OF APPROVAL OR REJECTION	CERTIFICATE OR DESIGNATION NO.	SIGNATURE OF AUTHORIZED INDIVIDUAL

FAA Form 337 (7-67) (8320)

Fig. 3-4. FAA form 337: Major Repair and Alteration—This document must be approved for the installation of a tow hitch on any aircraft.

91.169: Inspections. Unless the towing plane is used for other purposes, i.e. leaseback, carrying passengers for hire, flight instruction, etc., an annual inspection in accordance with the directives of this section will satisfy the FAR requirements for banner operations. However, the restrictive endorsements (special provisions) attachments to the Certificate of Waiver may otherwise specify additional inspection requirements (see item #6, Fig. 3-3). Maintenance records are detailed in FAR 91.173.

91.207: VFR fuel requirements. The time and distance range of your aircraft will undoubtedly be shortened by a significant factor with the drag load of a banner. It is *doubly* important to have adequate fuel reserves at destination in case problems are encountered in releasing the banner.

FAA Form 337. This important document (Fig. 3-4) must be completed and approved subsequent to the installation of a tow hitch on any aircraft. A copy should be kept with the other required documents to remain with the aircraft at all times. Samples of previously approved forms for various aircraft are available from Gasser Banners, Inc. and routinely provided for convenience.

AC 43.13-2A. The subject of the Advisory Circular 43.13-2A is the acceptable methods, techniques, and practices of aircraft alterations. Chapter 8 of 43.13-2A addresses the installation of the tow-hitch hardware. Earlier editions of this circular included sections on tow plane considerations and tow-line information. A recent change (6/1/79) has deleted that material from 43.13 publication. The information, however, is included in this book in other sections dealing with those subjects (Chapters 2 and 4). The content of AC 43.13-2A, Chapter 8 "Glider And Banner Tow-Hitch Installations" is reproduced for your convenience in Appendix E of this book.

As a supplementary comment, it seems obvious that the installation of a tow hitch to conform to glider specifications is the *only* way to travel. This clearly increases the utility of the aircraft without restrictive covenants that might someday haunt the owner.

EQuiPmeNT
A. TAYLOR

Chapter 4
Equipment

As a prelude to the discussion of banner towing operations, the equipment hardware will be introduced in some detail. Although other banner equipment is available, this discussion will be limited to that manufactured by Gasser Banners, Inc., of Nashville, Tennessee. Their merchandise is perhaps the best known and most widely used in this country and time has tested the quality of their products and workmanship. This endorsement, however, does not override the virtue of comparative shopping in a competitive market. *Extreme* caution must be exercised when contemplating the purchase of used equipment, occasionally seen in the various aviation classified ads. Always inspect *each* piece of equipment with obsessive-compulsive detail and never, that is, *never* purchase used equipment sight-unseen.

The subsequent narrative serves to highlight the individual components of the complete banner apparatus to include the tow-hitch, grapple hook assembly, towline, banner, and their respective connections. Remarks on assembly and operational tips will be interjected as appropriate to the immediate topic. The discussion shall begin with a description of the banner proper and progress forward to the tow-hitch and aircraft connections.

Figure 4-1 depicts a schematic of a typical banner in assembled form. Refer below for detailed descriptions of the labeled components.

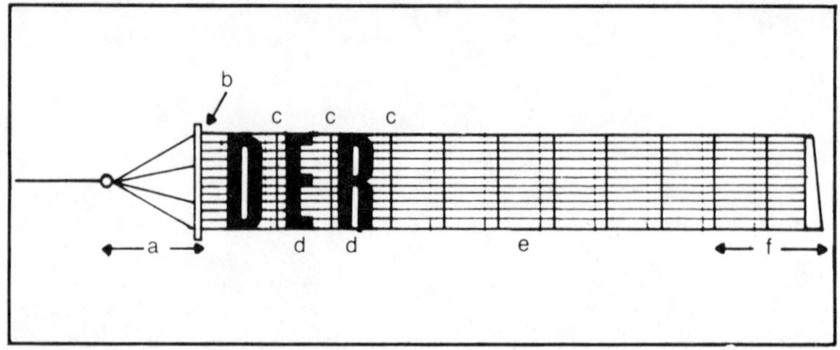

Fig. 4-1. Banner components. A. Mast assembly. B. Mast assembly connector rod. C. Connector rods. D. Letter character. E. Word spacers. F. Tail flag assembly.

THE MAST ASSEMBLY

Also called the *lead pole*, this unit (Figs. 4-2, 4-3) represents the heaviest single component in the banner. The bottom of the rigid metal portion is weighted to stabilize the attitude of the banner in flight; In other words, to keep the sign upright. A series of eight heavy-duty, interconnected nylon straps, each sewn into a loop on the trailing side, serves to connect the metal pole to a rigid fiberglass connecting rod. This, in turn, provides the connecting clips for the first character of the banner copy. Four metal rope guides on the leading side of the mast pole space the nylon bridle harness lines at their point of attachment. The heavy metal bridle ring at the terminal portion of the bridle harness is the point at which the towline joins the mast assembly. It is especially important to carefully preflight the bridle harness lines prior to each tow as they are subject to chafing where they pass through the bridle ring.

THE CONNECTING RODS

Connecting rods are flexible fiberglass units that are placed between each character (letter, space, punctuation mark, etc.) of the assembled banner. They serve to support the banner and give it vertical structure. Rods also provide the connector clips to which each character must attach. (As an aside, connector rods lend a means of handling the banner during ground operations and are useful for storage.)

Two popular styles of connector rods are shown in Fig. 4-4. The right rod is an "old-style" rod with "old-style" clips. It is a safe guess that the left rod and accompanying clips are of the "new-style" variety. While personal preference will ultimately

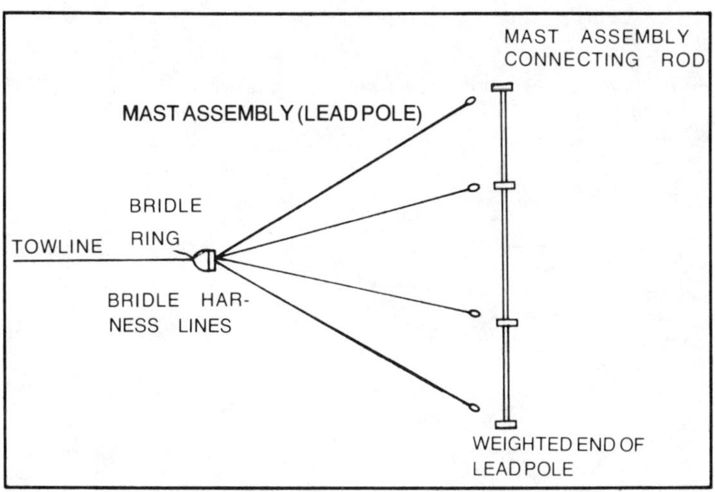

Fig. 4-2. Mast assembly (lead pole).

dictate the usage of either type, the distinction between them isn't entirely academic.

Old Style Connector Rods and Quick Disconnect Clips

Being the more rigid of the two types, this style connector rod is also heavier by weight. These rods are easy to handle on the ground and only two support points are required for rack storage.

Fig. 4-3. Mast assembly.

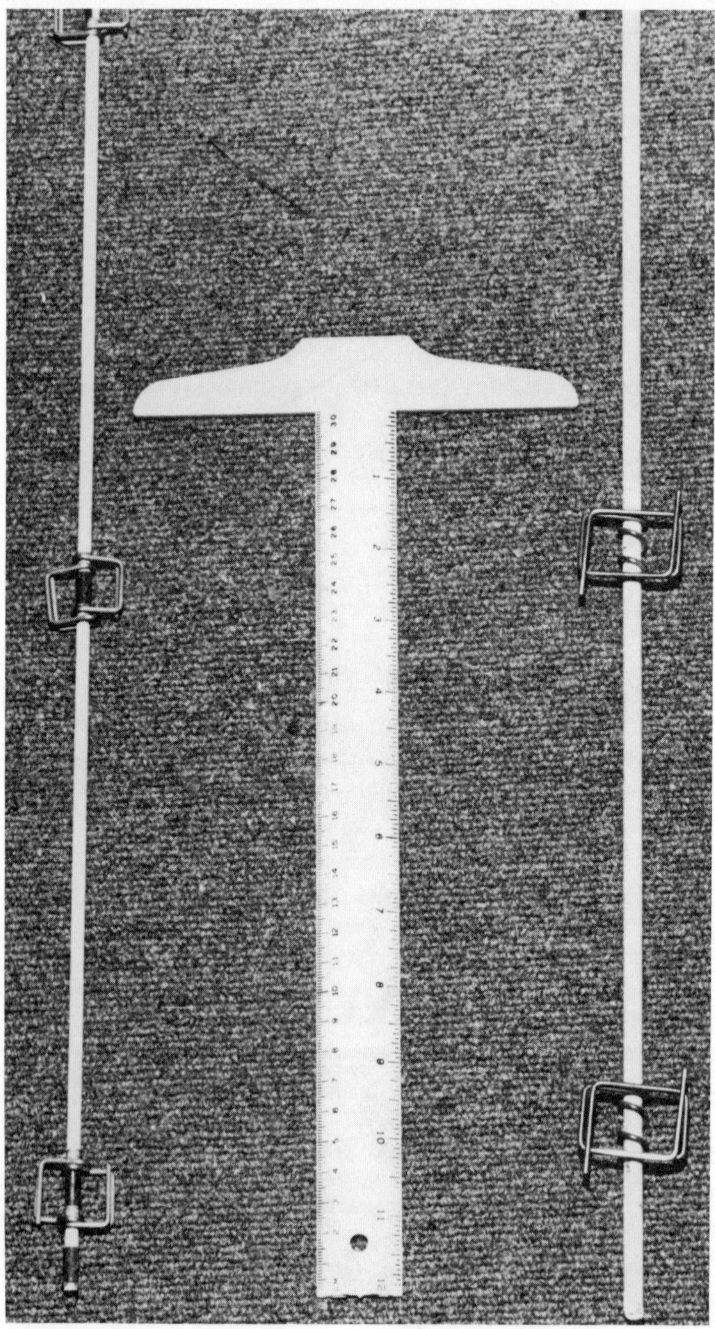

Fig. 4-4. Connector rods. Old-style is on right, new style on left.

The old-style quick disconnect clips (or QD clips) are single unit metal rigid wire configurations. Each QD clip is immobile once positioned on a connector rod, thereby providing constant spacing for each of the nylon straps vertically. Figure 4-5 depicts the old style QD clip in two views. Because of its single piece construction, it literally has a "front" and "back" side as shown. The "wings" connect the trailing and leading straps of successive letters. The "fuselage" section will always be positioned to the back of the banner. It must be noted that during launch, the banner usually undergoes some forward sliding motion across the ground before becoming completely airborne. It is the fuselage portion of the QD clip that should interface with the launch surface. This means that the banner should be positioned face up for launch with the old style QD clips and connector rods. While it is possible to launch with the winged surfaces against the ground, there is some increased risk of snagging a surface irregularity, thereby exposing the banner equipment to stress that might be foreign to design limits. Chapter 8 should be seen for warnings about the use of the old style clips.

New Style Connector Rods and Quick Disconnect Clips

The lighter, new style connector rod is less rigid than its bulkier cousin. Among the important consequences:

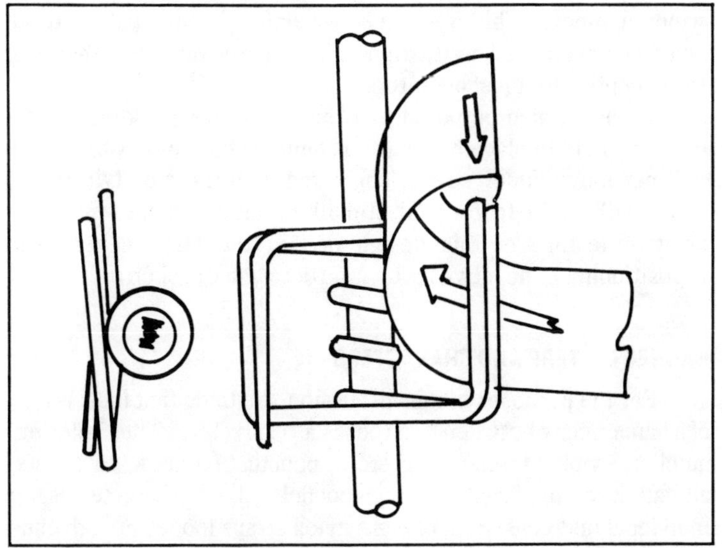

Fig. 4-5. Old-style quick disconnect (QD).

1) Decreased drag is a noticeable benefit of the lighter rods. A reduction in required cruise RPM is obvious in the crudest of comparative studies. The resultant decreased fuel flow will help to mitigate (however slightly) the impact of today's breath-taking fuel bills. Note that if your plane engine envelope has a middle RPM restriction, a very short banner with these light rods may put you right in the middle of the restricted RPM range in order to stay below 80 MPH. In such a case, the old style rods are better suited to the mission by virtue of their higher drag. (Decreased drag also means that a longer banner copy is possible with the new style components, all other factors being equal. If you charge by the job, this means more work; if you charge by the letter, it means more buckaroos!)

2) New style rods are somewhat more fragile than the old style variety. Some breakage is to be expected during normal use, although this is not a frequent or major problem.

3) Solo ground handling of the banner with new style rods is a bit trickier because the rods bow considerably when handled from the center. For the same reason, storage with the rods connected to the letters requires three or four support points to prevent bending and eventual warping.

4) Because of the clip design, the new style rods don't have a front or back side relative to the banner. Accordingly, the banner can be laid out either face-down or face up as needed for more ideal wind alignment. This is especially helpful if your final approach heading is restricted, a situation expected to be the rule rather than the exception for most operators.

It should also be noted that the new style quick-disconnect clips are held in place by central retaining clips (Fig. 4-6). These will not move under stress, compared with the old style types, which will yield to forceful stimuli by sliding somewhat. The positive feature of this design is that the QD's never need repositioning. A negative result is an occasional broken rod.

BANNER LETTERS AND CHARACTERS

For the purposes of this discussion, anything that flies as part of a banner copy between two rods is a "character." Thus, a letter, number, symbol (dollar sign, etc.) , punctuation mark, or a space all fall into the "character" pigeonhole. Each character is an individual unit consisting of eight nylon straps looped at both ends for union with the QD clip units of each rod. If nothing is sewn into

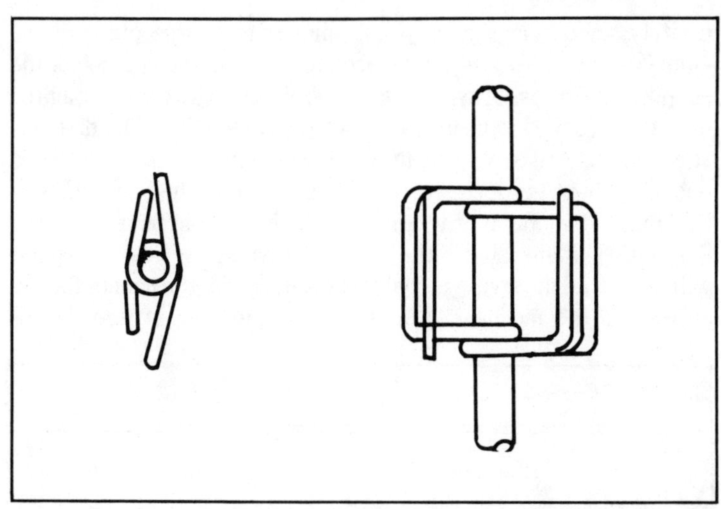

Fig. 4-6. New-style QD strap connections are identical to the old-style QD.

the straps, they represent a space or "word spacer." Figure 4-7 shows the basic 8-strap configuration between two old-style pods.

By keeping the spacing constant between the QD clips, we now have a symmetrical matrix from which to construct a numeral, letter, punctuation mark—virtually any symbol, provided the complexity isn't too outrageous. In the examples used in this book, the materials for the characters are of nylon fabric. The formed fabric is sewn into the strap matrix after being cut into the desired symbol shape and hemmed around the complete perimeter. Figure 4-8 shows a small punctuation mark (period) sewn into the bottom two straps of this character. Figure 4-9 depicts the front and back sides, respectively, of the letter "E". The straps are sewn in on the back side for cosmetic purposes, although this is hardly discernable at regular towing altitudes.

The straps interconnect with the QD clips as previously seen in the schematic of Fig. 4-5. There is only one way to correctly connect the straps to each type of QD clip, although Mr. Murphy will try to give you the benefit of perhaps a half dozen different approaches. The proper strap-QD clip interconnection is shown in Fig. 4-10 under tension. Note that the design is somewhat like the "Chinese finger trap"—the more tension within the system, the tighter the interface between the strap and the QD clip.

SPECIAL EQUIPMENT

Special logo or emblem panels can be constructed, again using the eight nylon strays as a basic unit in the banner. An appropriately sized heavy nylon panel (similar to a large piece of sail cloth) can be sewn into the horizontally oriented structure of the supporting straps. Any figure, symbol, or logo can be painted directly onto the cloth surface for a truly custom job. The distance at which this will be viewed must be kept in mind at all times when deciding about the inclusion of certain detail in the artwork. The recommended paint for this work is available under the Sherwin-Williams label in the "Bulletin Enamel" series. This paint will withstand the stress of folding or rolling with the banner fabric, as well as the mechanical duress of flapping the breeze during

Fig. 4-7. Basic eight strap configuration. This forms the matrix from which all banner characters are constructed. This is also used to provide word spacing in the banner copy.

46

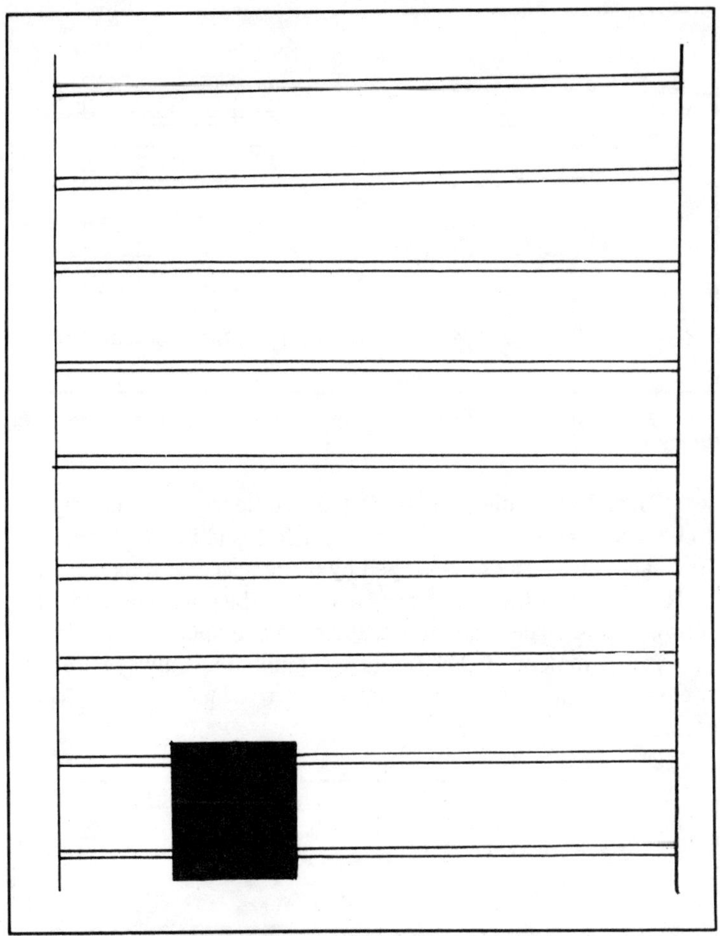

Fig. 4-8. Punctuation character (period).

flight. Figures 4-11 and 4-12 show such a panel and the logo original it represents.

Oddball characters such as fractions are made at the factory using a nylon netting as the background; the symbols are sewn into the fabric. This entire unit is affixed to the eight strap matrix appropriate to the banner size.

A variety of colors is available for the banner material. While it is best to stick with a single color for routine banner stock, it is nice to be able to offer such special effects as a green banner for St. Patrick's Day festivities and otherwise cater to special advertising needs.

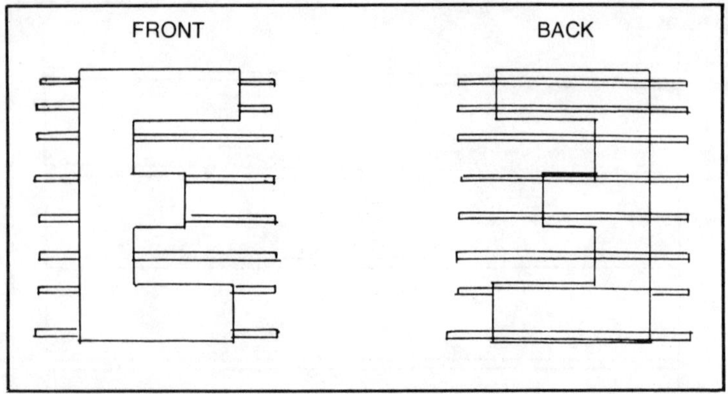

Fig. 4-9. The letter "E." Note that the straps are sewn into the back side of the character.

The adaptor pole (Fig. 4-13) allows different sized equipment to be flown on the same banner, i.e. 7 foot and 5 foot characters. The same effect can be managed by using spacer straps from a 7′ rod to a 5′ rod as shown. If a logo panel is to be placed immediately following the larger equipment, the sturdy support of the adaptor pole is recommended. The larger equipment must always precede the smaller portion of the banner for reasons of aerodynamic stability.

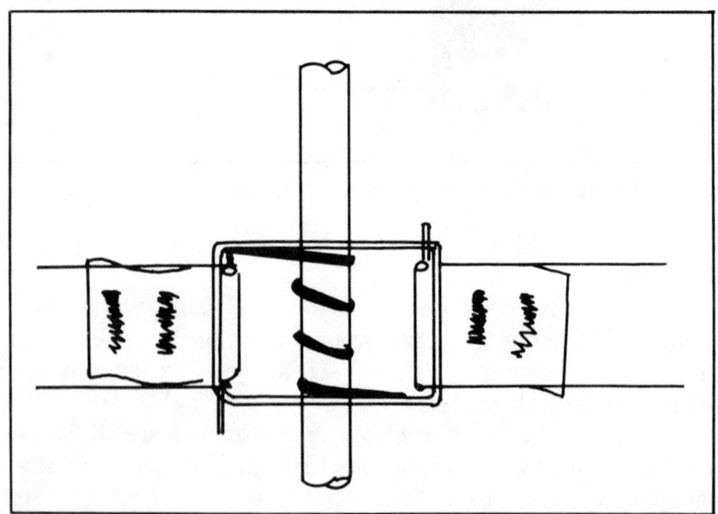

Fig. 4-10. Connector strap/QD interface. Note that this configuration becomes tighter as more force is applied.

Fig. 4-11. Custom logo panel for Gemcraft Homes, Inc.

Fig. 4-12. Original stationery emblem from which Gemcraft logo was drawn and painted.

Fig. 4-13. Adapter pole (7'/5').

THE TAIL ASSEMBLY

The final member of the assembled banner is the tail assembly (Fig. 4-14). This, in concert with the lead pole (mast assembly) helps to maintain the upright presentation of the banner by virtue of its aerodynamic characteristics. It further aids in keeping the banner in a more nearly horizontal (straight) attitude during flight by providing an upward vector from the aft end of a structure which otherwise has no lifting elements among its component parts. Without the tailpiece, the banner would be unstable posteriorly and the whipping effect of the aftmost letters would shred the sturdiest of materials in short order. The drag chute effect of the tail assembly further tends to stabilize the banner by reducing the waving action that would prevail in its absence. The tail assembly consists of a vertical drogue (drag chute) connected to an old-style rod with a set of spacer straps. The rod is threaded through the strap loops so that QD clips are not used in this particular component. The older drogue chute design is somewhat like a nylon pocket that catches the airstream. More recently manufactured equipment employs a design modification that permits airflow through the back of the chute, a feature that increases stability and reduces drag.

It should be clear that the drogue does have "top" and "bottom" sides. The consequences of installing this piece upside down might lead to a twisted banner, ruined letters, broken rods, and a red face. To avoid confusion, the "top" is so labeled.

BANNER CONSTRUCTION

The orderly assembly of a banner from the component parts is very much dependent upon the degree of organization present at the onset. For the purposes of this section, the banner assembly will be presented as if the components were taken right out of the box from the manufacturer.

Ideally, a waist-high table of sufficient length and width to accommodate the size of the equipment (i.e. 5′ or 7′) would probably produce the most comfortable working environment. Any flat surface, however, will do—including the floor. A hard surfaced floor can be devastating to the knees unless some protective cushion or knee pad is readily available and used. A well carpeted floor in front of the television has been the author's favorite spot despite the need for continued fielding of harassment from the lady of the house.

A sample of the banner copy should be written out and double checked for spelling, punctuation, and desired spacing. A slash mark between each letter helps to identify the position of each rod. For example:

D/E/R/R/Y/ /A/I/R/ /B/A/N/N/E/R/S/ /7/2/1/-/6/2/9/0/

A tally of the characters (character = anything between two rods on a banner) can be ascertained by summing the letters,

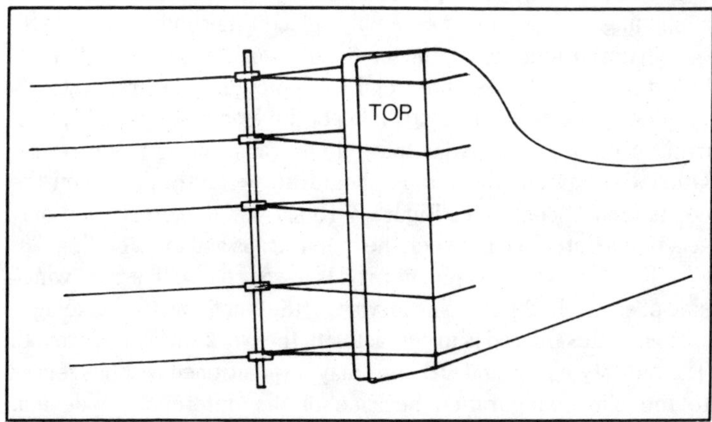

Fig. 4-14. Tail assembly. The "up" end is labeled "TOP."

numerals, spaces, and punctuation marks. This also provides a shopping list of characters to get from inventory.

A //	F	K	P	U	Z
B /	G	L	Q	V	SPACER ///
C	H	M	R ////	W	
D /	I /	N //	S /	X	
E //	J	O	T	Y /	

0 /	5	(.)
1 /	6 /	(,)
2 //	7 /	(-) /
3	8	(&)
4	9 /	($)

Totals		
	Letters	15
	Numbers	7
	Punctuation	1
	Spaces	3
		26

As a quick check, the connector rods pencilled in should be equal in number to the total for the characters, in this case, 26. (Note: the connector rod attached to the heavy webbing of the lead pole is not counted for this purpose. It does not uncouple quickly or easily from the mast assembly and therefore usually stays with it as a component part.)

With the assumption that all necessary rods and characters are available at arm's reach, along with the lead pole and tail assembly, you are now ready to begin the first steps of banner construction. Beginning with the mast assembly, place this rigid member with the vertical orientation to the left of your work area, as in Fig. 4-15, with the heavy or weighted end as depicted. The first letter is coupled to the respective QD clips of the lead pole assembly rod, using caution not to twist the nylon straps during the process. Otherwise stated, the straps should lie perfectly flat when the connection is complete. Figure 4-10 shows the correct and only acceptable interface between the nylon straps and the QD clips.

If an "old-style" connector rod is used, ensure that the "wing" side of each QD clip is facing forwards (the front) and the "fuselage" that encircles the rod is in contact with the work surface or ground. The new style rods and QD clips may be positioned without regard to the clip configuration because of the difference in design. However, it is recommended that you establish a routine for these

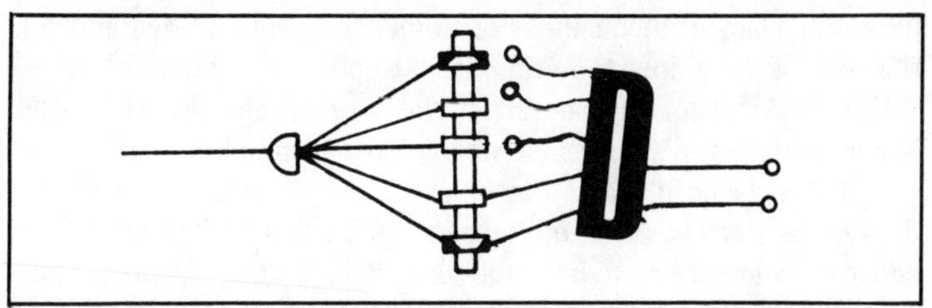

Fig. 4-15. The first letter to be assembled is attached to the connector rod that is part of the mast assembly unit. Be certain that the weighted end of the lead pole is at the bottom.

connections; for example, the trailing edge clip free prong always pointed "up".

A connector rod is now attached to the trailing edge of the "D" in a similar fashion, again using care to ensure that the straps aren't twisted. The subsequent letter ("E") can now be appropriated and connected as before. Note that the straps are sewn into the back of the letter. Unless you are very familiar with the equipment, you may be tempted to attach a symmetrical letter (e.g. "I", "M", "O", "H", etc.) with the straps facing the front of the banner. This would not affect the reading of the copy, but might interfere with your idea of cosmetic perfection.

The remainder of the banner is completed in this same monotonous fashion. To avoid the necessity for a workspace that is several hundred feet long, the banner can be folded accordian style as it is assembled. Or, it can simply be rolled up around the lead pole as Fig. 4-16 demonstrates.

The final step in banner assembly is heralded by the placement of the last letter in the copy, to which the tail assembly will be attached. With assurances that the labeled "top" is correctly oriented, the derriere of this aerial sign is thus complete. (Although the word "derriére" has commonly been used in an anatomical sense, it is more accurately translated to mean "the

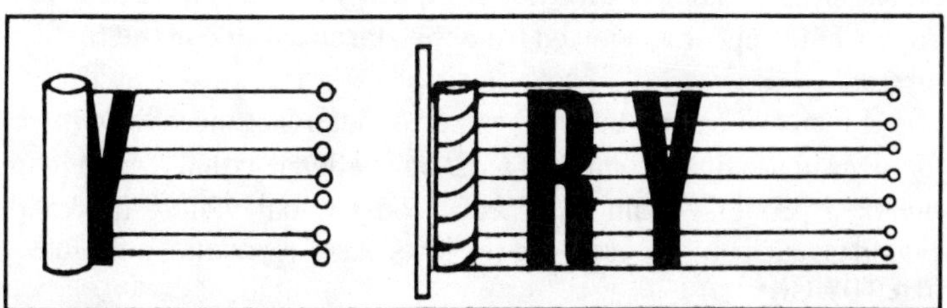

Fig. 4-16. As banner assembly progresses, the sign can either be rolled up or folded like an accordian.

back end". Since the business of banner towing is conducted from the rear of the plane, the temptation to claim the modified name, "DERRY AIR" for my operation was irresistible. For what it's worth, people really remember the company name.)

There is hardly any way around the tweedle-dee/tweedle-dum of banner construction because of the fact that there are 16 separate connections to be made for *each* character. As previously hinted, the amount of organization present from the onset will determine just how much of a dent a 26 character banner can take out of an afternoon. One way to cut down on assembly time is to have the letters stored on rods, with the rod attached to the trailing edges of the inventory characters. This essentially reduces by half the time required to connect successive units of the banner copy. The next obvious time-saver would be to have the inventory so arranged so that you don't have to hunt through a box full of nylon to find your next letter or symbol, whatever it may be. Some additional suggestions are to be found in Chapter 10.

Although "front-to-back" banner assembly is considered the norm, there are circumstances that would make "back-to-front" construction advantageous. Towing along the beach is a good example. With conventional assembly, the message reads from the left side of the tow plane which essentially limits exposure of the good beachfront audience to one pass. The process can be made more efficient by altering the sign construction in several ways.

One method is to construct a bi-directional sign as in Fig. 4-17 (A). This looks a little weird but it gets the job done because now the sign can be read from *either* side of the tow plane. The obvious disadvantage is that it doubles the length of the banner, a fact that may trespass on the towing limits of the bird.

A more aesthetically pleasing method comes in handy if suitable airports are situated at opposite ends of the beach. By alternating pickup sites and banner assembly, multiple clients can be satisfied in a single afternoon. Figure 4-17 (B) shows how the copy would appear as viewed from the starboard side of the towing aircraft.

Finally, it cannot be emphasized enough that there is *no room for error* in the final product. It *must* be spelled correctly, each loop *must* be properly connected, each rod without visible flaw and bound on each end according to the laws of aerodynamics and God.

THE TOWLINE

Prior to June, 1979, the specifications for the towline were listed in AC 43.13-2A, Chapter 8 (FAA) (*Glider and Banner*

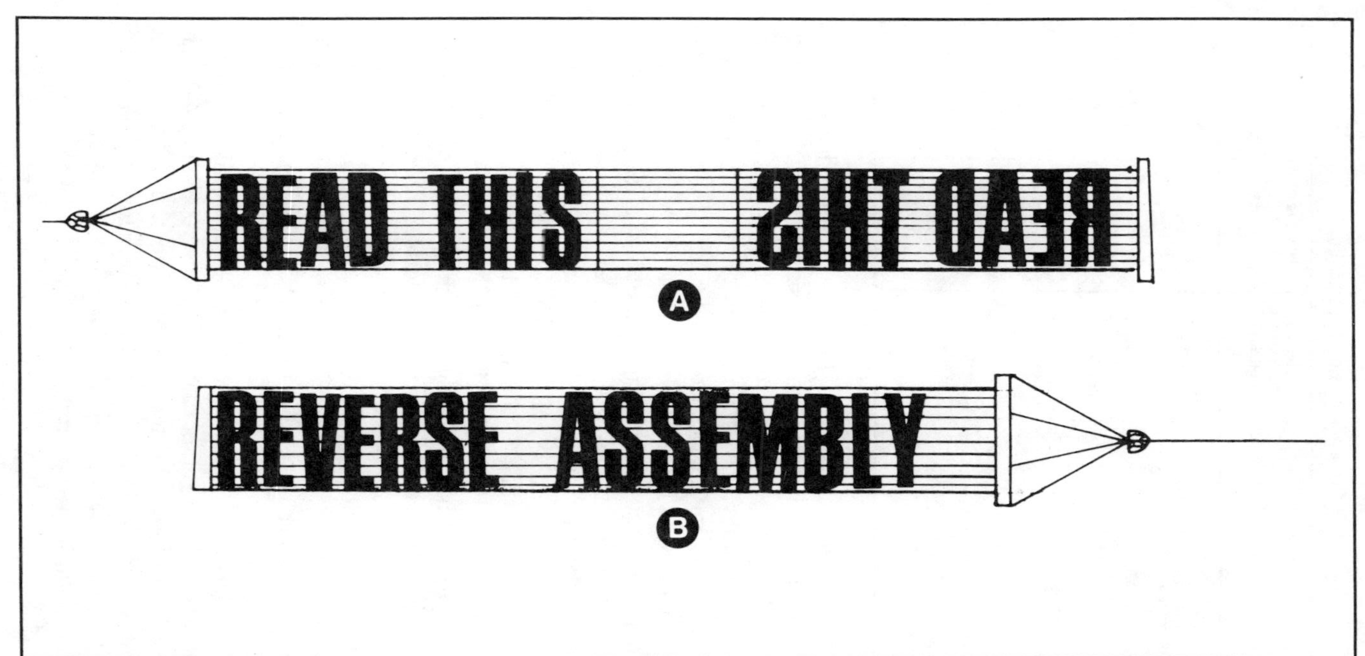

Fig. 4-17. A. A bidirectional sign can be used for short messages. The copy is legible from either side of the tow plane. B. Back-to-front assembly can be useful for a unidirectional tow where the viewing aucience is expected to be to the right of the plane, e.g., at the beach.

Table 4-1. Towline Specifications.

SIZE AND FIBER	BREAKING STRENGTH IN LBS.	WORKING % OF BREAKING STRENGTH	ELASTICITY
¼-inch Polypropelene (Hollow Braid)	1200	17%	Moderate
5/16-inch Polypropelene (Hollow Braid)	1650	17%	Moderate
⅜-inch Polypropelene (Hollow Braid)	2200	17%	Moderate
¼-inch Polyethelene	100	17%	Moderate
5/16-inch Polyethelene	1400	17%	Moderate
⅜-inch Polyethelene	1800	17%	Moderate
¼-inch Nylon (Twist)	1485	11%	High
5/16-inch Nylon (Twist)	2300	11%	High
⅜-inch Nylon (Twist)	3350	11%	High
¼-inch Nylon (Solid Braid)	1250	11%	High
5/16-inch Nylon (Solid Braid)	2500	11%	High
⅜-inch Nylon (Solid Braid)	3100	11%	High
¼-inch Dacron (Solid Braid)	1125	11%	Low
5/16-inch Dacron (Solid Braid)	2250	11%	Low
⅜-inch Dacron (Solid Braid)	2800	11%	Low
¼-inch Manila	600	5%	Low
5/16-inch Manila	1000	5%	Low
⅜-inch Manila	1350	5%	Low

Towhitch Installation). Due to the fact that the discussion of towlines wasn't exactly germane to the subject heading ("Aircraft Alterations") of this advisory circular, the entire section was deleted. Fortunately, the data has been retrieved from cracking, yellowed pages of the now outdated publication. The information is hereby revitalized (Table 4-1) with considerably less threat of being dethroned by a blank page. The data is reproduced for comparative purposes *only* and the reader is officially urged to consider his choice of towline material to be limited as follows:

1. ¼" Solid Braid Nylon
2. ¼" Solid Braid Nylon
3. ¼" Solid Braid Nylon

Elasticity and useful life score high among the reasons to choose nylon over other fibers. The former characteristic is important in attenuating the instantaneous loads experienced during the pickup that would otherwise be passed on to the aircraft. The latter just makes good sense. Solid braid lunges ahead of its twisted counterpart by a neck at the finish line by virtue of its ease in handling.

The rope connection between the banner and the towing airplane is referred to as the "towline" or "tow rope" (Fig. 4-18). In its most basic form, the tow line has two parts. The trailing segment is simply a straight rope length with a tow ring secured to one end. The leading segment, the *pickup loop*, passes through the tow ring and ultimately forms a geometric plane when draped on the uprights.

A 350′ length is available from the manufacturer for use with the takeoff method of launch. Gasser also supplies a 250′ tow rope as a stock item for the more popular aerial pickup method. There is no upper limit for the rope length (the matter of practicality excepted here.) However, there *is* a technical problem in photographing the banner in trail with the aircraft using tow lines of this length. It is permissible to shorten the cord in increments until the desired length is established, provided: 1). the banner should not be so close to the tow plane so as to have its flight characteristics influenced by the propwash or other aft airflow disturbances created by the aircraft, and, 2). there should be ample "zoom room"—the distance between the pickup assembly and the bridle harness. This is to allow adequate exchange of airspeed for altitude between pickup loop engagement and banner launch. This length varies accordingly with the type of towing aircraft, banner length and size, etc. Anything less than 100 feet is probably pushing the limits for most planes.

THE GRAPPLE HOOK AND CABLE ASSEMBLY

At first glance, this rather ominous appearing component (Fig. 4-19) seems better suited to dredging through the muck in

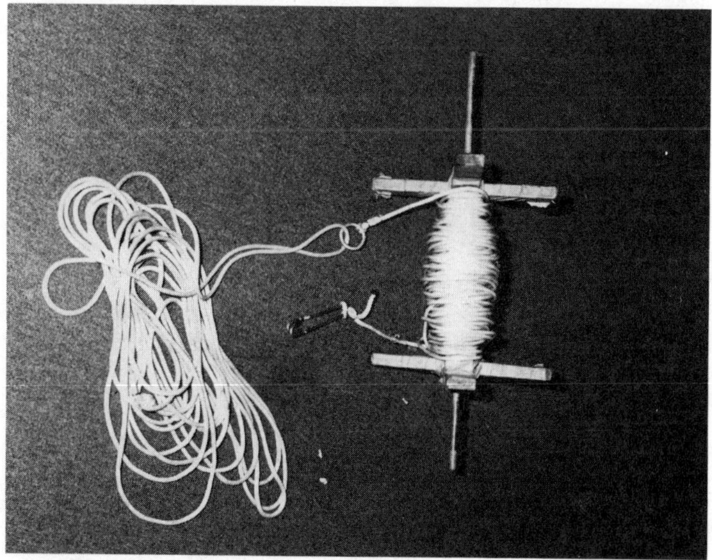

Fig. 4-18. The towline or tow rope consists of the pickup loop connected to the main towline through an "O" ring. The leading end of the towline has an adapter clip that connects to the bridle harness of the mast assembly.

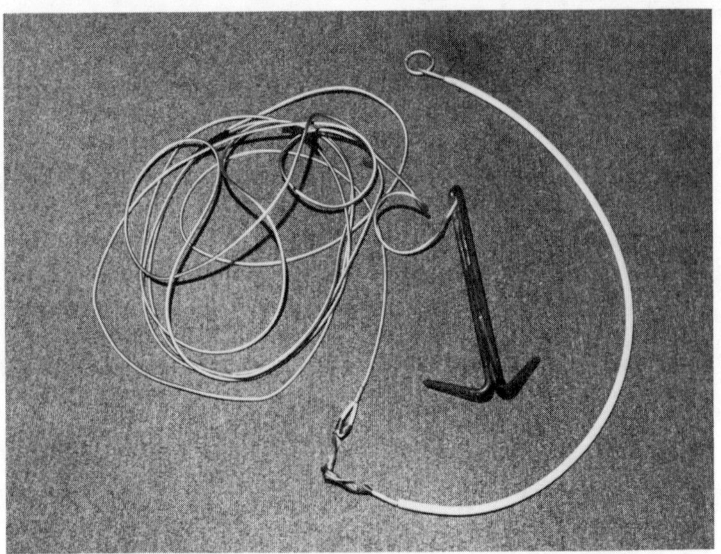

Fig. 4-19. The grapple assembly consists of the welded steel tow ring, a short forward section of cable, usually covered by a synthetic tubing, the safety link, a long section of vinyl-coated wire "tiller" cable, and the grapple hook.

lake bottoms in search of drowning victims than as a tool of commercial aviation. The grapple hook swings like a pendulum from a 30 foot vinyl-coated 3/16″ wire cable. The forward end of the cable is affixed to a steel tow ring that makes the connection with the aircraft tow hitch. The three words *least* used to describe this menacing device are *soft, subtle*, and *sexy*.

The cable itself is interrupted by a safety link near the leading end. The purpose of the safety link is to incorporate a "weak" link in the system. Its purpose is twofold:

1. To provide protection for the banner equipment should it be inadvertently flown through an unyielding object, such as a fence.

2. To provide protection for the towplane and pilot in the event of entanglement, as described in (1).

The safety link is actually a section of the same nylon material used in the straps of the letter characters. The term "weak" is indeed a relative one because this little booger has a minimum break strength specification of 500 pounds. There is a tendency to inspect this tiny connection with some degree of trepidation under the realization that this "two-bit" strap (literally) is the only thing between you and a potential multi-million dollar law suit. Any urge to eliminate this safety link should be *resisted unconditionally* (Fig.

4-20). Towing without the safety link is self-destructive behavior that requires follow-up care by a psychiatrist. It's also a violation of FAA standards.

While there is no regulatory mandate to change the safety link at specified intervals, it is recommended that this component be frequently changed. There is no practical way to assess wear and tear on the nylon strap. Indeed, the link may become attenuated without visible evidence of damage. For the minuscule price of replacement links, it is well worth the pilot's peace of mind to change the safety link for *each tow*. As a reference point, however, a single unscuffed link should be okay for several dozen tows, provided it is not subjected to stresses out of the ordinary. Replacement directions are shown in Fig. 4-21.

The grapple hook and cable must be stowed securely and properly to prevent inadvertent release and/or fouling during deployment. The placement of the "O" ring is a very critical step. Figure 4-22 portrays the potential problems to be avoided during

Fig. 4-20. Just one possible unpleasant consequence of not using the safety link.

this procedure (reproduced with permission from Gasser Banners, Inc. *Instruction Booklet for Gasser Banner Equipment.*)

The first three feet of the grapple cable runs through a plastic semi-rigid tube that prevents kinking in the cable or snagging on various posterior goodies like the rudder horn and tailwheel. The latter two possibilities should be carefully considered if applicable to your aircraft and all reasonable precautions taken to prevent them from happening.

The aft section of the grapple hook cable is then secured by whatever means available in the cockpit or cabin. Masking tape has the appeal of being easy to use and cheap. It is probably necessary to tape the cable in several places in order to hold the wire taut against the fuselage during takeoff. Another satisfactory method is depicted in Fig. 4-23, using Velcro as the adhesive factor. As a side benefit, the use of Velcro serves as a means of monitoring the length of the cable from the cabin back to the tailhitch hook. Any covert slippage or snagging of the line would be easily noticed.

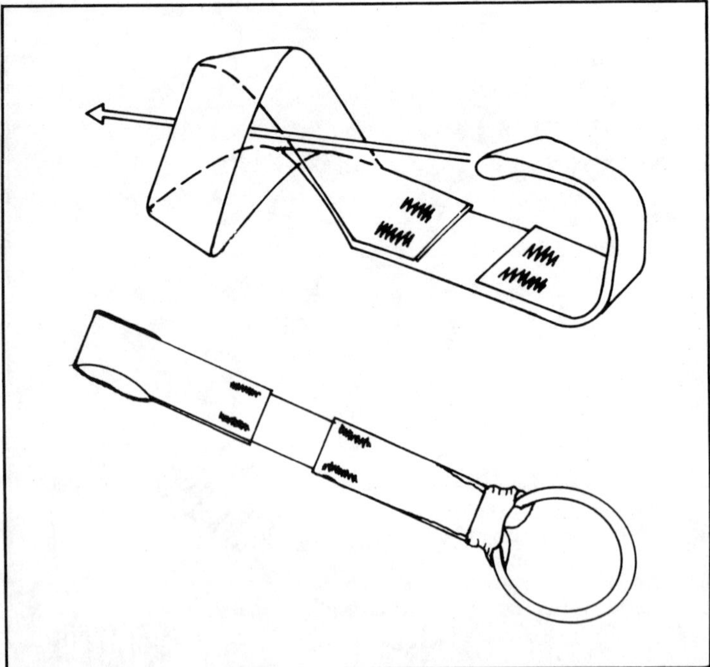

Fig. 4-21. To replace the safety link, remove it from the short (forward) section of cable first; then, uncouple the link from the long (aft) section. Reverse this order to connect the new safety link.

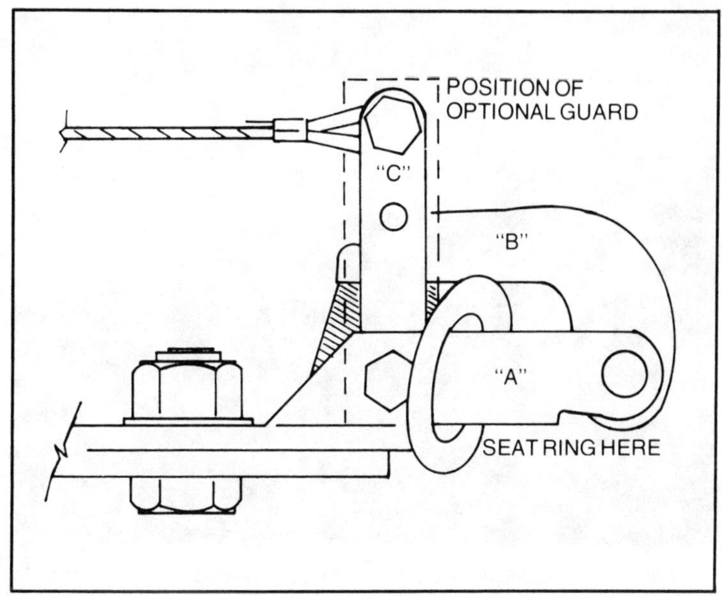

Fig. 4-22. The Schweizer tow release mechanism shown above is common to virtually every glider/banner tow hitch in use in the U.S. today. It is a simple and reliable device, but there is one precaution which must be taken in its use by those who employ the aerial pickup method of banner launching.

Whether the grapple hook is dropped from the cockpit or released from a point elsewhere on the plane, the hook's cable will have to be pulled forward and kept under tension until the hook is dropped. If the cable ring is placed around the "pelican hook" part of the mechanism (marked B in the diagram), the forward tension could conceivably nudge the latch arm, C, forward. If C is pushed far enough forward, the release will open, dropping the hook and cable on whatever is beneath the plane. Even if the arm is moved only slightly forward, this may result in the release opening under the sudden load of the pickup. Neither case is very likely, but there is no need to take any chance.

The hazard can be eliminated simply by being sure to place the ring on the A arm of the mechanism, as shown (for the sake of clarity, the grapple hook cable is not shown in the diagram, only the ring). If the release mechanism is mounted in an inverted position, often the case with nose-gear planes, enough slack in the cable could allow the ring to slip off A and onto B. If you think this is likely in your case, a little masking tape can be used to temporarily hold the ring onto arm A. The pickup and tow load will break the tape and allow the ring to ride in the correct tow position.

The problem can be eliminated permanently by adding a guard to the tow hitch which would prevent the release from being opened accidentally. The simplest form of such a guard would be a piece of steel strap, about ⅛" thick by ¾" wide, bent into a U-shape. The ends of the U may be welded to the main tow hitch bracket so that an arch is formed over the latch arm C, leaving enough clearance for it to move freely inside the guard. If the position of the tow hitch on the fuselage is such that it might be struck in the event of a tail-low landing, the guard should be a rigid, box-like structure which would protect the entire release mechanism against damage (courtesy Gasser Banners, Inc.).

Fig. 4-23. A narrow strip of Velcro can be wrapped around the grapple cable to hold it securely in the cockpit or cabin.

Stowage of the cable may be done according to the plane's cabin or cockpit configuration as required. This may mean simply coiling the cable on the empty seat. Many tow planes won't have this luxury, and a suitable alternate system must be improvised. A large-bore cardboard cylinder core is an excellent means of containing the cable prior to deployment. The cable is folded accordian-style and inserted as shown in Fig. 4-24. The core may be taped to any suitable fixture within easy reach of the pilot during flight, or even left loose if safety permits. Cable stowage inside the cockpit or cabin is a master key to Pandora's box without careful attention to the potential hazards of fouling on clothing, seat belts, or anything that might interfere with deployment. Folding the cable as described is probably preferable to coiling as it decreases the chances for inducing unwanted knots and subsequent cable kinking.

An alternative to stowing the cable inside the pilot's compartment would be to have a socket mounted on the forward end of the tow hitch bracket. One prong of the grapple hook can be inserted into the socket receptacle and held in place by tension on the cable. The remaining loop of cable can be snugly secured within reach of the pilot. Deploying the hook is then accomplished by feeding three or four inches of slack back to the grapple so that it

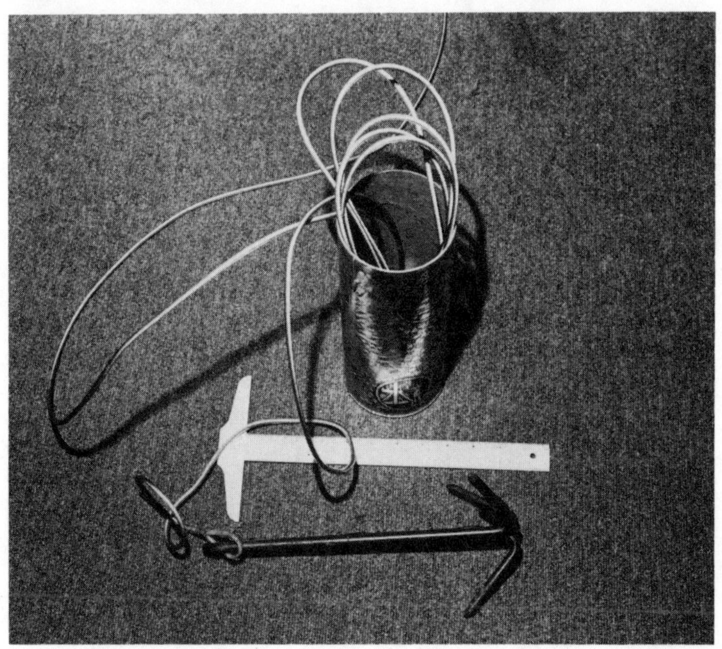

Fig. 4-24. Excess cable can be stored in the cockpit by folding it into a cardboard paper core.

Fig. 4-25. Another method of deploying the grapple utilizes a socket on the tow hitch. To release, simply feed a few inches of cable out and the grapple will fall out of the socket. The remaining cable is then thrown outward and downward to clear the aircraft.

falls out of the socket. The remaining cable can then be tossed out and away from the aircraft (Fig. 4-25).

TOW HITCH

The selection and installation of a tow hitch is serious business. It falls under the category of major repair and alteration to the airframe, thus requiring FAA Form 337 to be completed by a suitably credentialed craftsman. On the average, about six hours of elbow grease are required for the conversion if a standard tow hitch kit is used. Final inspection and approval may be required by the local Flight Standards District Office of the FAA. (Some factory installed or factory supplied tow hitch installations already have Supplemental Type Certification, thus avoiding the need for the FAA Form 337.)

Although you will have a qualified Airframe and Powerplant mechanic to complete the installation, your input concerning the location of the release handle may be helpful. The guidelines for installation are dictated by the FAA, but there may be several suitable locations for release handle placement. Remember, these directives were intended for *glider* operations. As far as the author knows, there are no known glider tow pilots that operate by throwing the glider out of the window after takeoff. So do yourself (and others) a favor by at least explaining to the installation expert that you are going to have hooks and cables *inside* the cabin and that you are going to be jacking with them during deployment. Together, you can work out an arrangement to preclude releasing the row ring before you've even deployed the damn grapple hook.

The Schweizer tow release mechanism is considered by most as the standard for tow hitches in the United States. Common to both glider and banner towing operations, this aircraft accessory line has withstood the test of time for reliability and quality. Specific stress analysis or test data is available for each type of aircraft on which installations have been approved, along with samples of previously approved FAA Forms 337 specific to aircraft types (Available from Gasser Banners, Inc., Nashville, Tennessee). Although the towhitch requirements are not as stringent for banner use as they are for glider towing, it seems unthinkable to implant an inferior quality towhitch and sacrifice the potential dual use of the aircraft. Furthermore, the guidelines for banner tow hitches listed in AC 43.13 (126 b.) provide an *inadequate* margin of safety and clearly are irreverent to the aircraft structure. A 28 character 5' banner with a 250' tow rope weighs approximately 30 pounds. If the tow hitch had only the capacity to support *twice* that load, it becomes apparent from the preceding discussions that the *towhitch* would become the weak link in the system. Keeping in mind that a cable tunnels somewhere through the fuselage to link the pilot with the release mechanism, one only need to imagine the resultant disembowelment catastrophe that would abruptly follow a relatively *minor* overload. The thought of altering the safety link tolerance to accommodate such a fiasco brings bile to the mouth.

The text of FAA AC 43.13-2A, Chapter 8 may be found in Appendix E. This is the most current issue as of the publication date of *Aerial Banner Towing*, but should not be referenced without searching for subsequent changes. Supplementary comments about paragraphs 126 and 127 are contained in the preceeding text. The text of Gasser's instruction pamphlet contains additional critical review that is paraphrased here for the reader's convenience and annotated by reference to paragraph number.

129. The installation in Figure 8.6 (43.13) [Editor's note: Appendix Fig. E-5 in this book] *lacks any mechanism to prevent swivelling as specificed in item (d)/. This could result in inadvertent release during turns.*

132. Although approved by the FAA, some factory installed tow hitch releases do not have the leverage-type release control specified in item a.5. The practice of using alternate release mechanisms is to be discouraged, in spite of official blessings.

GROUND OPERATIONS
A. TAYLOR

Chapter 5
Ground Operations

A unique sequence of procedures must be performed in order to get airborne with a banner safely in tow. Relinquishing control of the banner to Mother Earth is likewise a procedural discipline. This chapter is introduced with an overview of towing operations without regard to detail so that the dissection of the events to follow will hopefully come into focus with more perspective. It should be underscored that *flight safety* is the overwhelming consideration in this, and in any other flying endeavor for that matter. Although every precaution has been taken to avoid presenting material that may prove to be unsafe when applied to the particular aircraft or flying environment in your situation, the reader is expected to make appropriate adjustments and/or revisions to compensate and render his operation *completely safe* in all respects.

Figures 5-1 through 5-6 provide a pictorial essay of the salient events for an aerial pickup and release.

Figure 5-1: The banner is laid out on the ground surface with the towline trailing to the pickup assembly apparatus.

Figure 5-2: The pickup assembly is shown in close-up. The looped segment of the towline is draped over the uprights in a perpendicular plane to the direction of flight.

Figure 5-3: The tailhook is released after takeoff and trails behind the aircraft.

Figure 5-4: The tailhook is guided precisely between the uprights to snag the loop.

Fig. 5-1. The banner is spread out on the ground at an angle to the flight path of the tow plane. The towline trails to the pickup assembly (aerial pickup method) or directly to the towing aircraft (takeoff launch method).

Fig. 5-2. The leading segment of the towline is an endless loop of nylon rope that is draped over and supported by the vertical uprights. This forms a geometric plane into which the grapple hook will be guided (aerial pickup method only).

Fig. 5-3. After takeoff, the grapple assembly is manually deployed from the cockpit or cabin. Keeping good tension on the forward section of cable (right hand), toss the cable and hook out and down after making sure that nothing is fouled within the cabin area. The grapple assembly trails beneath and behind the towing aircraft.

Fig. 5-4. The grapple hook is guided into the area defined by the circumference of the pickup loop (aerial pickup method only).

Figure 5-5: The towing aircraft then "zooms" to exchange airspeed for altitude as the slack in the towline is taken up. The banner then peels off the ground in trail.

Figure 5-6: The aircraft and banner is maneuvered over a safe drop zone. The pilot releases the tailhitch mechanism and the banner plummets to the ground.

Each of these events will be examined in some detail throughout this chapter and the next. For the sake of clarity, the presentation will more or less follow a chronological sequence, keeping in mind that the ground operations have ample room for modification of the procedural order.

AERIAL PICKUP METHOD-PICKUP AREA

A suitable surface must be available for the aerial pickup method of launching a banner, as previously introduced in Chapter 2. A concrete or other hardtop surface is perhaps ideal for this purpose, assuming that it is completely flat. The area should be carefully inspected to assure no ledges are present, for example at the expansion seams, that might interfere with the sliding motion of the banner as it lifts off the ground. A reasonably groomed grassy surface is well suited for a pickup area and, in general, in greater supply than the artificial option. High weeds and other rough

Fig. 5-5. The towplane exchanges airspeed for altitude until the slack is taken up in the towline. At this point, the banner will peel off the ground to fall into trail behind the aircraft.

Fig. 5-6. At the end of the mission, the tow pilot manuevers for position over a safe drop zone and discards the banner by pulling the tow hitch release handle. The pilot will experience instantaneous acceleration at the moment of separation. The banner falls safely to the ground for later retrieval.

foliage can poke holes in the banner equipment as it is forcibly yanked over the surface for a short distance during launch. They are, therefore, a relative contraindication, and should be mowed off or another site sought.

An *adequate* clearway for both approach and departure is a commonsense must. The italics are used to emphasize the need for individual assessment of the local environment to include terrain, airspace, and aircraft limitations. As a guide, the approach corridor should be clear of any significant obstacles for at least ¼ mile with a ½ mile allowance on the departure course. You should have a prudent lateral clearance, too: say, 500 feet to either side of your flight path in case maneuvering is necessary. While it would clearly be a departure from routine to begin a lazy 8 while climbing out with a banner in tow, you always want to build some fat into your airspace, whenever possible, in case an evasive maneuver is necessary to avoid another aircraft. Obviously, this should never, ever, *ever* happen or become even close to being necessary. However, it highlights the often overlooked fact that even pilots make mistakes, not the least of which involve errors in clearing. Imagine, if you please, some hamburger with 50 hours total time (or 10,050 hours total time, for that matter) spiraling down over your pickup area, eyes glued to the airspeed indicator, one sweaty fist clutching the controls to overcome his multi-G turn, the other hand notching his belt in anticipation of the evening as he demonstrates to his girlfriend how aces enter the pattern from the inside. You missed him completely while clearing because of a high wing, relative altitude separation, or whatever—or you saw him and assumed he was going to fly a more orthodox pattern and proceeded with your pickup in good conscience. Now, after a storybook launch, your eyes return forward with dismay to find your windscreen filled with an alarming array of Dzus fasteners from our hero's machine as he struggles to make the best of the situation, now without regard for his belt. An uncomfortable situation, at best, for the tow pilot, and one further aggravated by the lack of lateral clearance if you hadn't planned ahead.

Relative isolation of the area is another concept swiftly learned by the tow pilot. Airports are saturated with all kinds of curious buggers, all with the common denominator of a flair for the unusual. This is not just limited to the minor aggravation of having to explain for the N-thousandth time what the letters are made of, how it will be picked up, how much does it cost for a banner, what do you do when you're ready to land, have you ever dropped this thing on somebody's stupid skull, not to mention fielding questions on general aviation potpourri such as, how fast can you go, how did you get into this business, and the one that always starts out "my brother wants to take flying lessons . . ." There is a matter of safety that is a real concern. How unnerving to be on final approach for a pickup only to find some local urchins playing skip-rope with your towline. No—better to set up in some semi-remote real estate location to discourage interference by those best left in the grandstands.

As a final consideration for the departure course, it is well to have an area suitable for a drop zone along the immediate climbout corridor. This is in case of some complication such as a twisted banner or a broken connector rod. Admittedly, these are infrequent problems but the concept of building fat into the system still applies.

BANNER LAYOUT

Unroll or unfold the banner, depending upon your storage technique. (If the banner is folded accordian-style, it may require the assistance of another person, especially in windy conditions. It also helps to hold the folded portion of the banner very low to the

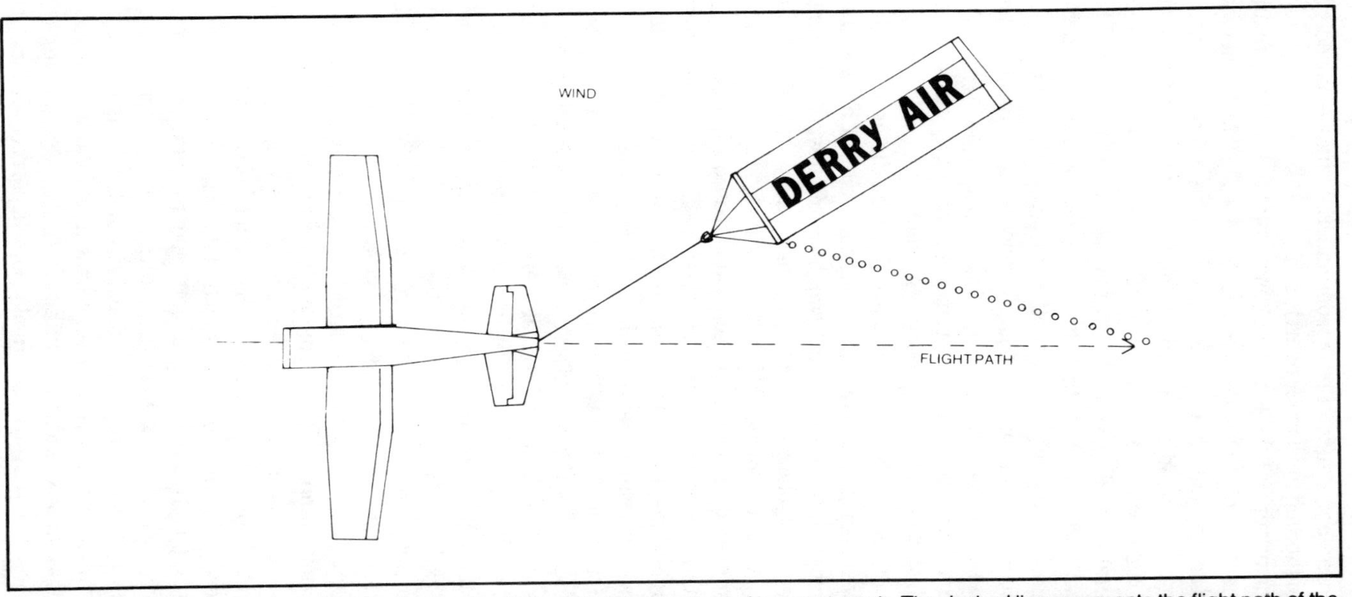

Fig. 5-7. Overhead schematic of the ground set up for the aerial pickup method of banner launch. The dashed line represents the flight path of the tow plane as it traverses the pickup environment. The dotted line shows the flight path of the mast assembly as the banner assumes the trail position behind the towing aircraft. Note that the heavy (longer) end of the lead pole points towards the flight path. The banner could also be spread out face down to the right of the flight path if needed to avoid unfavorable conditions (e.g. high weeds, water, terrain irregularities, etc.) on the left. Hold this illustration up to a mirror to visualize the correct "face down" configuration.

ground to prevent a sudden gust from undermining the unreeled portion and flipping the banner over.)

The heavy end of the lead pole must point towards the flight path of the tow plane as depicted in Fig. 5-7. The banner is positioned at an angle not to exceed 45° to the flight path and, ideally, into the wind as shown in the diagram. For practical purposes, light winds, say less than 10 knots, will not affect the pickup operation at all. Even a downwind pickup is not unthinkable if you are willing to concede the additional strain on the equipment with additional allowances for takeoff clearway, etc. As the winds get over 10 knots, plus or minus gusty conditions, the banner can loft up and even flip over during crosswind operations. It is best, then, to heed recommendations regarding wind direction when planning the layout of the banner on the ground.

The banner may be positioned face down on the ground if the winds and/or launch area favor this configuration. As before, the lead pole must be situated with the weighted end toward the flight path of the tow plane. For reasons previously introduced in Chapter 4, it is recommended that the "new style" rods be used for banner assembly if an inverted pickup is anticipated.

With the banner grossly spread out on the ground, it must now be thoroughly inspected from the bridle harness to the tail assembly with careful attention to detail (Fig. 5-8). Take any slack out of the banner that may be present by pulling on the tail assembly and sliding the banner across the ground surface for a few feet. (This also serves to ensure that there are no hidden twigs or other such obstructions that might interfere with the free gliding motion of the equipment during launch.) Beginning with the tail assembly, walk "forward" (towards the front of the copy) and visually inspect each loop/clip interface. Correct any and all discrepancies on the spot, including twisted straps, snagged clips, bent clips, etc. The integrity of each rod can also be evaluated with this inspection, although it is uncommon to experience problems. If the old style clips and rods are being used, the spacing needs to be checked and corrected as needed. As the lead pole assembly is approached and scrutinized, be certain that the bridle harness lines are not tangled around the lead pole and that the lines are furthermore not twisted among themselves. The banner should now be lying flat on the ground and in exactly in the configuration it will appear when airborne. It is wise, however, to take another trip to the tail assembly and back to re-check each of the items previously inspected as well as a final check on correct spelling of

Fig. 5-8. Close attention to detail on the preflight inspection is the best insurance for a safe aerial banner towing operation.

the banner copy and the overall condition of the letters. This inspection is an extremely important procedural routine that is not only a matter of consideration for your equipment but significantly pertains to *flying safety*.

THE TOWLINE

The towline may be connected to the bridle harness ring by using a metal clip such as the one shown in Fig. 5-9. Simply squeeze the spring clip to open the prongs and slip the harness ring into place. It is advisable to have this disconnect clip at the attachment of the bridle for several reasons. Firstly, it is easy and quick to attach and inspect. More importantly, it is easy and quick to disconnect. In the unlikely event that you may have to someday land with, God forbid, a banner in tow (Chapter 8) this feature will save you some very harassing moments as you struggle to evacuate your plane and equipment from the active runway while everyone else circles the pattern and titters on the radio about the situation.

Other than these minor points, there is little to contraindicate tying the towline directly to the bridle harness, using an anchor bend as shown in Fig. 5-10.

The towline is then trailed to the site of the pickup assembly at a slight angle towards the flight path so that the banner will not

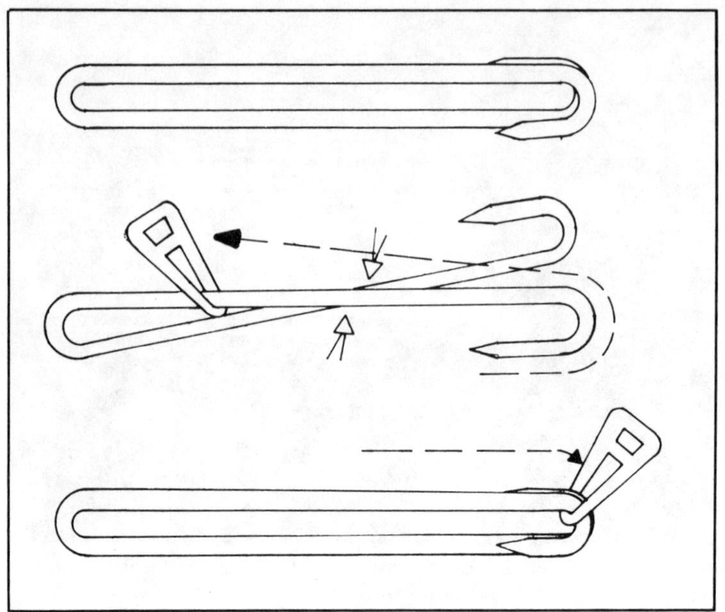

Fig. 5-9. Squeeze the towline connector clip to open the terminal prongs and slip the bridle harness ring into position.

underlie the flight path in any way. Inspection of the rope is properly accomplished as it is let out. Herniations (bulges) in the nylon ropes are a sign of stress that mandate testing prior to use, or replacement as needed.

The towline terminates with the "O" ring that connects the pickup loop with the main tow rope. With the slack taken up between the "O" ring and the banner, it is now time to set up the pickup assembly at the point where the "O" ring lies. Before doing so, be certain that the pickup loop is thoroughly inspected for the same stressful signs as the tow rope, including an evaluation of the integrity of the rings that complete the loop.

THE PICKUP ASSEMBLY

The pickup poles should be placed as widely apart as possible without causing the towline pickup loop to distort into an inverted teardrop shape. That is, the pickup loop should be roughly rectangular in shape when draped across the uprights, conforming to the imaginary plane defined by the visible portions of the poles. Figure 5-11 explains. The reason for this precaution is that if the towhook were to pass through the area between the vertical

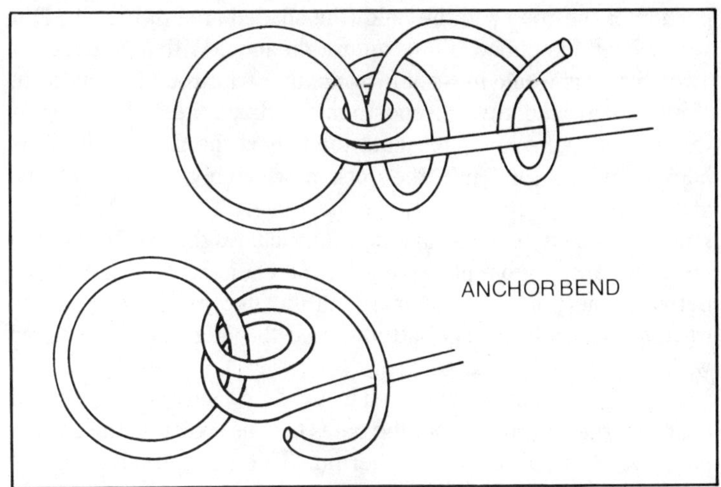

Fig. 5-10. The anchor bend is an effective knot for towing purposes. If this has been pulled tight while wet, it is extremely difficult to untie.

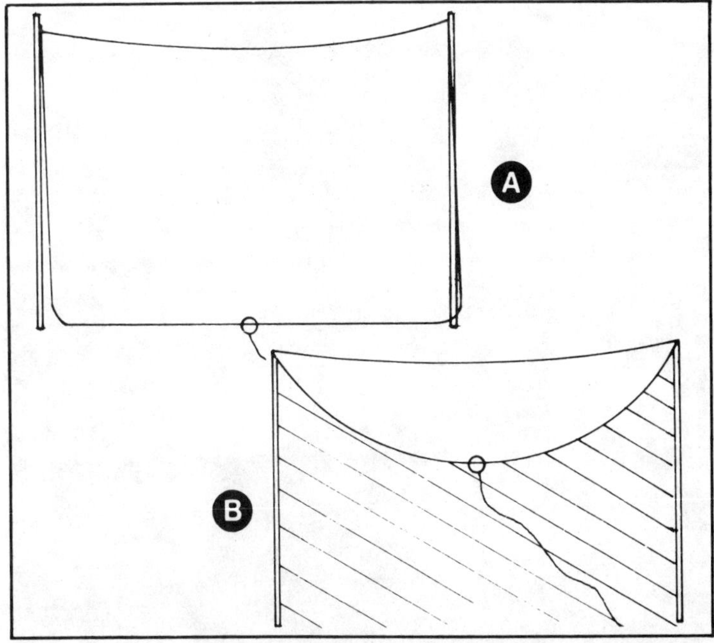

Fig. 5-11. A. Correct position of the pickup loop configuration. The pickup poles should be set apart widely enough to permit the loop to conform to the plane defined by the uprights. B. If the uprights are spaced too widely, there is an increased chance of missing the engagement should the grapple hook pass through the shaded area.

portion of the rope and the pole (the shaded area in Fig. 5-11) it would in all likelihood fail to capture the loop. With a groundcrew available, this would mean at least another trip around the pattern. Without a groundcrew to reposition and drape the pickup loop on the uprights, the tow pilot must then abort the pickup, drop the hook, land, reconfigure the pickup assembly, reconnect the grapple hook to the aircraft, and start over. This is a frustrating situation at best, and the easy solution includes the use of a ground crew and proper pole placement for the equipment. The distance between the poles, in summary, is dictated by the size of the towline pickup loop. The latter should therefore be constructed with an appropriately large girth.

The poles may be stabilized in any one of several ways. If the pickup surface is dirt or sod, the poles may be sunk into the earth if the ground is soft enough to permit. This has the advantage of

Fig. 5-12. These upright support cores, portable, sturdy, inexpensive, and easy to use, were made from ⅝" concrete reinforcing rods.

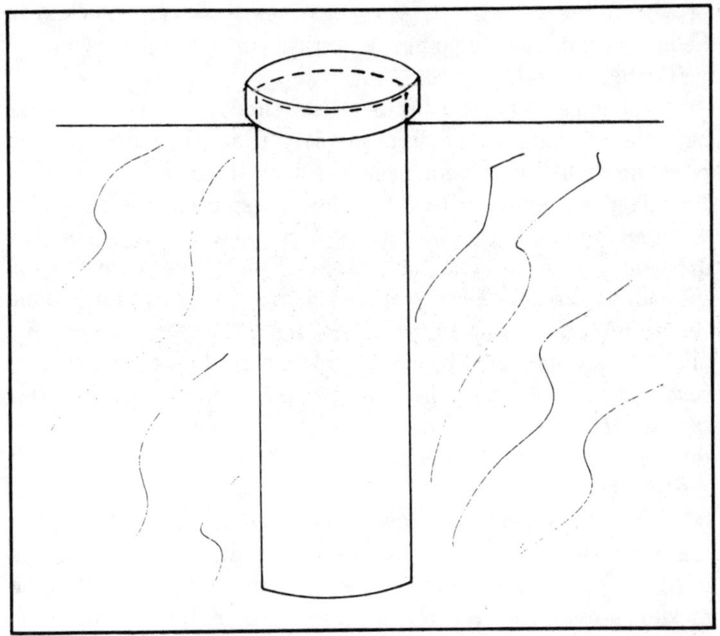

Fig. 5-13. If permanent base support receptacles are to be used, bury them flush with the ground and keep them capped when not in use.

reducing the amount of hardware needed to tote around. On the "con" side of the issue, it can be an excercise in frustration to locate the damn things from a distance, as you will have to on your pickup approach. Among the things that can be done to mitigate this problem are the use of flags and/or streamers on the uprights (also useful for wind information), ground crew communications, and the use of reliable landmarks for the placement of the pickup assembly.

Another satisfactory method for securing the uprights is depicted in Fig. 5-12. These pole support cores, for lack of a better name, were constructed from a single length of ⅝″ concrete reinforcing rod. The crossbar was welded from the same material, and serves to aid in the foot-assisted implantation of the support core into hard ground. The hollow upright fits snugly over the rod's exposed end. These transport quite easily and may be used in an emergency for grave markers, church services, and warding off vampires.

If you desire a more permanent arrangement, any suitable receptacle can be buried flush with the ground level (Fig. 5-13). These, of course, must be capped or plugged while not in use to prevent filling up with dirt, grass, or wildlife. Such receptacles can be made from plumbing pipes, electrical conduit, polyethylene or PVC pipe, etc.

The support bases (Fig. 5-14) supplied by the manufacturer have the advantage of being easily seen from a distance. Certain precautions are worth noting when utilizing them. The side view in the diagram demonstrates that the crosspieces can be folded together for ease in transporting the base supports (Fig. 5-15). The uppermost crossmember component is therefore not flush with the ground. If, perchance, this piece was situated perpendicular to the flight path of the towplane, underlying free space would potentially allow the grapple to engage the support base. *Position the support bases so that no daylight can be seen under the uppermost crossmember from the approach heading.* Otherwise stated, place the higher of the two crossmembers *parallel* to the approach path of the towplane.

The pickup loop of the towline is now ready to be draped over the horizontal support prongs on the uppermost ends of the uprights. Refer back to Fig. 5-11 for the correct appearance of the pickup assembly poised for launching a banner. It bears repeating that the pickup loop should more or less conform to the rectangular plane defined by the pickup poles. During windy conditions,

Fig. 5-14. Stock support bases have the advantage of being very visible from long distances.

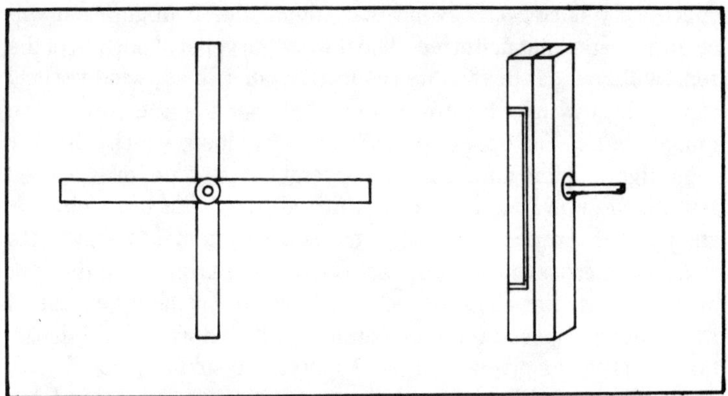

Fig. 5-15. Support bases fold for compact storage and easy transportation.

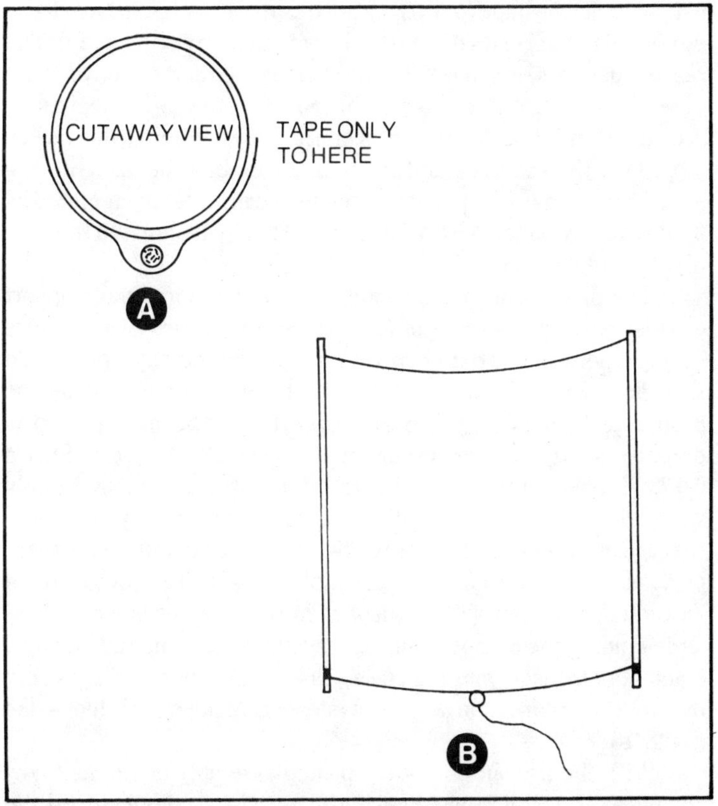

Fig. 5-16. A. If breakaway tape is used, do not wrap it all the way around the upright. Use just enough to hold the rope in place. B. The pickup loop may be lightly taped to the lower segment of each upright.

specifically during crosswind operations, this configuration will become somewhat distorted. The downwind vertical portion of the rope will bow out beyond its respective pole; the upwind vertical rope will bow inward into the zone belonging to the path of the grapple hook. The uppermost portion of the loop, which normally sags slightly in the middle from the weight of the rope, droops even lower and skews toward the downwind pole. The most obvious answer to this problem is to align the entire system into the wind to reduce the crosswind component causing the problem in the first place. This is not always practical, however, usually because of restrictions imposed on the clearance of the approach and departure courses at a given location. An alternate solution would be to use breakaway tape to stabilize the pickup loop against the uprights to minimize its distortion. Extreme caution is always in order when using breakaway tape for this purpose as it is possible to launch one or both of the uprights if the rope is fastened too securely. For this reason, use masking tape not wider than ½ inch and do not wrap the tape circumferentially around the pole. Figure 5-16 depicts an acceptable method for this procedure. After draping the pickup loop, tape the upwind side first near the base of the upright. The slack in the remaining portion of the loop can be easily depleted by pulling gently on the downwind side and finally taping it to the base of the upright as before.

The placement of the pickup loop on the horizontal support prongs must rank as one of *the* critical procedures of the entire banner operation. Ground crews must be thoroughly briefed and then brainwashed, re-brainwashed, and then braindried on the proper technique for this *very simple* task. The pickup loop is draped over the support prongs in a manner illustrated in Figure 5-17. Extreme care must be taken to preclude inadvertently wrapping the rope around either the support post (Fig. 5-17B) or the upright support itself (Fig. 5-17C-E). The potential consequences of an improper configuration relate to the towing of an unwanted pre-lead pole, suspended by tension alone. Notwithstanding the potential for lost revenue, annoying time delays, equipment damage, and a red face, this situation has obvious safety overtones of more than academic interest. Make sure it looks like Fig. 5-18.

With the preparation of the pickup assembly equipment now complete, the only remaining chore yet to be accomplished is the rigging of the grapple with its cable to the tailhook connection of the towplane. The operational details of the tow hitch hardware have already been introduced in Chapter 4 for your reference.

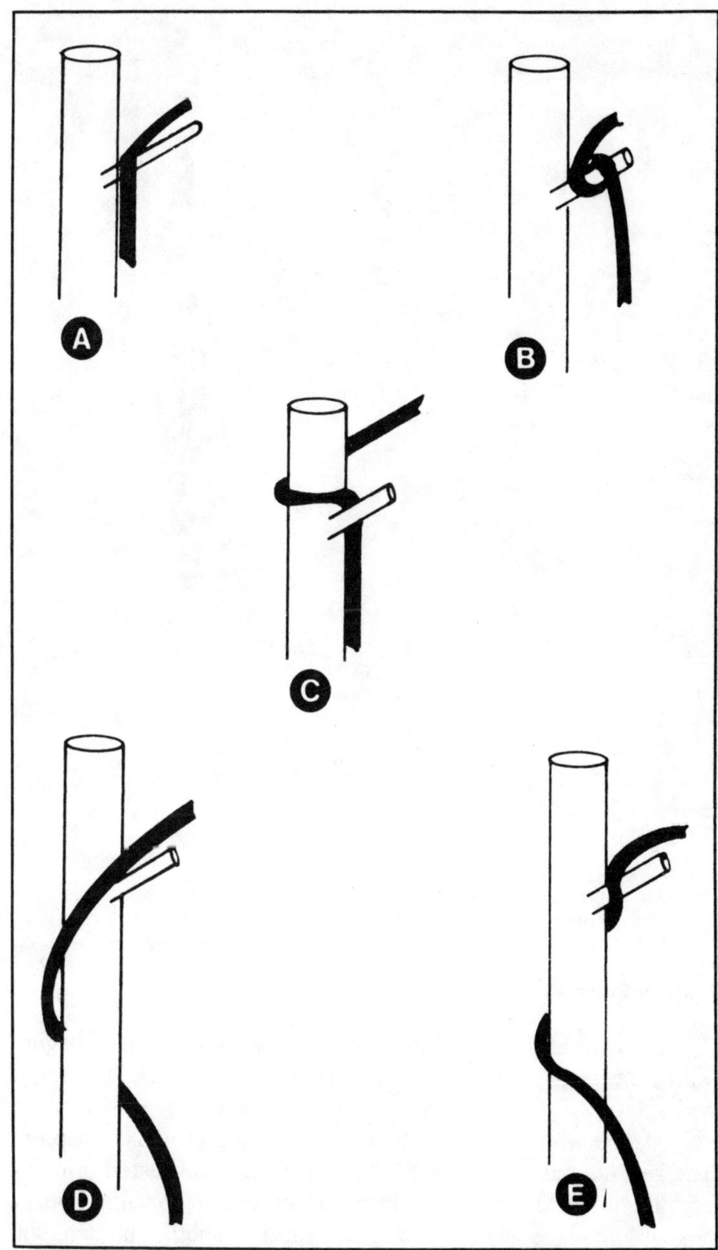

Fig. 5-17. A shows the only correct configuration of the pickup loop on an upright. If pickup loop is situated as shown in B—E, results may be damaged uprights or a launched pole.

Fig. 5-18. Pickup loop on upright will look like this, if correct.

As with all of this equipment, preparation for flight begins with a thorough inspection of the grapple hook, cables, "O" rings, and safety link. The latter should be replaced as needed or in accordance with the recommendations of the previous chapter. The release mechanism of the tow hitch must be tested prior to each flight in addition to visual scrutiny of the components. This is best accomplished with the kind assistance of another person, but can be done solo if necessary. The terminal "O" ring of the grapple cable should be routinely connected to the tailhook of the tow hitch as in Fig. 5-19. With firm pressure directed aft on the grapple

Fig. 5-19. Have an assistant pull the tow hitch release handle and check for proper release operation. This should be done with a simulation of towing pressure (30-50 pounds).

cable to simulate towing force (approximately 30-50#), an assistant should pull the release handle in the cockpit. Assuming no difficulty is encountered with the release mechanism, reconnect the cable as before and swivel the cable in a 10° cone around the

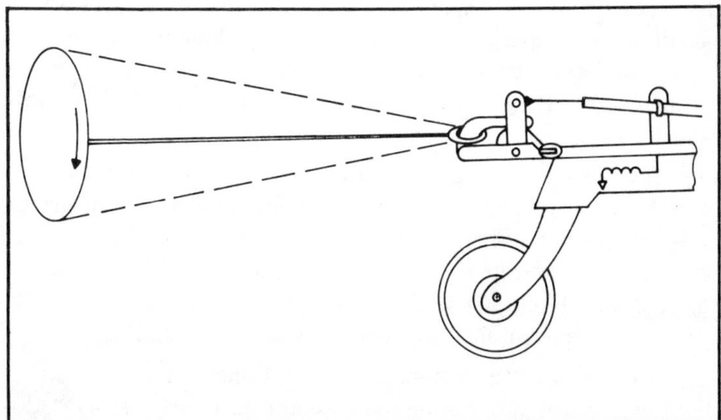

Fig. 5-20. The tow hitch should further be tested throughout its normal cone of displacement (10-20 degrees).

tailhook (still with aft pressure) to check for lateral and vertical stability of the tow hitch (Fig. 5-20).

The grapple hook and its attendant cable must be stowed for takeoff according to the method best suited to the pilot and aircraft. As this general subject has already been addressed in Chapter 4, its duplication here is unnecessary. Regardless of the method chosen for rigging the cable and hook, it is assumed that the operator will take appropriate precautions to ensure that no fouling of either component will occur during ground ops, taxi, take-off, or airborne maneuvering.

GROUND LAUNCH METHOD

Essentially a variation on a theme, the ground launch technique deserves some tailored comments. The ground layout is depicted in Fig. 5-21, virtually identical to the aerial launch schematic minus the pickup assembly paraphernalia. Previous remarks about the launch surface are applicable.

There is obviously no requirement for an approach clearway, but the departure clearway is necessarily longer for this method of banner deployment. Nonetheless, this shouldn't be a problem for a plane capable of this maneuver and *under no circumstances* should a takeoff launch be attempted if obstacle clearance is in doubt.

The banner liftoff can be assisted somewhat by propping the lead pole with a forked stick instead of leaving it flat on the ground (Fig. 5-22). This is helpful because the tow rope slides along the ground up until the point at which the slack is taken up. The resultant angle at which the tow rope engages the lead pole is very shallow in comparison with the aerial pickup technique. The resultant force may tend to torque the lead pole parallel to the ground rather than lofting it immediately. Stresses of this nature should be avoided.

The towline is connected directly to the safety link instead of the pickup loop. The short connection between the safety link and tow hitch may be adapted from the cable assembly or improvised from tow rope and appropriate metal rings. This is connected to the aircraft tow hitch just prior to takeoff. As before, the towhitch release mechanism should be tested for proper function.

Meticulous attention must be paid to the surface over which the towline will pass during the takeoff run. Leave *no* chance for entanglement with runway lights or other such obstacles that might dampen the cheeriest of spirits because of carelessness or haste.

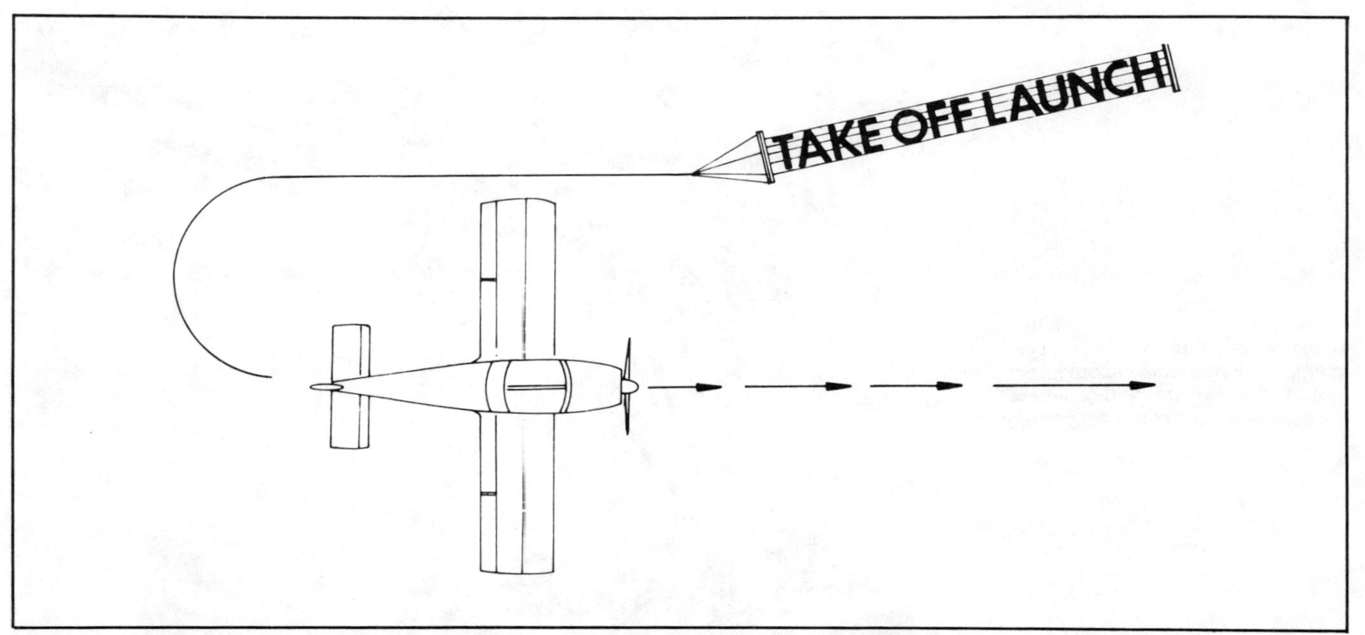

Fig. 5-21. Layout of the banner equipment for the takeoff launch is virtually identical to the aerial pickup method (except for absence of pickup assembly). It is especially important to angulate the banner to avoid the lead pole dragging over the remaining banner as it comes off the launch surface.

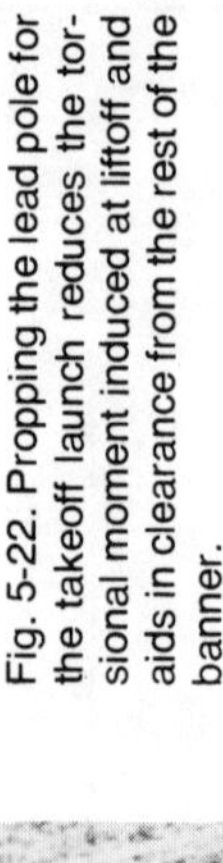

Fig. 5-22. Propping the lead pole for the takeoff launch reduces the torsional moment induced at liftoff and aids in clearance from the rest of the banner.

CHECKLIST

Regardless of the method used, it is wise to carry a checklist of operational procedures for reference on each flight. A sample of a typical checklist for both ground set up and aircraft operations has been provided in Appendix A as a guide. The ground crew member, if used, should have his own copy.

AERIAL OPERATIONS

A. TAYLOR

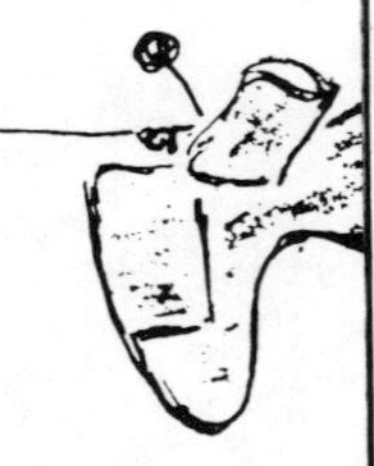

Chapter 6
Aerial Operations

The meat and potatoes of the operational aspect of banner towing is necessarily centered around the ability of the tow pilot to defile the words of John Gillespie Magee, Jr. by snagging one of those "surly bonds of earth" and actually *flying* with it. It is here that the product of all this labor gets displayed before the public. The preparation is complete, and hopefully perfect, for there is *no way* to modify it in the air. With an overview already introduced in the opening of the previous chapter, the challenge of the performance is now center stage.

AERIAL PICKUP METHOD

The emphasis in this book is strongly weighted towards the aerial pickup method of launching banners. Even though this technique is procedurally more complex, it is by far the most popular style among current operators. The majority of planes presently used for banner towing just don't have the short field takeoff capability required for the ground launch technique. Some airfields won't permit the ground launch method because it requires setting up in very close proximity to the runway. It is probably best, then, to learn the aerial pickup method as a backup technique even for those who prefer the alternative deployment.

Taxi/Takeoff

With the grapple and cable securely stowed according to whatever method is preferred, no extraordinary procedures need be accomplished for start, taxi, or takeoff. It seems that any law-abiding, safety conscious pilot would automatically pencil in some appropriate reminders at the end of each checklist sequence (i.e., Start, Before Taxi, Before Takeoff, etc.) to remind him of any peculiarities specific to the foreign configuration. Such items might include cable tension, cable stowage, grapple stowage, and towhook release handle. Repetitive entries must be lauded as a sound practice that serves to reduce the possibility of human error.

It is well to have a ground assistant eyeball the aircraft and tailhook configuration as the plane leaves the chocks. The propwash introduced at engine start could conceivably uncover or produce an unwanted aberration in the tailhook/cable assembly. This would be most easily observed from a vantage point outside the cabin.

After Takeoff - Grapple Deployment

Throwing a two pound grapple hook overboard during flight seems to violate all kinds of laws and general principles of common sense. Be that as it may, the alternatives seem less inviting, e.g. taking off with the hook already in trail scraping along the runway, etc. As experience matures your taste for the macabre, the unnatural overtones of this act will eventually wane into silent routine.

The deployment of the grapple hook has procedural variants inherent in the method of stowage (Chapter 4), so any attempt to cover all possibilities would surely prove to be an exercise in endurance. Some common elements and problems are worth mentioning.

The most basic principle for *any* method of deployment is that of tension on the grapple hook cable section that attaches to the tow hitch itself. A relaxation of tension, even for a moment, can allow the "forward" section of the tow cable to find all kinds of mischievous things to get into. (To name a few: fouling of the rudder horn, entrapment with aft flight control surfaces, wrapping around the tailwheel of conventional-type birds, tow ring displacement.) Proper tension can be held with one hand while the other is used to gather the loose cable excess and grapple hook. Throwing the hook and excess cable *first,* followed by releasing the taut end, provides the utmost assurance that you are in control as 'Murph' hovers overhead.

Climb to a safe altitude before tossing the hook. "Safe" means whatever your judgement dictates, *not* what you have the guts for.

Suppose an FAA check pilot was sitting beside you. What altitude would you choose? 1,000 feet? 600 feet? The flying environment will probably influence a prudent choice, but chances are anything less than 400 feet AGL would invoke instantaneous "OFF" flags in both eyes of any official observer.

Trim the aircraft for a slight climb. Ordinarily a brief routine, grapple hook deployment can get messy in a hurry when things don't go according to Hoyle and you're busier than a one-legged man in a butt kicking contest.

It bears repeating that the cable and hook *must not* be fouled on loose or fixed items in the aircraft—pilot included!

Deploy the tailhook assembly over an unpopulated area whenever possible. If for no other reason, it makes you feel better and reduces anxiety.

Summarizing this section is easy: Keep tension on the "forward" section of the cable with one hand, gather up the towhook and loose cable with the other, and throw everything out and down to clear the aircraft. (I heard a story about some unfortunate that managed to throw a grapple hook *through* the fabric on a low-wing aircraft. It's most likely a distortion of the truth, but I can't help but to think of that every time I give the old heave-ho.)

A final thought: if there is *any* question about the configuration of the towhook, cable, or tow hitch release mechanism after deployment, the *only* way to find out for sure is by inspection. If you have a ground crew observer, he *may* be able to assist you during a fly-by. Otherwise—"TS"—drop it in a safe area and start over rather than to risk the unknown.

THE PATTERN

Any pattern consistant with sound flying safety practice is acceptable. The unique elements of a pattern for a banner pickup must include a safe straightaway for hook deployment and provisions for entering and departing a suitable pickup area. In many cases, entering the "normal" traffic flow will suffice with the pickup area offset in the runway periphery. See Fig. 6-1.

THE APPROACH FOR PICKUP

The final approach to the threshold of the pickup assembly has a lot of room for technique. Ultimately, two basic styles emerge.

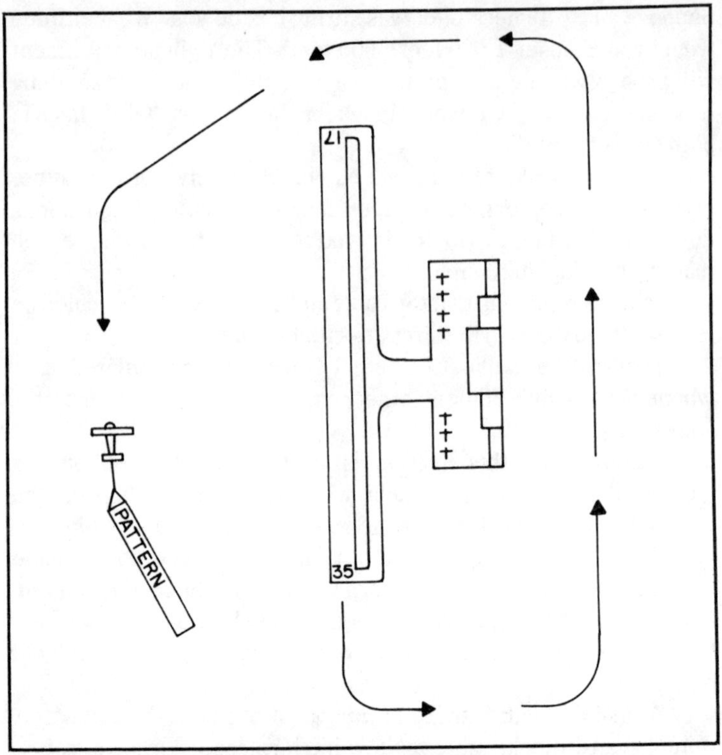

Fig. 6-1. Conforming to the local traffic pattern is probably the easiest technique to set up for an aerial banner launch. Controlled airports may dictate otherwise. Any pattern consistent with flight safety is acceptable.

Fig. 6-2. Serial photographs of glide path technique. Approach has just been started.

Fig. 6-3. Ground crewman uses hand signals to show relative position of grapple hook (not visible in these photos).

The Glide Path Approach

The descending, or glide path approach is a worthwhile technique to master. It shortens the final approach clearway requirements so that the ground equipment can be set up closer to obstacles, within reason. The sequential photographs (Figs. 6-2 through 6-6) portray the technique from the ground observer's view. The "disadvantage" of the descending pickup pivots on the need to reverse vertical direction at a critical moment.

A suggested technique for this method is to use the pickup assembly as an "aim point" for a glide path. Whatever style is

Fig. 6-4. Pilot has made final corrections for clean pickup. Pickup assembly is just coming into view in right of photo.

Fig. 6-5. Looks like it will be a good pickup. Note stable, slightly nose-high attitude of aircraft.

chosen, the end point of the approach is to guide the grapple hook somewhere inside the area defined by the pickup loop.

The Level Approach

The advantage of a level approach translates into a better zoom capability after pickup loop engagement. This is probably the best method for smaller birds that are a little shy in the excess power department, particularly if the banner is of considerable

Fig. 6-6. As forward-and-down visibility from Stearman is marginal, pilot must be on correct tracking path long before this point.

length relative to the limits of the tow plane. The terrain along the approach path must be sufficiently flat and free of obstacles in order to accommodate this method. It takes a bit more practice to master this approach so as to get a feel for the position of the grapple hook. It should "fly" well clear of the ground below (say, five to eight feet) and yet retain a vertical margin below the level of the pickup loop horizontal (approximately three to four feet). Once the judgement for this has been developed, the approach is simply reduced to guiding the aircraft over the uprights (Fig. 6-7).

Airspeed

There are no magic numbers for the approach airspeed. Higher speeds offer better zoom capability but decrease the trailing angle of the grapple assembly (Fig. 6-8). Gasser recommends the optimum approach speed along the order of 1.1 to 1.2 times the best rate of climb speed for the towing aircraft. Whatever your choice, the airspeed should remain constant, especially in the last few seconds prior to engagement.

Power

Figure 6-9 demonstrates what happens to the trailing angle of the grapple assembly when power is abruptly added or chopped. Jinking around with the throttle on short final is poor technique and will shave your batting average at the pickup assembly. Crossing the pickup threshold at or near idle is to be discouraged. The temptation is there, especially on the glide path approach technique. The response time of the engine is slower and this is no time to induce a sputtering powerplant because of lousy throttle technique. Carry enough power to stay out of carburetor icing range, and make smooth power changes, if needed.

Groundtrack

The flight path of the approaching tow plane should be perpendicular to the uprights. Crosswinds are to be corrected for with crab angle. Favor the upwind side, as the cable tends to trail slightly displaced to the downwind side. **Do not attempt a wing low slip to maintain the desired groundtrack.**

There should be enough length to the final approach groundtracks to allow sufficient time to establish *perfectly* stable flight parameters, including directional control, airspeed control, and altitude control. This is especially critical during the final 50 yards or so.

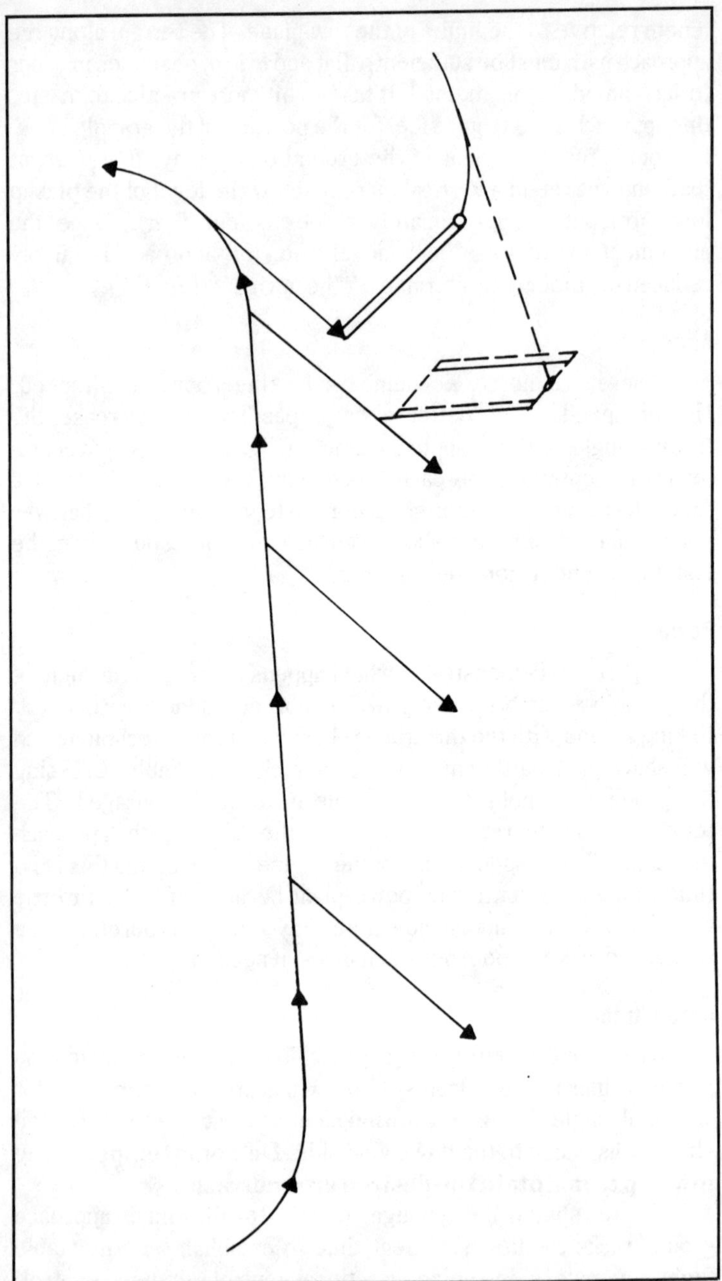

Fig. 6-7. The level approach keeps the grapple hook at a constant height above the ground during the last phase of approach to the uprights.

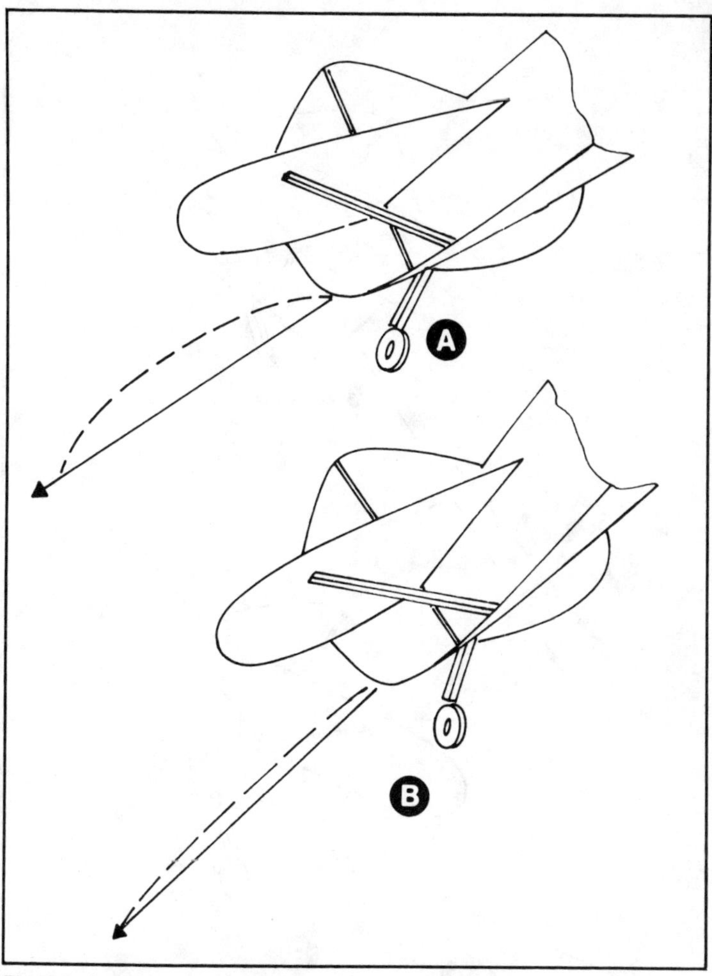

Fig. 6-8. A. At slower speeds, the grapple assembly trails lower beneath the aircraft (greater trailing angle). B. At higher speeds, the trailing angle decreases, the bowing phenomenon is reduced, and the grapple hook flies comparatively higher relative to the tow plane and the ground.

A change about any axis of the aircraft transmits to the three pronged parasite in the rear with subsequent effect on its ever oscillating path. *Stability*, therefore, is the key to success with the aerial pickup method of banner towing.

THE ENGAGEMENT

The only requirement for successful engagement is that the tow hook enter somewhere into the area defined by the pickup

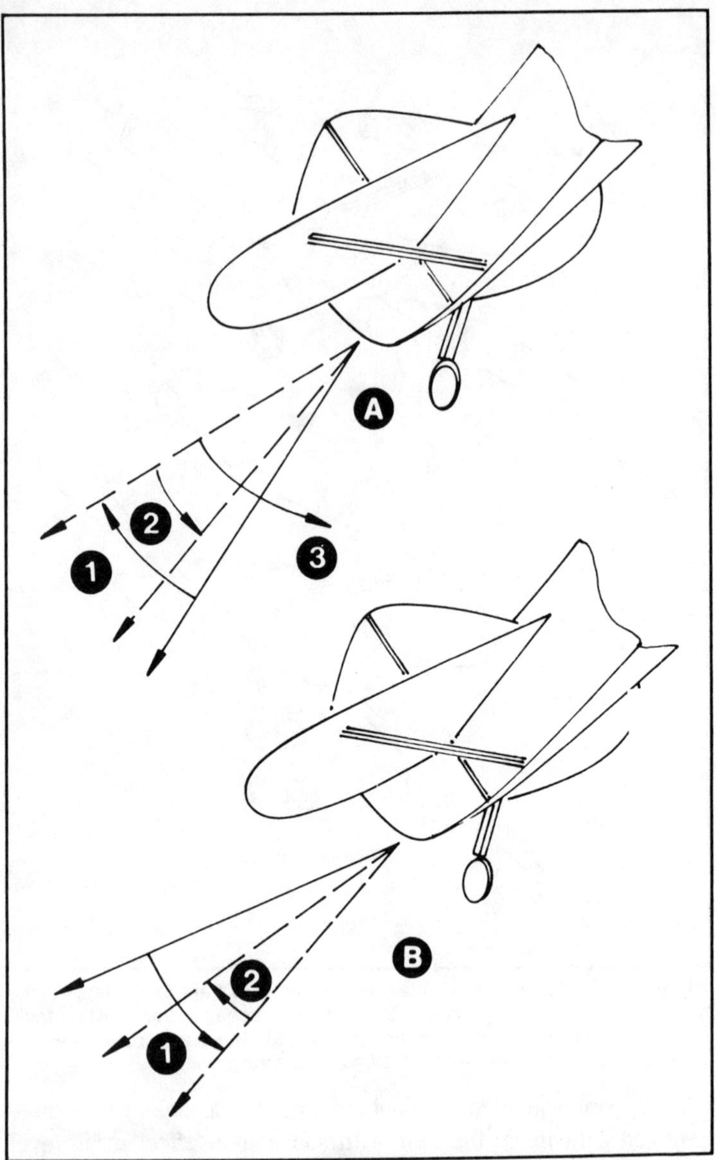

Fig. 6-9. A. As power is added, the grapple assembly first swings up (1) and then returns to stabilize (2) at the reduced trailing angle compatable with the increased speed. If the tow plane is "zoomed" at the same time power is added, the return oscillation of the grapple assembly (3) will be of greater magnitude as the grapple assembly orients to the new trailing angle relative to the plane. B. Abrupt reduction of power will cause an increase in the trailing angle of the grapple assembly (1) before it stabilizes (2).

loop. The cable will knock the horizontal portion of the loop off its suspension prongs and guide the rope down to the grapple.

Full power may be added to commence the zoom maneuver once the aircraft and the grapple hook are past the uprights. Adding power too soon will cause the grapple to swing up, possibly converting an otherwise dead ringer into a missed approach. Tardiness with the throttles compromises the magnitude of the zoom. The balance between the two must be found by practice.

Another technique may be used that takes advantage of the phenomenon illustrated in Fig. 6-9. By beginning the zoom maneuver just prior to passing over the uprights, the hook will swing up as power is added but then arc downward in a pendulous track. By coordinating this with the lag time for aircraft altitude gain, the cable should strike the pickup loop more nearly perpendicular and enhance the percentage of engagements. Furthermore, it adds a few seconds to the zoom time by getting an earlier start. This highlights the old adage that the two most useless things in the world are runway in back of you and sky above you. This takes considerable practice to perfect. A few dry runs with a ground observer's feedback will help enormously.

THE ZOOM MANEUVER

Exchanging kinetic energy (airspeed) for potential energy (altitude) is the final step of launching the banner. The objective is to gain as much altitude as possible (within the safe limits of the aircraft) prior to moving the banner. This needn't be a *violent* maneuver but is rather best described as a very *positive* one. The sequence of photos (Figs. 6-10 through 6-16) show an acceptable technique. Too shallow a climb angle will induce a torsional moment in the lead pole that may even cause it to swing over into the first or second letters of the sign as it comes off the ground. There is also a higher chance of breaking one or more rods. An over-enthusiastic pull up is to be discouraged by the danger of running out of airspeed and ideas at the same time.

THE ABORTED PICKUP

If the pickup loop is missed, intentionally or otherwise, the first order of business is to maintain aircraft control and continue to clear for other aircraft in the area. If another attempt is appropriate re-establish a position in the pattern and try again. Otherwise, the hook must be dropped as a separate maneuver before returning to land.

Fig. 6-10. Zoom maneuver. Pickup has just taken place and banner is starting to move.

Fig. 6-11. The zoom maneuver involves exchanging airspeed for altitude. Note climbing attitude of aircraft.

Fig. 6-12. Torsional moment of lead pole may be induced if climb angle is too shallow. This can cause lead pole to swing over into first character or two as sign comes off ground.

Fig. 6-13. This photo sequence shows that proper climb angle is being flown. Note that lead pole and first characters have already assumed the proper vertical orientation, even though half the banner is still on the ground.

Fig. 6-14. Note that this banner utilizes an adapter pole; first part of banner is 7′ characters, back section has 5′ letters.

Fig. 6-15. When zoom maneuver is performed correctly, the banner comes smoothly off the ground, just like peeling a banana.

Fig. 6-16. With the banner now in the air, pilot is off on way to another successful mission.

If a launch must be aborted *after* banner engagement, it is usually best to do so as quickly as possible by actuating the release handle. Continued flight with a cripple banner can be devastating to the equipment, not to mention the safety implications.

BANNER LIFT-OFF AND DEPARTURE

As viewed from the aircraft, the banner is stripped from the ground in a very graceful motion. If visibility permits, the pilot should crosscheck the launch and be prepared to pull the tow release handle to abort a faulty pickup, for whatever reason.

It is assumed that the departure course in front of the launch area is a suitable drop zone and clear of persons, objects, etc., as it should be. Reasons for immediate abort include twisted banner, pole pickup, or signal from the ground crew to do so.

Assuming a clean pickup, adjust power and climb attitude as required for the local area and depart in compliance with local procedures.

TOWING

Cruising the skies with a banner is surprisingly easy. The aircraft tends to have greater stability about the vertical and horizontal axes. Less rudder is required to counter torque during climbout. A few items peculiar to towing are worth noting.

Total drag is increased during towing operations for two reasons: parasite drag is increased by virtue of the banner in trail, and induced drag is increased because of the lower airspeeds used for towing and subsequent higher angles of attack. The important consequences of the increase in total drag coupled with the lower towing speeds might be best expressed through equalities: Increased Total Drag = Increased Thrust Required = Increased Fuel Consumption = Decreased Range and Endurance. In other words, it's harder on the engine, operating costs per hour are increased, and you can't go as far or stay up as long as you could without dragging the rag.

Here's another series: Slower airspeeds = Increased Angle of Attack for Level Flight = Decreased visibility (for some aircraft). Lowering the flaps partially helps to reduce the angle of attack required for level flight at a given airspeed. This may also promote better cooling of the engine secondary to better airflow patterns. The liberal use of cowl flaps is advisable to keep engine temperatures within limits.

The addition of parasite drag shifts the total drag curve up and to the left. The increment in induced drag shifts the curve up and to the right. This may change the best angle and/or best rate of climb airspeeds, a point the reader should at least be aware of.

It's more than an aerodynamic rule of thumb that drag increases directly with the square of the airspeed—it's a law (Assuming subsonic airflow, according to the drag equation: $D = C_d \frac{1}{2} \rho V^2 S$).

Therefore an increase in towing speed of 10% will result in an increase in drag of slightly over 20%. The lowest towing speed that still provides a safe margin above the stalling angle of attack will produce the optimum tow speed in most cases. Higher speeds have increasingly detrimental effects on the banner equipment. 80 mph (70 KIAS) should be used as a *maximum* acceptable speed in consideration of the banner. It won't come unglued at higher speeds for short periods of time, but the effects of wind-whipping accelerate equipment depreciation.

You don't have to be Captain Golden Gloves to keep the banner stable during flight. It just takes *smooth* aircraft control. Even though the banner is designed to "fly" right side up, it will swirl if given the opportunity to do so. It is *particularly* vulnerable to this as descents are initiated. While there is no detrimental effect to the equipment, it must be remembered that your client is paying you to *display* his sign, *not* to screw it through the air. If encountered, the swirling motion will dampen itself out as the descent stabilizes or you can level off (or begin a slight climb) to halt the aberration.

The banner trails about 50—70 feet below the level flight path of the towplane. During descents, it may stabilize at or slightly *above* the aircraft's altitude. This latter characteristic may become useful if a landing with the banner in tow becomes necessary (see Chapter 8).

DISPLAYING THE BANNER

If the folks on the ground are to read the sign, they've got to *see* it first. This is not so much a problem along the beaches when the towplane audibly announces its arrival well in advance of the advertising display. But flying alongside the highways—that's a different story. Any amateur people-watcher can testify that well over ninety percent of American drivers are too busy picking their noses to pay much attention to anything else. They look straight down the highway with an occasional glimpse in the rearview mirror to verify that no one has sneaked up to invade their private ritual. Those who dare to scan their eyes more than one degree off centerline are indeed in the minority, and even then seem restricted to the two dimensional perspective along the front/back - left/right axes. And with the windows up, and the air conditioner blasting, and the radio blaring, Mr. & Mrs. American Driver won't tune in to the distinctive sound of an approaching tow plane. No, friends, it's got to be *shoved* in front of their eyes!

Towing along the highways can be super-effective, especially during the big city rush hour traffic jams. As previously suggested, merely paralleling the highway won't produce the desired audience. A few lazy turns *across* the highway at intervals will alert the victims of the traffic snarl that there will soon be something *else* to look at besides the engine temperature gauge and the clock (Fig. 6-17). As long as they know it's coming, they'll keep tabs on it because that's the way folks are. They can't stand not knowing what that sign says.

From an advertising standpoint for highway towing, it's probably best to fly a circular or racetrack pattern over a defined segment of the road, especially an area of heavy traffic congestion (Fig. 6-18). In this manner, virtually every observer transiting this portion of the freeway will make visual contact with the banner.

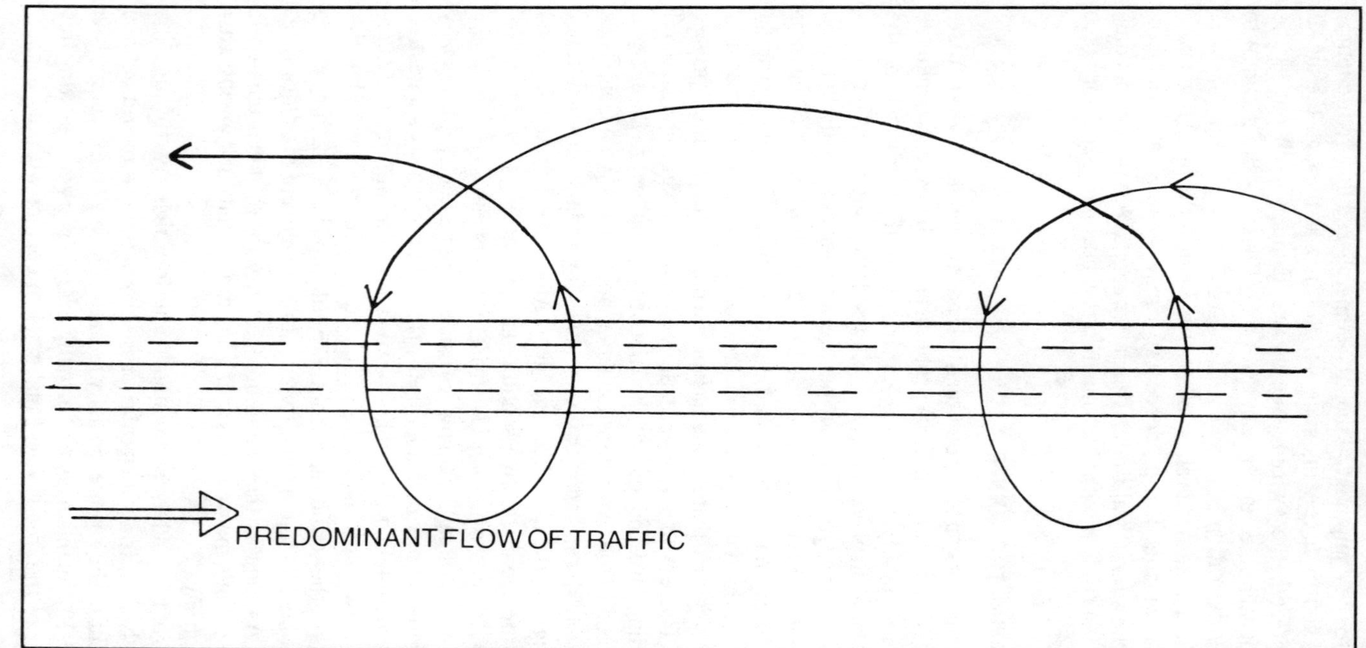

Fig. 6-17. When displaying the banner to the highway audience, maximize exposure by working against the predominant flow of traffic. Brief circular excursions across the highway every few minutes will call attention to your presence to traffic in both directions.

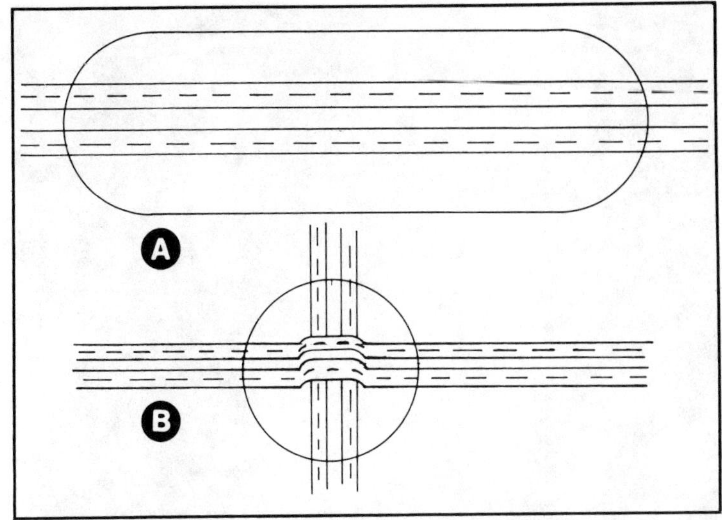

Fig. 6-18. Two other patterns that are effective for highway displays. A. Racetrack pattern. B. Circular pattern. This is particularly useful at an intersection, overpass, or cloverleaf, as sign is exposed to drivers travelling in all four directions.

From the pilot's standpoint, however, this technique of displaying is fraught with the danger of pilot fatigue, even for relatively short missions. Some variation is necessary to overcome the monotony. Flyers are reminded of the need to enforce quality clearing discipline, as the temptation to relax standards is accentuated with boredom.

It is important to maneuver into a position from which the sign can be easily seen. The observer should not have to look up higher than 45° from the horizon to read the message. Basic trigonometry says, then, that the towplane should be displaced at least as far away from the observer as is the height of the towplane in feet AGL. The pictures in Figs. 6-19 through 6-24 were taken simultaneously by the ground observer and a photographer crewmember at varying distances and angles.

An easy way to select a lateral displacement distance from the viewer is assess how well you can read ground-level signs in the vicinity.

Banking the aircraft in a coordinated turn will cause the banner to bank also. This tilting action enhances the readability of the message from the spectator's viewpoint as Fig. 6-25 shows. All turns should be made to the left (assuming the customary left-to-right sign construction) whenever possible.

Fig. 6-19. The next six photos were taken simultaneously from tow plane and ground observer, and will give an idea of legibility of banner from different distances and positions. Ground observer is at pickup truck with partial banner laid out behind it.

Monitor the banner at least every two to three minutes throughout the duration of flight. If rear visibility is restricted, a steep turn (45° bank angle) will bring the sign into view. Any aberration from a "normal" appearing banner in flight is sufficient justification to terminate the flight to investigate the problem.

BANNER RELEASE

Releasing the banner is a truly critical phase of towing, second only to the launch, and a close second at that. If the drop is to be made at an airport, you will want to announce your arrival well in advance, especially if there is no control tower:

"Jetsetter Unicom, Clothmoth 007, 5 miles North with a banner in tow. Will enter left hand traffic for a banner drop to the West of Runway 03."

The drop zone should be clear by prior arrangement but this must be affirmed *visually* before the actual release. The optimum surface for dropping a banner is a well groomed grass field. Concrete, or any other unyielding surface could damage the lead pole.

The descent should be smooth and unhurried. As always, monitor the position of the banner, especially as your altitude brings the aircraft and banner into the proximity of obstacles. Guide the aircraft over the drop zone and plan to arrive there at approximately 150 feet AGL.

Fig. 6-20. With tow plane almost directly overhead (see Fig. 6-19) the banner is fairly readable, but note how "HOUSTON 721-6290" is hard to see due to slight twist. Aircraft is 60° above horizon.

Fig. 6-21. Aircraft is now slightly higher and more to the side of the road.

Fig. 6-22. Banner is now completely legible. This is very close to optimum viewing for ground observer. Aircraft is at 1000 ft AGL, and 45° angle to horizon.

Fig. 6-23. Tow plane is now nearly half a mile from the audience area.

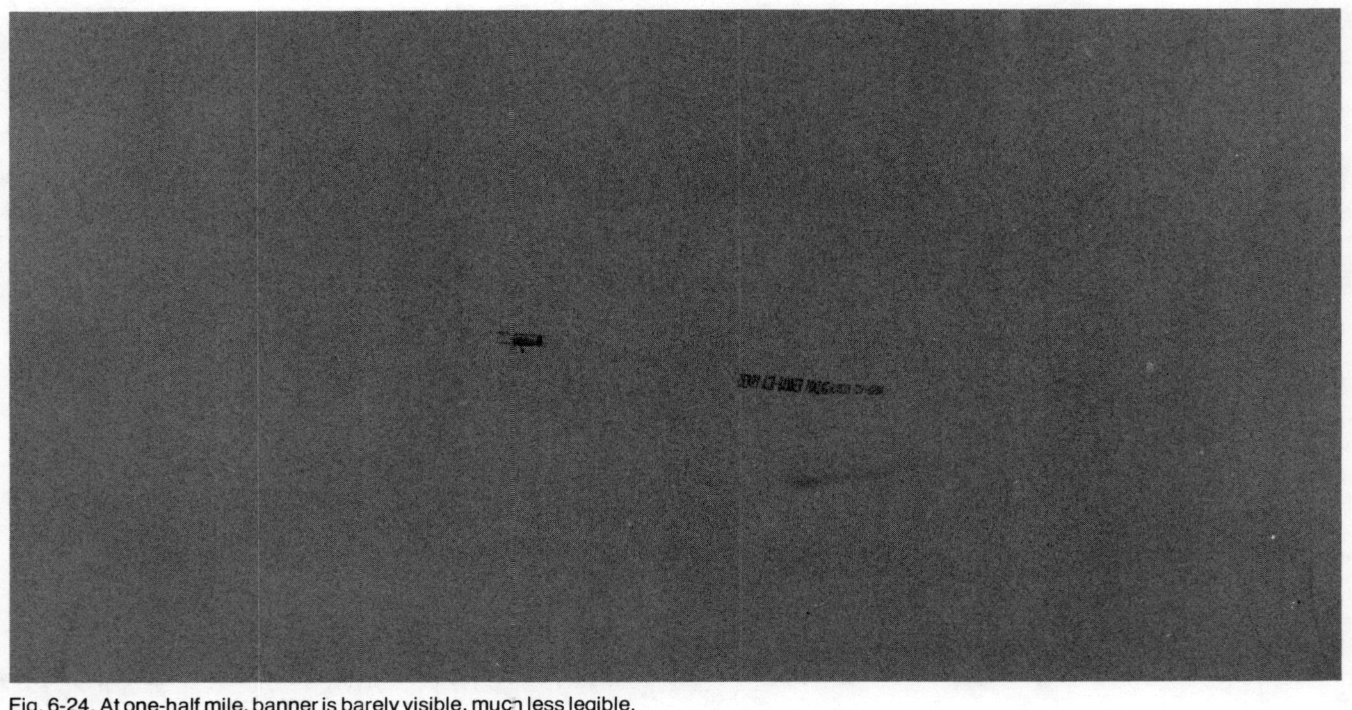

Fig. 6-24. At one-half mile, banner is barely visible, much less legible.

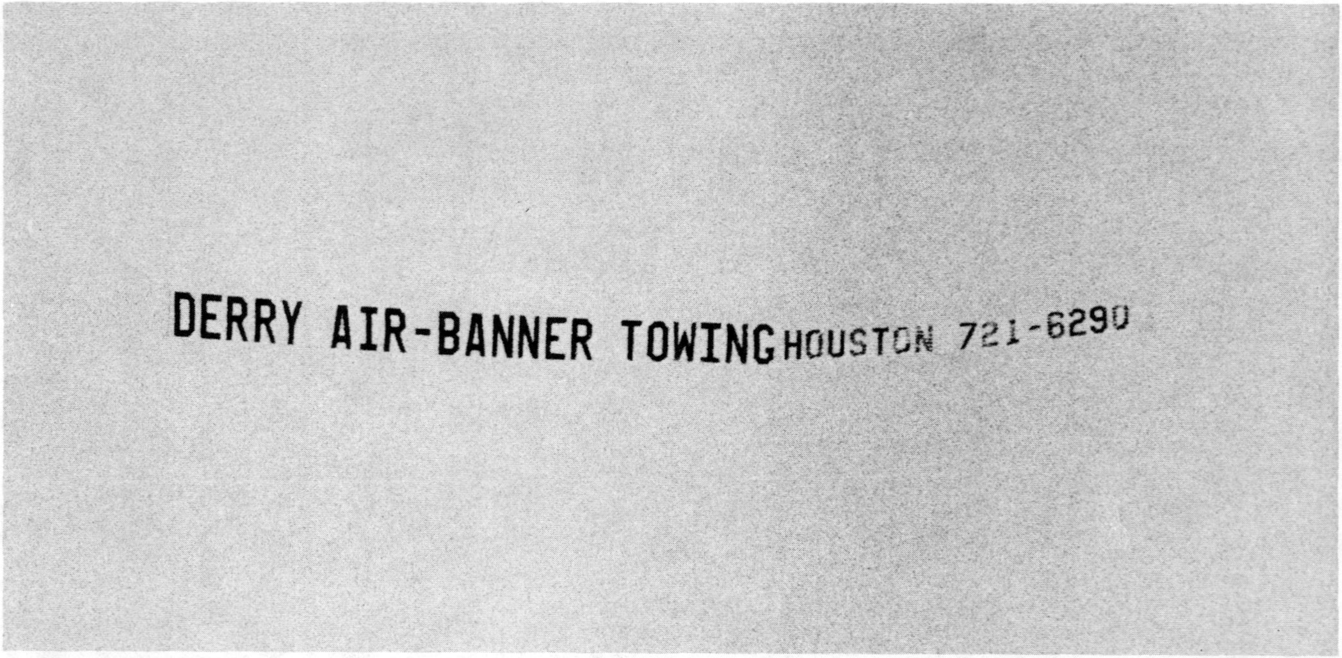

Fig. 6-25. Banking the tow plane will also cause the banner to bank. This aligns the vertical axis of the banner closer to the line-of-sight perpendicular. The net effect is a more legible copy.

Fig. 6-26. As the banner is released, the lead pole plummets sharply downward, with the rest of the banner falling in trail behind it. The lead pole will often stick in the ground in an upright position.

Fig. 6-26. As the banner is released, the lead pole plummets sharply downward, with the rest of the banner falling in trail behind it. The lead pole will often stick in the ground in an upright position. (Continued from page 117).

The banner will be trailing slightly downwind of the tow plane prior to release. After jettison, the wind will significantly affect the groundtrack of the sign as it tumbles earthward. The leadpole will only displace slightly downwind during its trajectory. The trailing banner is at the mercy of the crosswind component and the altitude from which the release is effected. The sign may therefore splay out and angulate as much as 90 degrees from the ground track of the tow plane. With a long banner copy, this might translate into more than 100 feet of lateral displacement from the point at which the lead pole strikes the ground. This is why the drop zone *must* have adequate clearance from parked aircraft, taxiways, etc. Compensate for effects of the crosswind by approaching the drop zone on the upwind side. By aligning the drop directly into the wind, the phenomenon is altogether eliminated (Fig. 6-26).

RETRIEVAL

The banner should be gathered up as soon as possible after the drop. The lead pole often sticks in the ground vertically, thus exposing the fabric area of the first few letters to the surface winds. The resultant stress could snap a rod, although experience shows this to be a rare occurrence.

THE GROUND LAUNCH METHOD

If you have a plane that can get airborne and attain an altitude of at least 100 feet within 700 feet of brake release, the option of the ground launching the banner is yours. The setup, previously described, is quite simple. Once on the runway, configure the aircraft for a minimum run takeoff and conform to the recommended procedures for that maneuver. Once airborne, attain the best climb angle speed at least until the banner has lifted off the ground and all obstacles are cleared. Since this maneuver is done along the runway environment, you must have ground personnel available to clear the banner should it drop or the takeoff is otherwise aborted. If the takeoff is ground aborted, be certain to release the tow rope as early during the rollout as is practical to prevent damage to the banner.

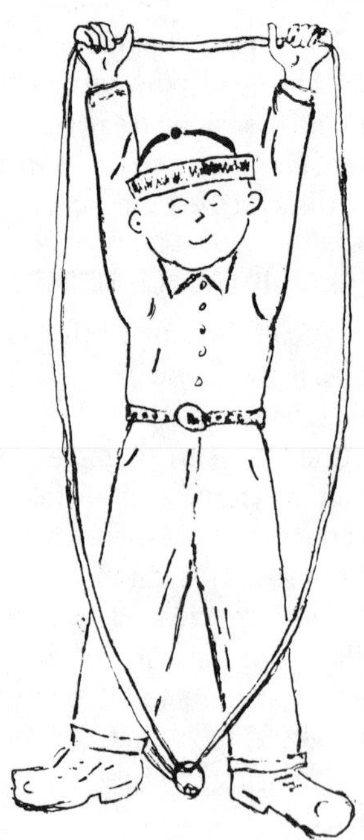

THE GROUND CREW
A. TAYLOR

The ground crew is such an important, albeit optional, complement to any organized banner towing operation that the subject deserves its own chapter. This handy assistant not only serves as a helper to reduce the time factor and work load, but also provides a much needed link between the tow pilot and his equipment, both on the ground and in the air.

The drawbacks to solo operation are many. You've got to lug all of the equipment out to the pickup area. The banner needs to be unfolded and spread out properly (no small trick in a 15 knot wind). The pickup assembly must be set up and rigged. Finally, the plane must be configured and tested as discussed in Chapter 5. Without anyone to check your plane over as you taxi out, you take off with full power and a lot of trust that everything will be just the way you left it when the moment of truth arrives. How would you handle these problems without a ground crew?

Problem #1: As you are taxiing away from the chocks, you notice some slack in the grapple hook cable.

Answer: Taxi back to a safe location, shut down, get out, and re-inspect the tailhook/tow hitch assembly to be certain that the "O" ring hasn't been displaced from its proper position or that the cable isn't fouled on the tailwheel (if applicable)

Problem #2: After takeoff, you throw the grapple hook overboard and the cable momentarily hangs up on your sleeve. You aren't certain that the cable isn't fouled on the tailwheel. (Tricycle jocks may go on to the next problem.)

Answer: Fly over a safe drop zone and pull the cable release handle. If the cable drops, land and start over; if it doesn't drop, land and start over.

Problem #3: After courteously fitting into the flow of traffic on a busy day, you now have ample spacing to complete your pickup safely. Although running slightly behind schedule, you have just enough time left to travel to your target for a 2:00 show *if* there are no problems with this pickup. As you roll onto final approach for the banner pickup, you notice three living bodies gazing at the pickup assembly with their backs to you and oblivious to your approach. You chide yourself for attempting to honk your horn in a machine not so equipped.

You go around and try again, confident that they will have cleared the area as you bargain for position and spacing in the busy pattern. On final approach for second time, the situation has changed. You are now *very* late for your aerial appointment and the mindless jerks are now facing you between the uprights wondering what in the hell you're doing flying so low to the ground with a fripping hook hanging off your plane. (Hint: The very first thought that pops into mind probably isn't the correct answer.)

Answer: Without any latitude in your target time, this mission is doomed to an abort. In any event, you must land, if only to hurl vulgarities at the now fleeing intruders. If time still permits, a re-inspection of the entire banner apparatus is needed in the interest of flying safety.

Problem #4: As you approach the pickup assembly on final approach, you notice that the wind has knocked the looped portion of the towline off one of the uprights.

Answer: Drop the cable and grapple in a safe area, return for landing, reconfigure the pickup assembly, re-rig the grapple hook and cable, and start all over.

Problem #5: As you pass over the uprights during an otherwise perfect pickup approach, you zoom in your familiar style to snatch a message off the ground. Something is different, as you anticipation of the tell-tale tug that usually alerts you to the banner snag is never fulfilled. Alas, a "missed approach" and your eyes, looking back, also mist to learn that the pickup loop was either blown off, knocked off, or prop-washed off. Whatever, it's *off* and you're hacked off because you already know the only solution to the problem.

Answer: Same as Problem #4.

Problem #6: You have just picked up a banner, as usual on your first pass. Such precision and grace; the crowds must be cheering! As you routinely look back to check the banner in flight after pickup, you notice something doesn't look quite right. It's nothing you can readily identify as a problem (like a twisted banner) but you are not convinced at the 100% level that there isn't a broken strap or something of that nature. You climb to a safe altitude and check it over visually with a steep banked turn. Everything looks o.k.

Answer: Return to a safe drop zone and release the banner. Then return for landing and start over.

Such a pity - this all could have been prevented, avoided, or solved with a ground crew assistant. At the risk of belaboring the point, let us briefly recapitulate each of these scenarios as they might present with a ground crew assistant added to the list of characters.

Problem #1 would have been satisfactorily handled, if not discovered, by a trustworthy ground crew member with adequate training. A loose cable is virtually a "thumbs up/thumbs down" situation when the slack is taken up from the cockpit. Any adjustment to the tow hitch mechanism and/or tail hook assembly can be achieved externally by the assistant with minimal delay.

Problem #2 is simply solved by a flyby as a prearranged procedure to confirm a safe configuration. Air/ground communication (either radio or via flag signals, to be discussed later) provides the necessary link for this service, which obviates the need to drop the hook, land, re-rig, etc.

Problem #3 probably never would have existed in the first place. The ground crew can readily inform folks to keep their distance, in accordance with common sense and any Special Provisions directives that may be on the Certificate of Waiver.

Problem #4 and problem #5 become minor annoyances rather than major, time-consuming, aggravating-as-hell disasters. The pilot need only set up for another swat at the rope while his earthbound assistant, in the meantime, has already resurrected the fallen cord to its lofty supports.

Problem #6 gets solved much as the solution to problem #2. The best possible view from the aircraft cabin can't begin to achieve the perspective afforded to the ground observer.

The preceding discussion grants *no* prerogative to push the judgement override switch. The ground crew assistant is merely an aide with primarily a two-fold purpose: 1) to render service during routine procedures in order to minimize time delays and maximize safety and, 2) to provide the tow pilot with information he otherwise would not have available to him for an appropriate decision process. It should be unnecessary to state that the final responsibility for the safe operation of aerial banners *exclusively* rests on the already bruised shoulders of the pilot in command.

With all this as a cruelly lengthy introductory note, the spotlight now turns toward the duties, responsibilities, and training of the ground crew assistant.

SELECTION

Your ground crew assistant needs better credentials than twelve to sixteen breaths per minute. He (or she) needs to have a *brain* with the ON/OFF switch safety wired to the ON position. Additionally, *common sense* is the output most needed and receives the most number of points on the qualification scoreboard. With little else required of this employee besides, perhaps, the willingness to participate in a conscientious manner, it seems that any further probing into this subject is unjustified. It is assumed that this person's availability, access to transportation, reliability, and motivation will be taken into consideration for compatibility with the mission at hand.

As a final comment, it is well to tag someone with a tenacious character—that stick-to-it-iveness quality—who will become a familiar part of your operation. This avoids the annoyance of having to train somebody out of the woodpile each week and risk finding one who can't find his rear end with both hands and a flashlight.

DUTIES AND RESPONSIBILITIES

Defining the precise role of the ground crew in a banner towing operation depends largely on local circumstances that are unique to each operator. Indeed, the assistant may fit into a wide spectrum of roles:

- ☐ Procurement
- ☐ Banner assembly
- ☐ Aircraft crew chief
- ☐ Banner setup
- ☐ Pickup and launch observer
- ☐ Banner retrieval
- ☐ Equipment storage
- ☐ Maintenance & repair

While these are listed for the sake of completeness, the reader may put to rest any fears that a detailed dissection of each item might be forthcoming. Aside from the redundancy that would ensue because of the material covered elsewhere in this book, even the author would suffer from irreversible boredom. Rather, we will focus in on the activities surrounding the banner pickup and how the ground crew integrates into this brief albeit critical phase of operation.

The ground crew is charged with two basic responsibilities: 1) baby-sitting the banner, launch apparatus, and pickup environment while the pilot takes off and maneuvers to set up for snagging the towline, and, 2) communicating useful information to the tow pilot.

The first of these responsibilities covers a wide gamut of trifling details—if the wind flips the tail assembly chute: straighten it out; if the wind blows the pickup loop off: put it back up; if somebody walks into the area: get him the hell out of there; if somebody throws a baby between the uprights: get it off the flight path and recheck the loop—and so on in the same, sing-song monotony for every possible problem imaginable.

The second, and no less important, category of services provided by the ground observer is inherent in his ability to make observations that the pilot cannot, When given the tools to relay such information to the tow pilot, an air/ground communications system is born, the purpose of which is to provide the pilot with useful information from a different vantage point on an as-needed basis.

What kind of information is "useful" to the tow pilot? The answer is *any* information that will contribute to the safety and success of banner launch and recovery. Admittedly, this is a pretty loose statement and the ensuing comments are offered to alloy any frustration that it may have generated.

The aerial pickup method of banner launch has consistently received a disproportionate share of attention throughout this book. No exception is to be expected here. For it is in the brief moments (more accurately, *seconds*) prior to and just after the actual engagement of the pickup loop that the ground assistant earns his keep. The style with which this proceeds will vary according to sophistication of the communication system furnished to the users.

AIR/GROUND COMMUNICATIONS

Experience has proven the effectiveness of two entirely different methods of communication.

Flag Signals

Air/ground communication using flag signals presents a primitive albeit useful and economical method of information transfer. For a modest investment, which may run as high as several pennies, a signal flag may be constructed to serve the purpose well. For the budget watchers, a sturdy stick and a T-shirt, towel, or rag will suffice. Endless volumes could surely be compiled here on this subject alone, including such topics as size and shape of the handle, custom handgrips, staining techniques for exposed wooden parts, material selection, cost comparison data, psychological studies showing the correlation between banner pickup success and flag color, how to deal with cost overruns, field repairs, and so on. Considering such to be beyond the scope of this text, the author and the reader alike will be spared such agony by proceeding to more germane topics.

Five basic flag signals are all that are needed to make this system workable. These can be easily learned and assimilated by the dullest of minds in as much time as it takes the educator to spit out the words. Note that Figs. 7-1 through 7-3 depict signals for use only during the approach phase of the aerial pickup just prior to

Fig. 7-1. TOO HIGH: The grapple appears to be trailing too high for pickup loop engagement.

Fig. 7-2. RIGHT ON: The grapple appears to be trailing at a satisfactory height for proper engagement.

the actual engagement (or miss) of the pickup loop. The ground observer simply judges the position of the grapple in trail behind the tow plane and feeds this information back to the pilot. The pilot, in turn, may utilize this information to fine-tune his approach and gain familiarity with the vertical perspective for this precision maneuver.

Figure 7-4 titled "UNSAFE CONDITION" might alternately be titled "BANNER SHOULD BE ON THE GROUND." Depending upon the phase of launch in which it is used, the implications of this signal are at variance with any single definition. If, for example, the ground observer noted that the pickup assembly was improperly configured with the tow plane on final approach (a bad time to find this out, but admittedly better than not at all) he should issue the "UNSAFE CONDITION" signal, more accurately interpreted by the pilot as "go around". The same flag signal when used while the banner is airborne, i.e. immediately after launch, should be defined to mean, "Something is wrong; return to the drop zone and release the banner." Examples of this sort would be a twisted banner, snagged pole, or a ripped or otherwise damaged letter, etc.

The "OKAY" signal (Fig. 7-5) would generally be accepted as a "green light" to proceed as planned. Suppose the tow pilot were waived off by an "UNSAFE CONDITION" sign on his first attempt

Fig. 7-3. TOO LOW: This signal alerts the tow pilot that the grapple hook is in danger of contacting the ground.

Fig. 7-4. UNSAFE CONDITION: During aerial launch, this should be interpreted as a command to abort the pickup, at least temporarily. The ground observer may be able to remedy the situation to allow another attempt. If this signal is given while the banner is being towed, it suggests that something is wrong, e.g. damaged or twisted. In this case, the tow pilot should return to drop the banner and correct the problem.

at picking up the banner. He would execute a go-around and try to make observations as to the reason for the abort command. It may be obvious from a single glance at the pickup assembly that an adjustment is needed. He would then hold nearby or re-establish himself in the flow of traffic to set up for another swat at the line when he receives the "OKAY" signal (confirming that the pickup apparatus is now properly configured.) After launch, it is always a good idea to receive confirmation from the ground that all is well. The final appearance of the banner is, after all, your product.

Although this system is more or less a one-way street in terms of information flow from the ground to the towplane, there are a few things the pilot might consider beforehand should he wish to elicit a response from his partner on the ground. A prearranged signal,

Fig. 7-5. OKAY: Prior to banner engagement, this signal is generally given after the tow plane has been waved off (Fig. 7-4) for a minor problem that can be corrected by the ground crew. The OKAY signal tells the tow pilot that the launch apparatus is ready for another attempt. After the engagement of the banner, the OKAY signal confirms that the banner has an acceptable appearance from the ground.

such as dipping a wing, can be used to alert the ground observer to check the banner status and/or the tow hitch, cable, and grapple for any suspected deviations from the norm. He can then reply according to his observations, thereby minimizing the guesswork on the part of the stick-and-rudder man. A set of binoculars is a handy luxury, if provided to the observer.

The flag and its attendant must obviously be in a position so as to be in clear view of the pilot throughout his approach course to the pickup threshold. For most planes with reasonable forward and side visibility, this poses no problem and the flagman can stand to either side of the uprights at a safe distance, say, 50 to 75 feet laterally. In those aircraft where the visibility ain't so hot (again, using the Stearman as a 'ferinstance,' and the problem further accentuated by the high angle of attack and crab correction) it is important to situate the ground observer on the *downwind* side of the pickup assembly. This keeps the crosscheck between the aircraft aim point and the flag signal input on the same side of the plane. It does not, however, reduce the need for a systematic vigil towards the opposite side as a matter of good clearing technique.

It surely must be against the law not to mention safety as the concluding note to this topic. If for no other reason, this is why you want to choose a ground crew member who has both oars in the water at all times. No one or no thing should be in alignment with the tow plane once it has turned onto its final approach course, the banner and its attendant equipment excepted. Implement and enforce sound practices in the safety department and allow *no exceptions* — there is no other way.

Radio Communications

By far and away, the easiest and best method for air/ground communications is via two-way radio circuit. There is no need to teach and rehearse signals, no ambiguity over specific meanings, and no need to incorporate extra visual cues into a crosscheck. The drawbacks are essentially limited to the price tag and the potential for equipment failure.

The observer, in this case, need only to have the appropriate visual acuity, the talent to squeeze a mike button, a command of the English language, and an FCC Restricted Radio-Telephone Operator Permit. The latter is one of the few free-for-nothing good deals left in this world (at least as of the time this book went to press) and applications are available for the asking at any FCC office.

How simple, by comparison, to relay specific information:

"Ragpuller Zero-Niner Alpha, you've got a knot on your tow cable."

or,

"Dirty-Duty, Ragpuller here—will you check out my tow hook cable and make sure it's not wrapped around the tailwheel? I'll make a low pass to the West of the pickup zone."

The picture in Fig. 7-6 shows a mobile VHF radio setup that is powered from the truck's electrical system, adapted through the cigarette lighter socket. A wide range of radios may be selected for this purpose for a wide range of prices through the usual sources for this merchandise. A more portable system than the one portrayed here might be advantageous, specifically the hand-held, battery powered models.

GROUND CREW BRIEFING

Before the tow pilot straps an airplane onto his butt, he should *always* brief the ground crew on the upcoming events. Nothing should be taken for granted, regardless of how seasoned and experienced the assistant may be. A checklist for such a preflight briefing would look something like this:

1. Final inspection of the taxiing aircraft: the ground assistant should check key items as the aircraft pulls away from the chocks, e.g. position of the tow cable, tow hitch release mechanism, etc., depending on the particular configuration of the tow plane. A "thumbs up" to the pilot confirms that all is well.

2. Pattern direction and ground track.

3. Specific location for the ground observer to stand.

4. A/G Communication procedures (flag or radio)

5. Identification and correction of potential problems:

a) Persons wandering into the pickup area and/or approach/departure course.

b) Fallen loop

c) Configuration of pickup assembly

6. Aborted pickup plans.

7. Confirmation after pickup.

8. Drop zone and banner release procedure.

It is expected that local revisions of this guide to meet individual circumstances will be commonplace.

GROUND LAUNCH DUTIES

If the takeoff launch is employed, the ground crew assumes an altered role. A truly trusted assistant can help test and attach the

Fig. 7-6. A VHF radio unit serves well for air/ground communications. This one is powered by the electrical system of the truck through a cigarette lighter adapter.

towline to the tow hitch prior to takeoff with the acknowledgement that the responsibility for its correctness still rests with the tow pilot. This accomplished, the only remaining task is to stand by to assist in whatever way necessary should any off-routine problems arise during launch. An example of this would be clearing a dropped banner from the runway.

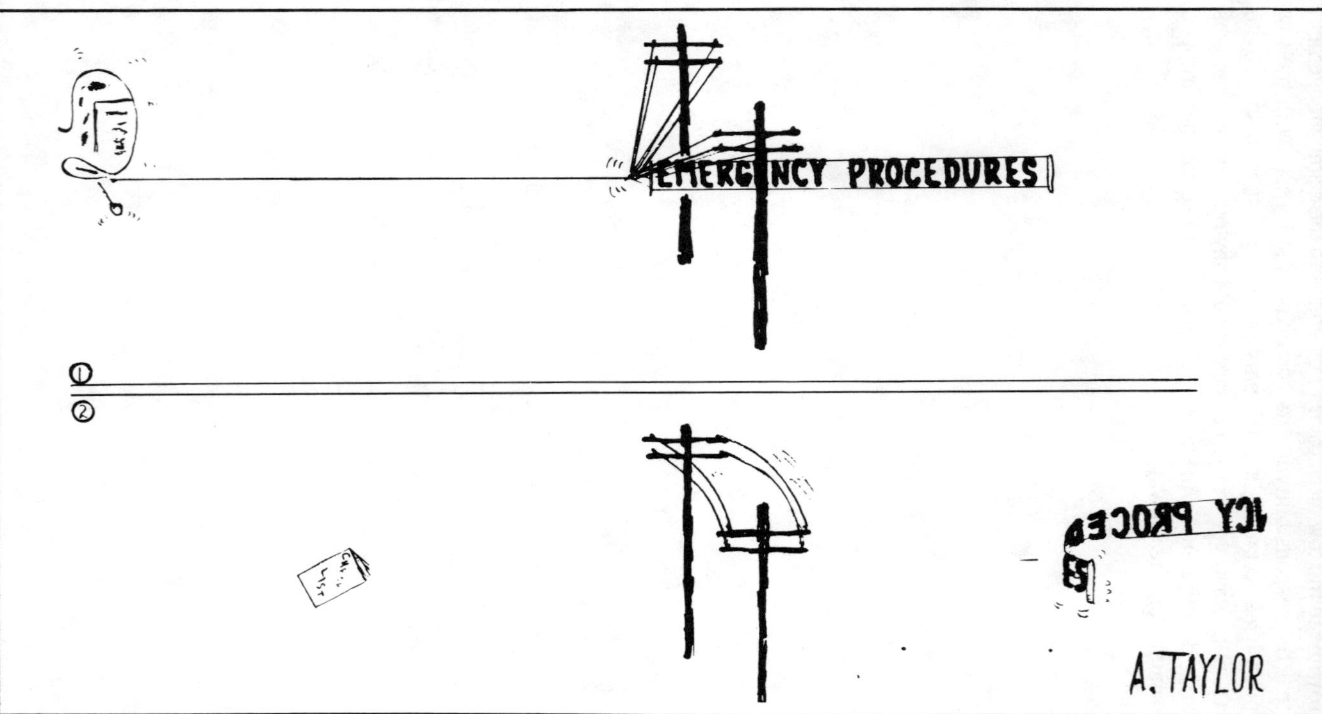

EMERGENCY PROCEDURES
A. TAYLOR

Chapter 8
Emergency Procedures

At a recent flight instructor refresher course, a rhetorical question was tossed to the audience: "What do you teach your students about night forced landings in a single engine airplane?" A voice popped up from the crowd amidst the generally unsettled chatter that ensued: "If you lose an engine at night, maintain aircraft control and guide the plane at final approach speed to around 150 feet AGL and turn on the landing light. If you like what you see, go ahead and land. If you don't like what you see, turn off the landing light."

There is little else in life's experience to parallel the "Gee, I wish I were doing something else right now" feeling that escorts the unexpected situation in aviation. By the same token, the satisfaction that emerges from the prompt and safe management of critical situations always yields a blissful euphoria, doubly underscored by the fact that a pulse and blood pressure can still be measured. This chapter is dedicated to the exposure of some adverse situations that might surface during the course of banner towing operations, with suggested plans for coping with these problems.

The topic of what constitutes an emergency in the first place is, of itself, fuel for endless discussion and stalemate debate. An "emergency" to pilot A might be a routine irritant to pilot B and vise versa. Emergencies don't always come packaged as variants of textbook situations, as many pilots will testify.

Here's an example of the truly unexpected! Despite a multi-thousand hour repertoire of jaded 'There I was, face-to-face with death' stories which time has a way of distorting at each re-telling, I can think of no situation that unglued me more than this

still vivid recollection. It was during my instructor days in the Air
Force in a T-38. After descent from a routine high altitude training
mission, a bit of condensation began to form on the windscreen as
often was the case. After turning up the canopy defog to alleviate
the problem, I was dismayed to learn that a heretofore silent
stowaway, in the body of a wasp, did not in fact appreciate the
200-some-odd degrees Fahrenheit just delivered by yours truly
through the piccolo tube he was cuddled against. So there I was, in
a closed compartment, interacting with a wasp that apparently still
harbored some ill feelings, judging from its disruptive behavior. As
it became apparent that this little creature wasn't going to let
bygones be bygones, I seriously considered ejection as the first and
only course of action at my disposal, since I was certain that my
trusty yellow checklist did not have a 'Bee in cockpit' procedure.
Ironically, the checklist ultimately became an integral part of the
solution as I upheld military tradition and met the now thoroughly
enraged enemy, face to proboscis. As the intruder turned base to
final in alignment with the gap between my sun visor and face mask,
I squashed 'im—with the checklist.

Knowledge and experience are the prime ingredients of the
decision making process, not only when the sirens are wailing, but
at all times. Failure to take proper action in response to a routine
problem can lead to a *real* emergency. Likewise, a mindless
knee-jerk reflex course of action can turn a non-emergency into a
true crisis. Sadly, no-no's such as these are all too common and can
further be eliminated by not just a little, but *extensive* pre-flight
thought. Hopefully, the framework for such thought will be
outlined in crude form as this chapter unfolds.

The basic guidelines for emergencies during banner towing
operations are no different than any other flying situation crisis:

    1. Maintain aircraft control.

    2. Analyze the situation and take the appropriate action.

    3. Land as soon as conditions permit.

       Above all, don't

```
PPPPPPPP   AAAAAAAA  NNN  NNN IIIIIIII   CCCCCCCC
PPPPPPPP   AAAAAAAA  NNNN NNN IIIIIIII   CCCCCCCC
PPP  PPP   AAA  AAA  NNNN NNN   II       CCC
PPP  PPP   AAA  AAA  NNNNNNNN   II       CCC
PPPPPPPP   AAAAAAAA  NNNNNNNN   II       CCC
PPPPPPPP   AAAAAAAA  NNN NNNN   III      CCC
PPP        AAA  AAA  NNN  NNN IIIIIIII   CCCCCCCC
PPP        AAA  AAA  NNN  NNN IIIIIIII   CCCCCCCC
```

If there is sky above and ground below and the flight controls still work, there isn't too much that can be called a true emergency *if potential problems have been well thought out in advance*. In this spirit, the remaining sections of this chapter are offered as a skeletal framework around which an approach to the management of towing emergencies can be constructed.

EMERGENCY SITUATIONS

Engine Failure During Flight

In keeping with the spirit of the three general rules for emergencies stated above, it isn't appropriate to rank-order a procedural format for this dread circumstance. This mnemonic might serve as a springboard for developing individual checklist plans:

FLIGHT (*Maintain aircraft control, best glide speed*)
FUEL (*Selector switch position*)
FIRE (*Battery, magnetos*)
FROST (*Carburetor heat*)
FREE (*Drop the banner*)

The loss of power-assisted flight takes on a added stressful dimension when normal glide characteristics are distorted by the drag of a banner. Few would argue with the virtue of dropping the load as a *first priority* procedural item with engine loss along the beach out over the ocean. But what about the unthinkable—losing the engine over a congested area? Should the pilot retain the drag liability while maneuvering to execute a forced landing? Or, should he abandon this shackle and deliver it into the hands of fate and Lady Luck? The dilemma is solved, of course, by always flying within gliding range of an emergency landing site with allowances made for the banner tagging along. If that means flying higher, then so be it; the pilot should *never* compromise the safety of himself and the folks below. Make it clear to your customers that some areas aren't amenable to towing at the usual altitudes and the pilot in command has the final say-so on this issue.

The glide ratio will vary inversely with the total drag, which can and should be demonstrated with a practice descent for judgement development.

Should the pilot be inadvertently stranded without power in a position from which he cannot execute a glide to safety with the banner, any procrastination about discarding the load is unlikely to improve the situation. The increased glide capacity resulting from dropping the sign should enable the aircraft to maneuver to a more favorable bargaining position with the ground. The only heavy portion of the banner is the lead pole. The odds are better than even that it won't strike some unfortunate pedestrian beneath.

On the other hand, the reduced landing rollout distance required with touching down on a prepared surface with the banner might enable the pilot to select an otherwise marginal forced

Fig. 8-1. Plan ahead. Don't count on a miracle to save your bacon.

Fig. 8-2. Stalling on launch, while a wonderful stimulus for the adrenal glands, need not be fatal. Ditch the banner at once.

landing site. Such an option would hopefully be relegated to the bottom-of-the-barrel-last-ditch-effort status and never construed as proper planning (Fig. 8-1).

Stalls During Launch Maneuvers

The cause of an aerodynamic stall is exceeding the aircraft's stalling angle of attack—*period*. Although drag has nothing to do with the genesis of the stall, the reduction of drag has important consequences in correcting the situation, especially at critical altitudes. The zoom maneuver execution undeniably paves the way for a stall situation to develop. If the pilot should become distracted, however briefly, during the relatively intense moment of the aerial pickup, the following steps should be initiated at the instant an incipient stall is recognized:

1. Relax back pressure to reduce the angle of attack.
2. Add full power (throttle should already be "to the wall" for the zoom).
3. Release the banner without delay to maximize excess thrust.
4. Utilize available altitude to regain aerodynamic control.

A decision to retain the banner under such adverse circumstances could result in the banner falling back to the ground, thereby further increasing drag, aggravating the stall, subjecting the sign to significant damage, and jeopardizing the pilot and plane.

Any temptation to salvage the towing mission by retaining the banner must be dismissed as folly (Fig. 8-2).

The remainder of this chapter is devoted to situations considered to be less critical by virtue of the available time for analysis between onset and resolution. Many are simply just things to watch out for, so that the learning process can take place through educationl as means vis-a-vis the trial and error method. As nearly as possible, the problem areas will be identified in the order in which they may be expected, by grouping them into the appropriate phase of banner towing operations.

PREFLIGHT SETUP

Banner Fouling in High Winds

Proper QD clip configuration must be assured prior to flight. Although visual preflight inspection is adequate under most circumstances, high winds can introduce wave motion in the banner fabric that can foul the clips, especially the old style types (Fig. 8-3). This *can* occur in the interim period between inspection and pickup. Straps that are snagged in this manner will distort the QD clips, resulting in bending or even breakage. (This warning is the result of an actual situation that resulted in a partial banner separation during flight. After nearly two hours of routine towing operations, the strained QD clips failed without warning and 18 letters with rods plummeted earthward into a benign location. In reconstructing the situation, the winds were noted to be gusting to 20 mph and parallel to the alignment of the banner layout.) *Always* have a ground assistant monitor the banner during strong or gusty wind conditions while the pilot maneuvers for pickup.

The fabric panels used for logos are more susceptible to the wind effects on the ground than are the stock characters. An alternative remedy for this problem is to use a special clip (Fig. 8-4) to connect the logo panel to the lead pole (instead of using the stock old-style connector rod). There is virtually no overhang by the prongs and consequently no chance for entrapment due to the mechanism previously described. Unfortunately, these are antiquated clips and in short supply. Should you run across these clips while scouting around for used equipment, they are well worth saving.

Base Placement

Although unlikely, engagement of the pickup bases is possible on a low approach. Always align the upper crossmember of the

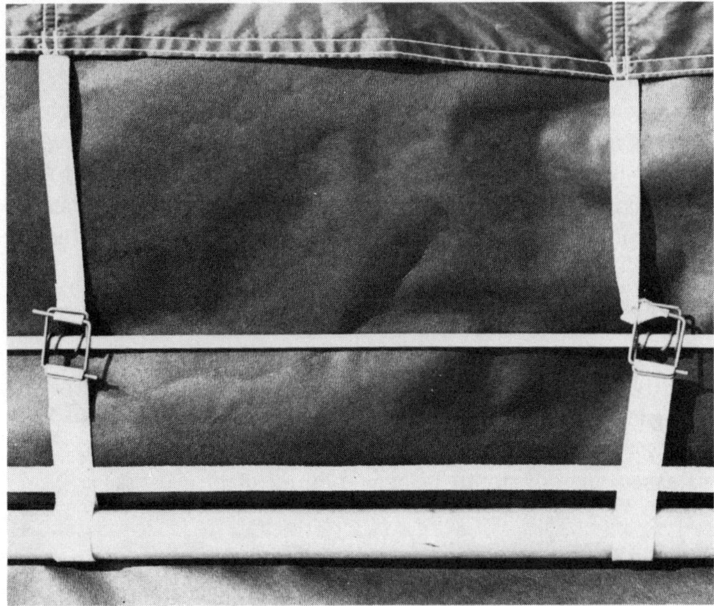

Fig. 8-3. Careful attention to detail must be given during the banner preflight. A fouled connector strap, like the one at right can be easily overlooked. If flown in this configuration, the QD will bend and can break.

bases parallel to the flight path of the tow plane to avoid this problem.

Taxi Precautions (Tailgear Aircraft)

Pivoting turns on other than paved surfaces may cause the grapple hook cable to be caught in the tailwheel strut of conventional aircraft. If this situation were allowed to go uncorrected, the cable could wedge in the strut mechanism and the banner might not release as advertised. That problem is addressed in a later section of this chapter. The message here is not to let it happen in the first place. Avoid pivoting turns in high grass or on rough surfaces.

PICKUP (AERIAL LAUNCH METHOD)

Engagement of the Pickup Loop

It is possible to engage the pickup loop with the main gear struts, tailwheel, or nosewheel. Of these, the latter would create the most serious situation by inducing a pitch-down moment from forward of the center of gravity thereby decreasing stability about the lateral axis of the aircraft. Engagement by the maingear struts or the tailwheel isn't nearly so serious as the cause (poor pilot judgement).

To compensate for such disasters, maintain an increased angle of attack and decreased airspeed in order to keep the towline against the fuselage. Announce the problem and intentions to the tower controller or over Unicomm as the case may be and land in accordance with the procedures stated at the end of this chapter.

Pole Pickup

Should a pickup pole happen to tag along for the ride, avoid overflying populated areas as much as possible and return to a designated drop zone for release. If the towplane has excellent rearward visibility, the problem can be recognized and the tow release actuated prior to the banner lifting off the ground.

Twisted Banner

Experience recalls that a banner will infrequently twist on itself during launch, usually because of a broken rod. Other than the observation that this induces gyrations and spiraling tendencies, this is generally a benign problem. The crippled banner can be safely guided to the drop zone with the added precaution of being alert for vertical oscillations that may cause the sign to deviate lower than its normal position relative to the tow plane.

GRAPPLE HOOK DEPLOYMENT

General Considerations

Every reasonable precaution must be taken to avoid unwanted entanglement of the grapple hook and its attendant cable during deployment from the cabin or cockpit. The pilot's immediate environment contains a variety of protruding knobs, handles, levers, and controls that tend to lure loose loops of cable. The towhitch release handle, for one, needs no introduction as the top-o'-the-list contraption to stay away from. Be aware that the hook and/or cable can snag on loose clothing or wrap around arms and wrists. In the event of entrapment, maintain aircraft control as the number one priority while continuing to clear for other aircraft in the area. It is advisable to retreat to remote airspace at a safe altitude before attempting to disentangle entrapped cable.

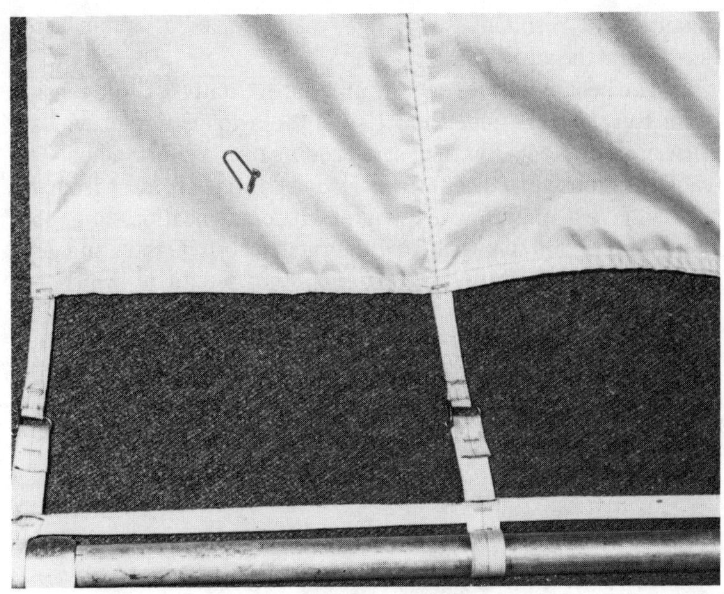

Fig. 8-4. An alternative type of clip can be used on the lead assembly, instead of a connector rod.

Conventional Aircraft

There are at least two problems unique to taildraggers:

1) Grapple hook cable entrapment by the tailwheel strut can occur unbeknownst to the pilot. This is to be avoided by following the recommended procedures outline in Chapter 6 (i.e. tension on the cable and use of the semi-rigid plastic tube around the proximal tow cable.) If fouling is confirmed or suspected, *do not* attempt banner pickup but rather return for a landing on a hard surface runway. (Landing on a soft surface is acceptable, if necessary, but engagement of the grapple hook in the turf is to be expected with subsequent separation at the safety link.) An attempt could be made to discard the cable in a designated drop zone *only* if continuation of flight to a landing can be completed without overflying populated areas.

2) Certain aircraft models (e.g. Citabria, Supercub) have rudder horns so situated as to invite the grapple hook cable to drape *over* the control actuator (Fig. 8-5). This, too, may go unnoticed at first because it doesn't require much rudder power to overcome the drag of the grapple hook assembly. Engagement of the banner, however, will slam that rudder to the opposite side in a New York

minute. The resultant yaw is best characterized as an unwanted surprise at the very least.

The best medicine, once again, is preventative. Cable tension prior to release is important. Use of the nylon casing previously described prevents the cable from coiling over a short radius. The cable and hook should be thrown out and down to arc away from the aircraft fuselage. Following Chapter 4 recommendations, the cable should be rigged to pass below the rudder horn. Gasser suggests two systems as further safeguards, the details of which are described in Figs. 8-6 and 8-7.

Fig. 8-5. Unprotected rudder horns, such as the one on this Citabria, are vulnerable to cable fouling during deployment of the grapple assembly. See text.

144

Fouling of the rudder horn is best remedied prior to banner pickup by unloading the offending cable in the drop zone or landing with the hook in trail as described in the previous section.

Should either of these problems be present after banner engagement, and release attempts are unsuccessful, follow the procedures for landing with a banner in tow outlined at the end of this chapter.

CRUISE

Overheating

The perils of operating aircraft are common knowledge to all aviators. The varied susceptibility among different aircraft to overheating problems during slow flight conditions leaves the burden of investigation up the reader. Preventative measures might include:
- [ ] Don't fly.
- [ ] Installation of an oil cooler.
- [ ] Use of cowl flaps.
- [ ] Cowling removal.
- [ ] Use of flaps to reduce the pitch attitude necessary for level flight, thus improving airflow characteristics.

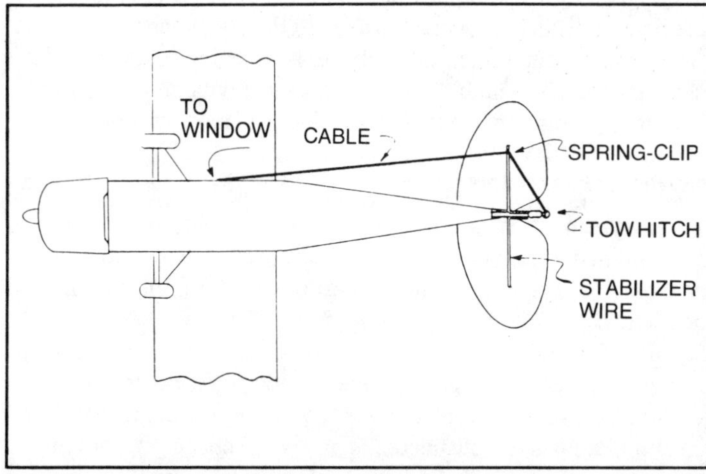

Fig. 8-6. A spring-action clip serves to hold the cable away from the fuselage until the hook has been dropped. This should be mounted on the lower stabilizer wire near its junction with the bottom surface of the stabilizer. The cable is routed from the release mechanism to the clip and then forward to the cockpit (courtesy Gasser Banners, Inc.).

□ Matching banner length with the aircraft's capabilities and flight conditions.

□ Use of recommended fuel grades.

□ Use of proper fuel mixture management.

Should an overheating situation arise during a towing performance that does not respond to whatever corrective means are available to the pilot, the mission should be terminated *at once* at the nearest suitable airfield. If travelling to an air facility poses any imminent danger to the continuation of flight, the banner will have to be abandoned in the interest of flight safety. It would be a nice gesture to tip the farmer whose field you selected to dump in.

It should be noted that cylinder head temperatures may become elevated without corresponding increases in oil temperature readings.

Weather

It will be assumed that none among the readership is interested in flying into thunderstorms, hurricanes, or tornadoes. Consequently, no admonitions will be tossed out regarding those subjects. Instead, a more insidious adversary will be mentioned in hopes of enlightening the unaware and reminding the experienced. During conditions of low temperature-dewpoint spread (less than 5°) the visibility can rapidly deteriorate from VFR to IFR literally in minutes. This is especially true with the coexistant calm winds. Adjust your mental flight computer to low towing speeds and don't stray far from acceptable recovery zones. Flying in the goo while towing is not only bad style, it is a violation of most waivers.

Inadvertent Banner Release

Little mention is needed about the precautions required to prevent inadvertent banner release during towing operations. Any unexplained acceleration should alert the pilot that this may indeed be a long day.

If it does happen, notify the appropriate controlling agency or contact the nearest Flight Service Station to report the incident. Collect as much information as possible, including time and location, known or suspected injury or damage to persons or property on the ground. An account of the details should be written out at the earliest convenience for future reference and investigation. Inspect the plane and towhitch after landing and take pictures, if possible. The cause of this serious problem *must* be identified *and corrected* prior to commencing another towing performance.

Failure To Release

If, for any reason, the towhitch release mechanism should malfunction over the drop zone, execute an immediate go-around

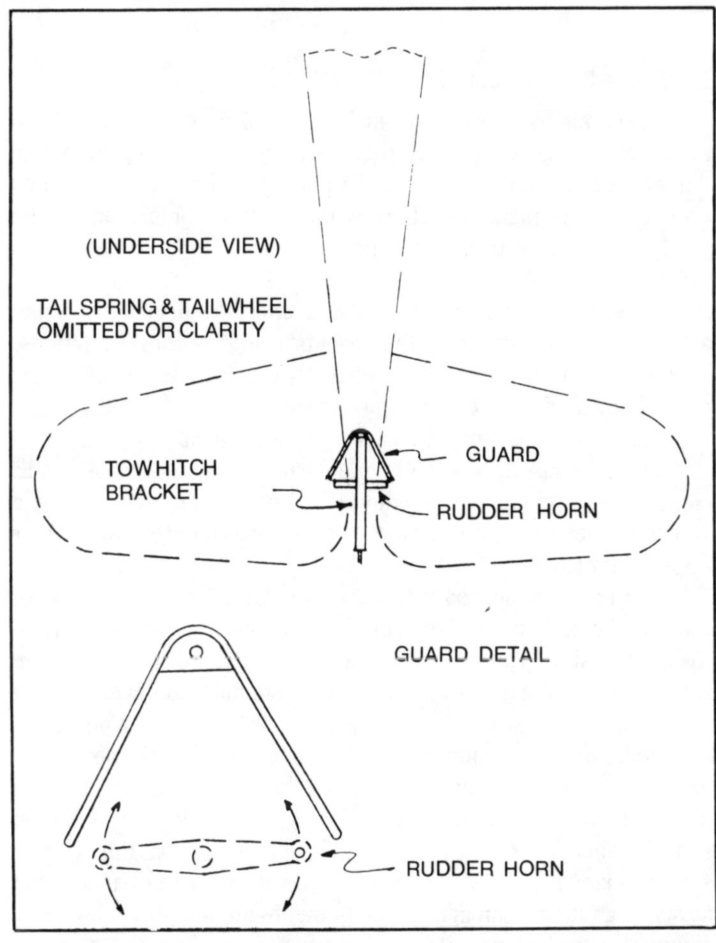

Fig. 8-7. A guard projecting outward and aft from either side of the fuselage serves as protection from cable fouling. This can be fabricated from a steel rod ¼" to ⅜" in diameter, bent at the center to form a "V" and attached just forward of the rudder horn. A small plate can be welded to the rod at the bend to provide a base which can be attached to the fuselage lower members. The length and spread of the "V" should be designed so that a cable sliding along the side of the fuselage toward the rear would be deflected outward around the rudder horn and clear of the steering arms and springs (courtesy Gasser Banners, Inc.).

and attempt to maintain a ground track over sparsely populated terrain. Attempt to ascertain the cause of the malfunction (tower fly-by, ground crew observer, etc.) if at all possible. Fuel permitting, it is probably worthwhile to attempt another release, but in all likelihood, submission to the inevitable ending of landing with the banner in tow will be necessary.

Landing With a Banner In Tow

It sounds like aerial hara-kiri. In actuality, it is not nearly so ominous, though admittedly there are some harassing moments during the first such experience. By abiding with the recommended procedure, the plane and its now unwelcome companion can be safely transitioned back to the ground with little or no damage to either.

The basic technique is to fly a short field landing, using airspeeds consistant with the manufacurer's recommendations. The approach should be planned with careful consideration for obstacle clearance for the trailing banner.

Announce your intentions on the radio; include the information that the runway will be momentarily closed to enable you to clear the equipment after landing. Fuel permitting, it would be a courteous gesture to allow others in the pattern to terminate their flights prior to this.

With a steep approach to keep the banner above the level of the tow plane, select an "aim point" far enough down the runway on the upwind side to permit the banner to also touch down beyond the threshhold and clear obstacles. The flare should be delayed slightly as the airspeed will fall off precipitously when the power is chopped. A firm touchdown ideally occurs just before the banner lead pole strikes the runway. Positive breaking at the point of an insipient skid will provide a surprisingly short rollout to a full stop straight ahead. The banner can then be pulled off the landing surface manually. *Don't* attempt to taxi off with the banner still attached as it will damage the equipment and probably take a few runway lights along with it.

A hard surfaced runway offers the least resistance to the dragging banner. A properly executed approach and landing will even spare the safety link, although it should be replaced after such stress. Grass or other rough surfaces offer an acceptable second choice, but the banner will probably suffer greater damage, especially at the loop-clip interfaces.

This problem also needs identification of cause and appropriate remedy prior to subsequent tows. Repeat performances are unwarranted.

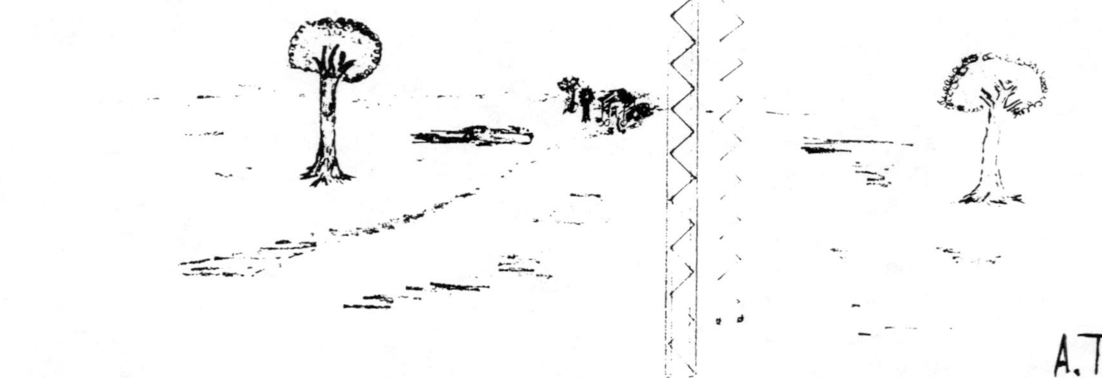
EQUIPMENT REPAIR
BANNER REPAIRS
NO WRITING
A. TAYLOR

Chapter 9
Equipment Repair

The repair and maintenance of banner equipment is a necessary function that makes good practical and economic sense. Little is required in the way of technical skills with the possible exception of the need to develop some finesse with the sewing machine. Otherwise, the only major ingredient needed for proper care of the banner components is time. This can usually be scheduled for more leisure moments with idle hands. Prompt repair and conscientious maintenance clearly extends the life of the equipment with important consequences. It reduces inventory purchases, increases inventory useable stock, and keeps the banner at its very best condition from both a cosmetic and functional standpoint.

SPECTRUM OF REPAIR

Not all of the banner equipment is amenable to home repairs. Problems with the tow hitch and/or its installation should be tended to by a qualified and licensed A & P man. Defective towropes should be replaced vis-a-vis repaired in most instances. Furthermore, the very heavy nylon webbing used on the mast and tail assembly elements requires the use of an industrial quality sewing machine. The latter represents a substantial investment by anyone's standards so that repairs of these areas are best left up to the factory.

Fortunately, most of the equipment problems that routinely surface are relatively minor and are easily corrected with a

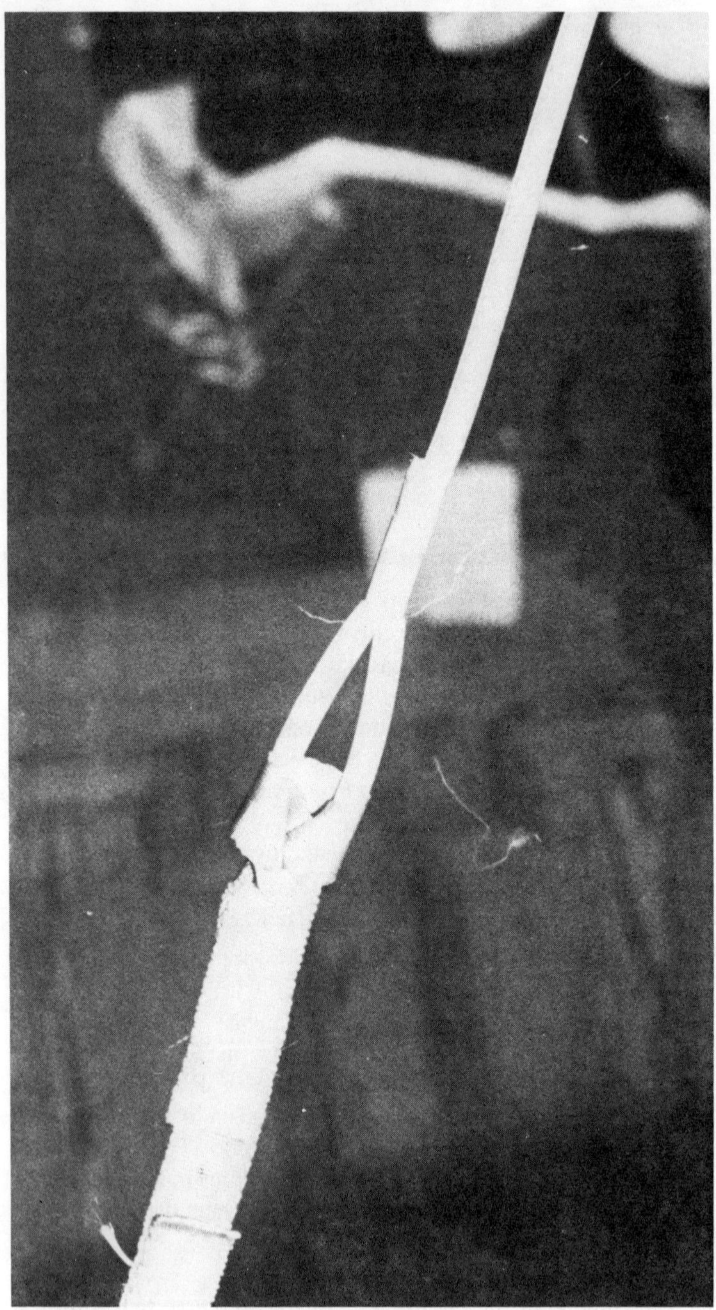

Fig. 9-1. Cable loop union with metal sleeve secured with Nicopress tool.

minimum of do-it-yourself skills. Without itemizing each and every little malady that may afflict the system, the subsequent list outlines the general areas to which repair and maintenance techniques may apply.

1. The grapple hook cable
2. The pickup assembly
3. Tow rope connections
4. Bridle Harness connections
5. Connector rod clips
6. Nylon fabric defects
7. Nylon strap webbing defects
8. Special equipment

REPAIR SUGGESTIONS

The Grapple Hook Cable

Periodically, it may become necessary to replace the cable that suspends the grapple hook. While the cost for a replacement cable direct from the factory (and complete with the grapple hook) is still modest by today's standards, it is possible to cut some financial corners here without sacrificing safety in any way.

Cable replacement on the grapple hook requires the use of a "Nicopress" tool. This device affixes the metal sleeve to secure the opposition of cable at the terminal loops (Fig. 9-1). Although the local hardware shops have shelves full of substitute gimmicks for this purpose, the awesome importance of this procedure *cannot* be overstated. It follows that this repair item be accomplished with the appropriate tool or otherwise delivered into the hands of competent professionals.

It shouldn't be too difficult to find local assistance for this specialized procedure, since aircraft control cables are secured with the "Nicopress" tool. One is likely to be available for hire at a nominal fee at any aerodrome where airframe structural arts are practiced. The cable is commercially available at many hardware stores. You will need about 30 feet of 3/32" vinyl coated "tiller" cable.

The Pickup Assembly

For all practical purposes, the only components of the pickup assembly susceptible to damage are the uprights. If either pole is inadvertently struck by the towhook during pickup, it will bend emphatically and irreversibly. Attempts to straighten deformed conduits invariably terminate in metal fatigue and subsequent separation at the point of maximum stress.

New uprights can be quickly improvised from 6′ lengths of aluminum conduit available at most any hardware, building supply,

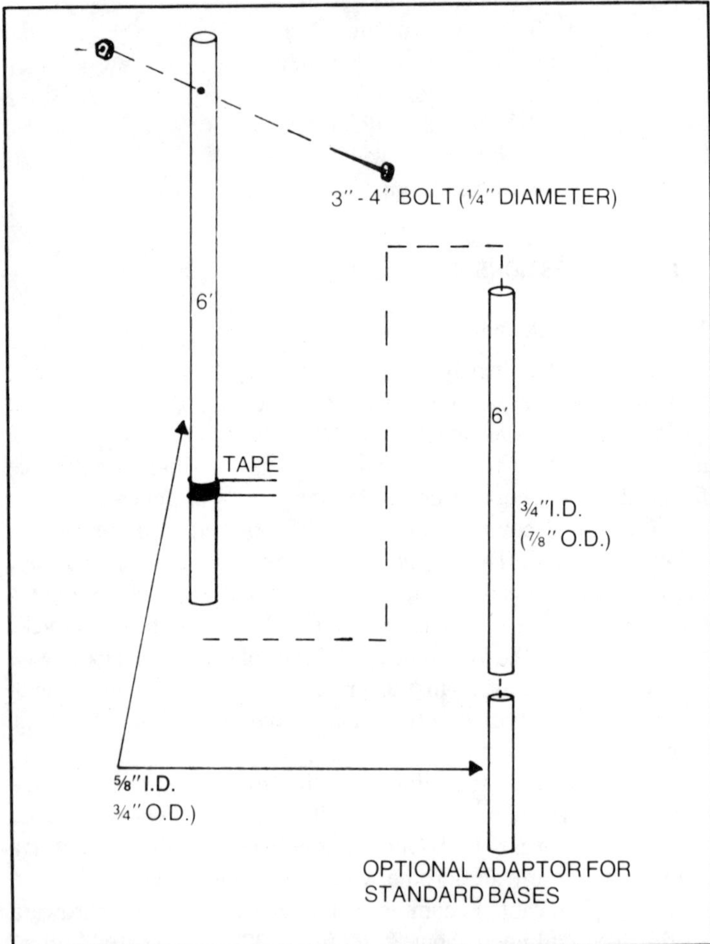

Fig. 9-2. Pickup poles can be quickly constructed using aluminum conduits or similar tubing. Drill a ¼″ hole near the top of the upper segment into which a 3″ or 4″ bolt can be secured. The ⅝″ tubing has an outside diameter of ¾″ which is the inside diameter of the larger lower segment. A small length of ⅝″ tubing can be inserted in the lower end of the bottom segment to adapt to standard base receptacles.

or home improvement center. Two lengths are needed for each upright. The outside diameter of the section destined to be the top of the upright should snugly conform to the inside diameter of the lowermost segment. ⅝″ and ⅞″ conduits interconnect nicely for this purpose. A bolt is needed for each upright to be constructed.

Following the schematic of Fig. 9-2, drill a ¼″ hole approximately 4 inches from the end of the smaller diameter tube, *slightly* angulated (not more than 5° - 10°) off of the perpendicular. The bolt is then inserted into the passageway just drilled with the protruding portion bordering the lesser angle toward the near tip of the pole. Securely torque the nut without exerting undue force that might crush the conduit form. Covering the threads of the bolt shaft with plastic or vinyl tape is an optional finishing touch directed at reducing frictional forces.

Approximately six to eight inches from the opposite end of the same pole segment, wrap eight to ten layers of masking tape around the tube to act as a laminar stop collar. This accomplished, the two segments need only to be joined to complete the task. (The inside diameter of the stock bases is less than the outside diameter of the recommended ⅞″ pole of the lower segment. A small segment (less than one foot) of ⅝″ tubing may be inserted into the lower segment as an adapter for this purpose.)

Rope Connections

Any time there is an interface between rope and metal, the potential for accelerated wear and abrasion exists. There are several such connections within the banner towing elements.

1. The bridle harness attachment points to the lead pole.
2. The focus of the bridle harness where it slips through the bridle harness ring.
3. The towline connector clip.
4. The attachment of the towline to the "O" ring.
5. The pickup loop interface with the "O" ring.

Of these, the bridle harness ring suffers the most wear because the harness ropes chafe against the metal guide at a single point.

The knots that secure the rope connections at other points are exposed to abrasive wear, especially if the takeoff method of launch is employed. Covering these vulnerable points with a protective wrapping, such as electrical tape, is worthwhile preventative maintenance.

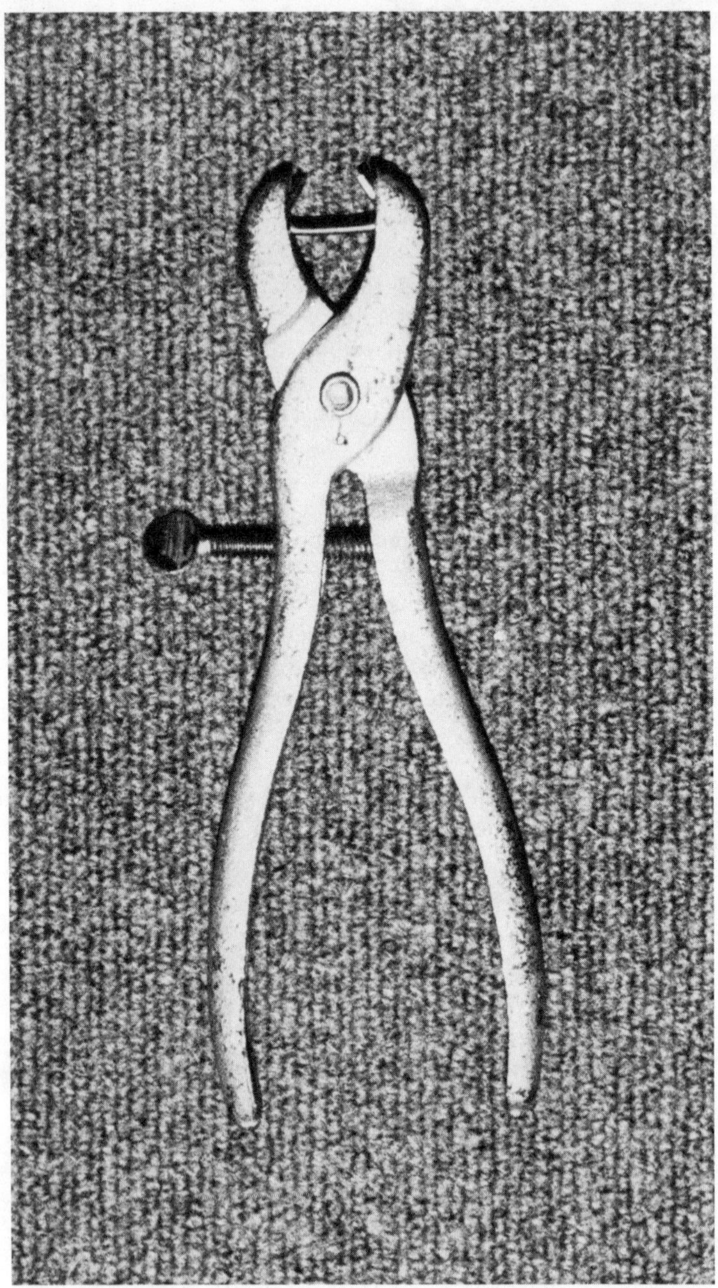

Fig. 9-3. Rope connections are secured with "pig rings" that can be set with pliers or this inexpensive tool (hog pliers).

Repair and replacement of rope ends requires an inexpensive tool from the local hardware known as "hog pliers" (Fig. 9-3). The tool is used to crimp the "pig rings" around redundant strands of rope for steadfastness. Suggestions for acceptable connections at different points are depicted in Figs. 9-4 through 9-6.

Rods & QD Clips

Broken rods are somewhat hazardous to have lying about due to the sharp fiberglass splinters present at the break. These can be sawed off and discarded. The segments remaining can be fitted with old bent clips and seated in a power drill for mixing paint.

Bent or broken clips should be replaced before using the rod for any sign assembly. Any aberrant configuration might put undue stress on the character loops and increase wear, possibly causing failure. New clips are cheap and replacement is easy, so there is no sense in cutting corners in this department.

Either style rod should be marked off according to the dimensions shown in Fig. 9-7. Keep one rod around as a "standard" so there won't be a need to measure each time a repair is necessary. The technique for each style rod is different and will be discussed separately.

Old Style Rods and QD Clips

Refer to Fig. 9-8 for the technique for removal and replacement of old style equipment. These are really quite easy to repair, even if replacement of one of the center clips is required. A bench vise makes the job much easier. The "L" shaped persuader is available from the manufacturer or at any hardware store.

New Style Rods and QD Clips

Inasmuch as these dudes don't slide on and off, the new-style QD clips take more time and effort to replace. Again, a good bench vice is the central piece of equipment. Following the sequence in Fig. 9-9, secure the NS rod in the vice with the clip to be removed just outside the grips. With a hammer and screwdriver (or other suitable persuader) give the retaining clip in the center a whack, remembering that the expert is the guy with the biggest hammer. Once the retaining clip has been ejected, the QD clip segments are free to slip off the ends of the rod. Unfortunately, this has to be done for each successive clip to get at and replace a faulty one. Fortunately, they don't bend very often.

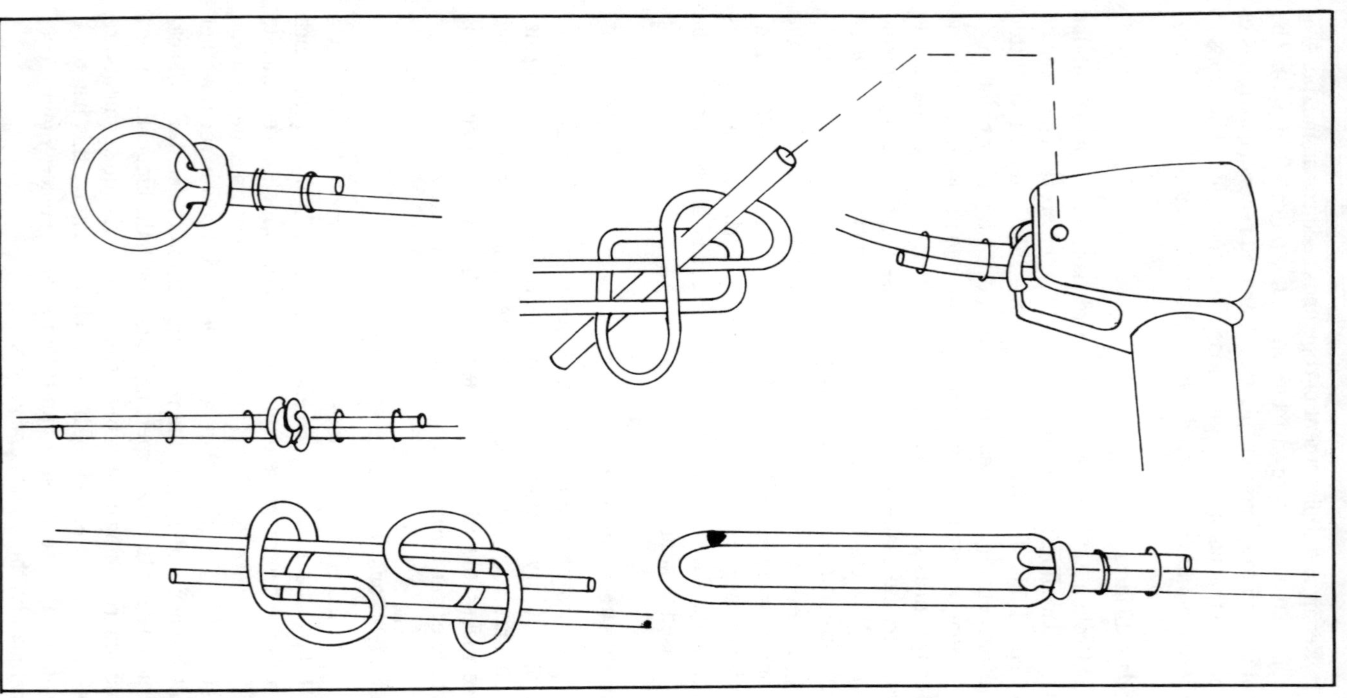

Fig. 9-4. Terminal connection of the pickup loop.

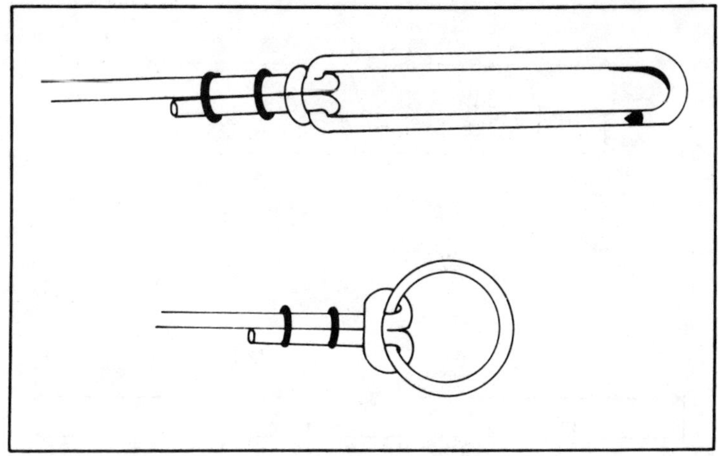

Fig. 9-5. Towline to "O" Ring and towline to rope connector.

Installation is logically a reverse of the removal procedure and considerably easier.

Fabric Repair

Prompt repair of even minor defects prevents progressive damage and clearly prolongs the useful life of textile components. When the airstream gets ahold of the tiniest defect, the repeated whipping action results in cumulative deterioration of the fabric.

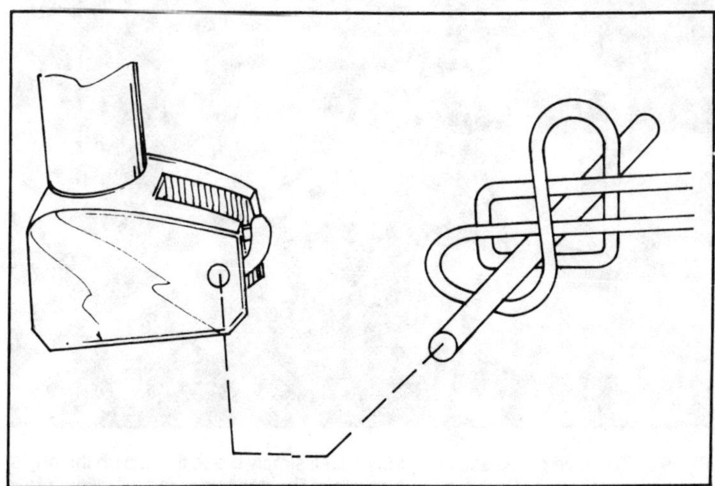

Fig. 9-6. Bridle harness to mast guides.

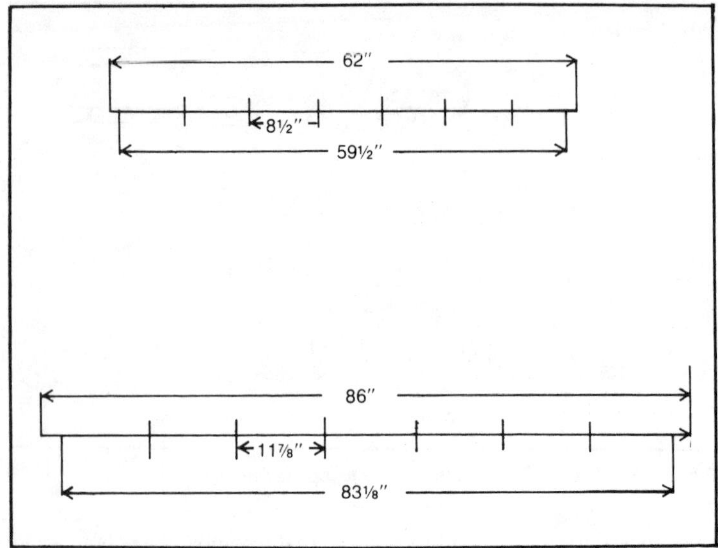

Fig. 9-7. Spacing for QD clips on connector rods.

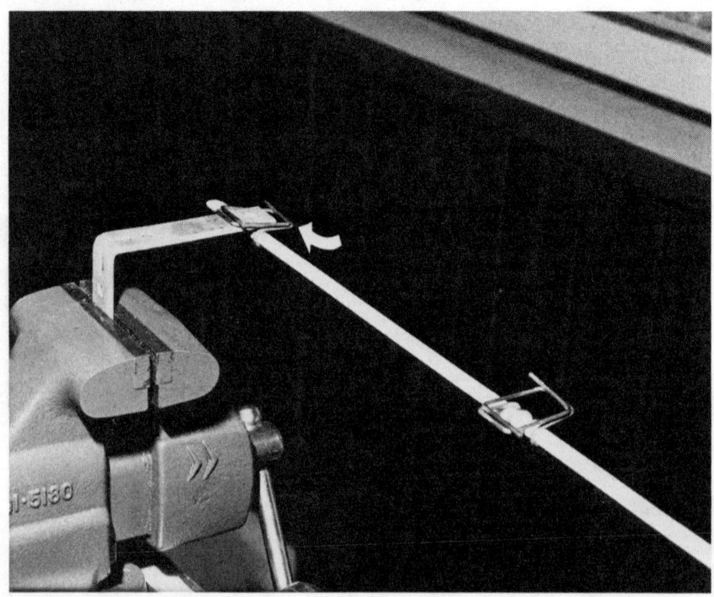

Fig. 9-8. To move or replace an old-style QD, simply brace the clip on the angle bar as shown and push up on the free wing. The connector rod will slide freely within the spring core.

This is costly and it looks bad. When taken to extremes it can be dangerous. A three hour flight is long enough without the added worry over the condition of a marginal loop noted prior to flight.

Of course, the best preventive maintenance is to avoid the harsh conditions that may predispose to fabric damage in the first place. A smooth launch area free of twigs, rough weeds, and the like will minimize the need for repair. Slower towing speeds, in keeping with the capability of the tow plane and pilot, also reduce wear and tear on the banner equipment. But even the most optimal conditions won't entirely prevent the miscellaneous rips, frays, frazzles and frizzes that crop up from time to time. Repair, therefore, is inevitable and makes good economic sense.

You may be fortunate enough to have a family member, neighbor, friend, or nice old lady down the street with a passion for nylon fabric repair. And, for a small fee, damaged letters can be

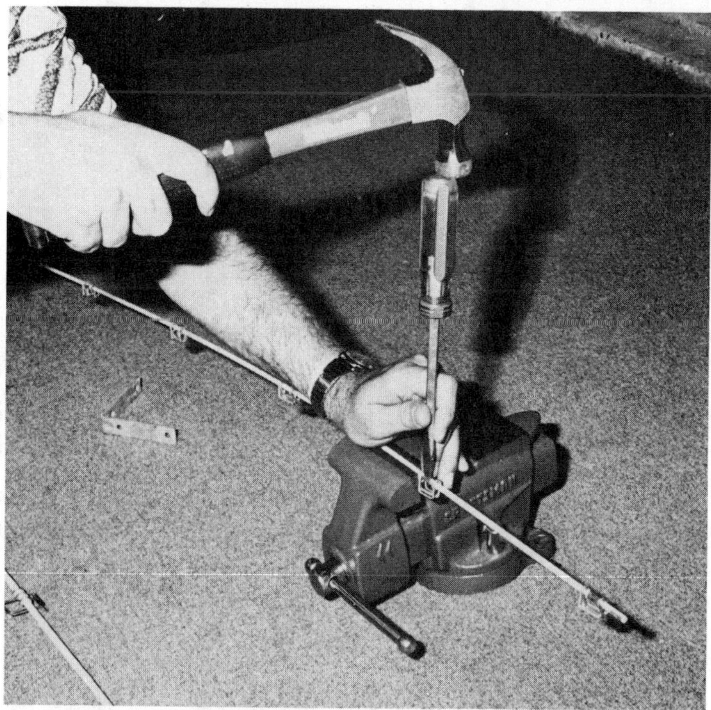

Fig. 9-9. The new-style QD's will not move until the central retaining clip has been forcibly removed. Secure the rod in the vise with the clip aligned concave up. Using a hammer and screwdriver as persuaders, extract the central clip. The QD is now free to slide in either direction.

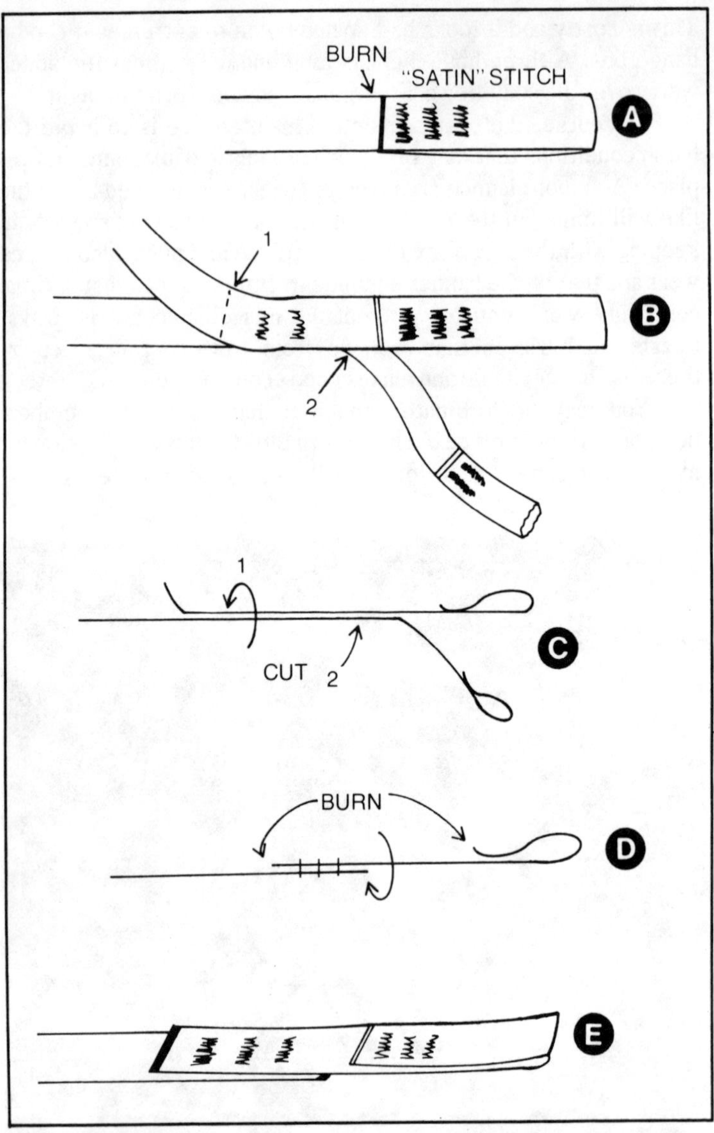

Fig. 9-10. Worn or broken connector loops should be replaced immediately. (A) Begin by folding a nylon strap on itself and secure the doubled portion with either a multiple zig-zag stitch or a "satin" switch. Match the "new" strap to the length of the defective one and unite the fabric as shown in (B) with a similar stitch. The stitching will define the excess straps, to be cut away (points 1 and 2 on B and C). Loose frays are always present after cutting and should be fused by lightly candling with a match (D). The finished product (E) will be strong and dependable. Triple stitching is recommended.

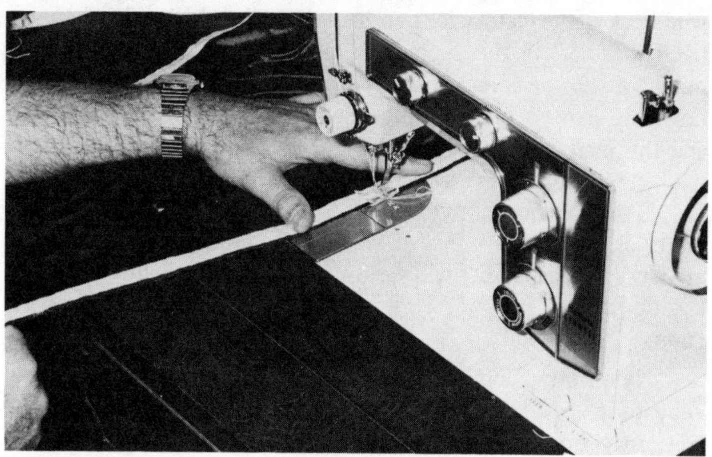

Fig. 9-11. As an alternative to loop repair, the entire strap can be removed from the character and a new strap sewn in. Use at least two rows of straight stitches.

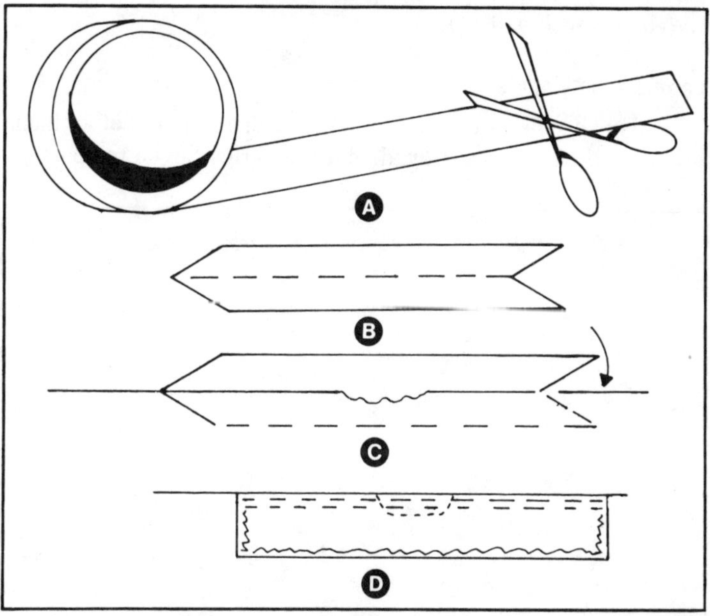

Fig. 9-12. Repairing fabric borders: (A) Cut a sufficient length of Ripair tape. (B) Fold the repair fabric along its longitudinal axis and remove the sticky-back. (C) Slip the fabric border into the crease of the repair material, avoiding air bubbles and creases. (D) Close the remaining flap of repair tape, again using care to avoid an irregular surface in the finished product. The repair is made permanent by straight stitching the lateral seam border and a zig-zag around the remaining perimeter.

sent back to the factory for overhaul. But what do you do when you need that "S" in the repair box and your family is sick, your neighbor is on vacation, your friend is out of town, the nice old lady down the street is in the hospital having her gall bladder ripped out, and the mail carriers are on strike? If you can get to a sewing machine, the problem is solved.

Given a choice, nylon thread is the best choice of repair material, though any thread will do. A sewing machine with a zig-zag stitch capability will yield stylish, quality repairs.

Loops

Nylon strap loop ends will show wear after prolonged and repeated use. Excessive stress, such as landing with a banner in tow, may even rip the fabric. New loops can be constructed in seconds using the technique shown in Fig. 9-10 from an old strap. If preferred, the damaged strap can be completely removed and replaced by a new strap by completely sewing the new strap across the character (Fig. 9-11).

Letters and Symbols

Many problems can be restrained with a match or other flame source. Loose threads, frayed edges, and rips can be temporarily

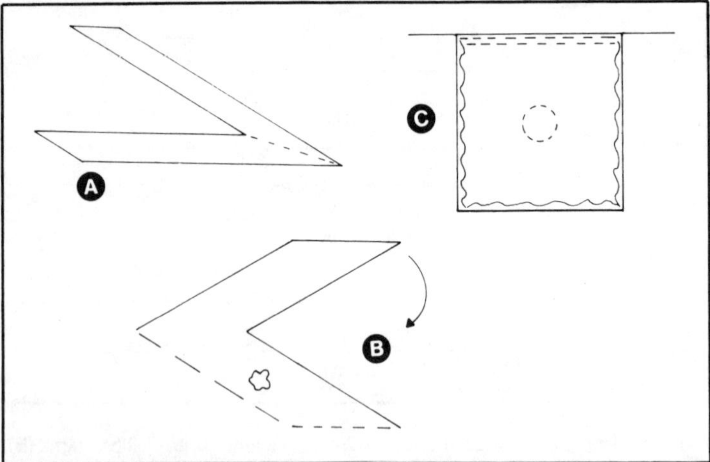

Fig. 9-13. Defects close to the border, but not on the edge of the fabric can be fixed in a manner similar to that shown in Fig. 9-12. The only difference is that the Ripair tape should be folded along its width to produce a symmetrical patch. The technique is otherwise identical.

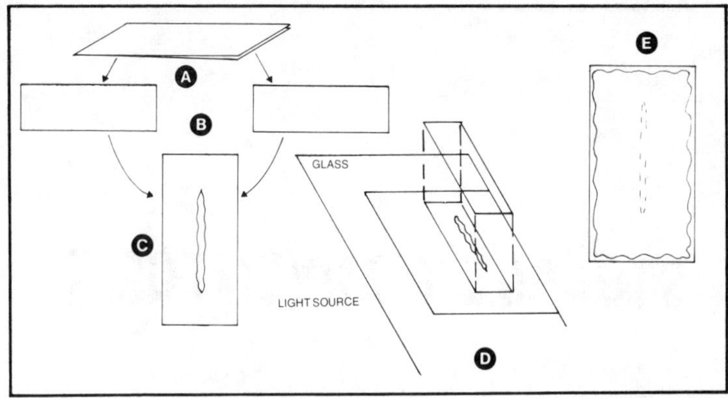

Fig. 9-14. For centrally located defects or large rips not amenable to previous techniques, it is necessary to cut the folded Ripair tape (A) into two halves (B). It is helpful to trim or burn loose threads and frays from the center of the defect before placing one of the patch halves into position (C) over the rip. The fabric can now be turned over (patch down) and placed on a glass-top table (D) with a source of illumination from beneath. Align the companion patch with its mate underneath and press it firmly in place. A zig-zag stitch (E) around the repair border completes the job.

halted from progressing by singeing the loose fuzz off the periphery, using care not to burn healthy fabric. Use caution—burning nylon melts and drips and burns and hurts.

Other field repairs can be accomplished with the use of Ripair tape, a sticky-back nylon fabric available from the manufacturer. Such repairs are only temporary, and need to be sewn later for continued integrity.

Rips are usually accompanied by modest stretching of the nylon fabric around the defect. Be certain to compensate for this during repair to avoid uneven spacing between the straps when the job is complete. A small pleat is usually all that is needed under the repair material.

Refer to Figs. 9-12 through 9-14 for fabric repair suggestions.

Special Equipment

Logo panels and other special equipment items are repaired much in the same manner as the routine fabric parts. Being of heavier material, however, the repair demands may outclass the average home sewing machine. If so, help may be locally available at a sail repair shop, for example. The option of sending the equipment back to the factory for repair is always open, if shipping delays are tolerable.

STORAGE & INVENTORY

Chapter 10
Storage And Inventory

Organization of the banner equipment is an absolute *must* if one wishes to avoid countless hours of sorting through boxes and boxes of twisted nylon looking for a particular letter or character. Proper storage likewise facilitates banner assembly along with the added benefit of increasing the useful life of the equipment. This chapter addresses some of the problems associated with the logistics of handling, storing, and cataloging a banner inventory and offers some suggestions for coping with these issues.

INVENTORY

Any operator will want to have an adequate supply of the right kind of characters for his inventory. This will, of course, depend upon the intended use corresponding to the proposed or actual type of operational requirements. In an attempt to generalize the ungeneralizable, the collection of paragraphs to follow may lend some idea as to where to start for the beginner.

1. *Anticipated business volume* is an obvious factor that will dictate the number and assortment of characters needed in stock. During peak periods, it is not uncommon to launch up to 5 or more different banner copies in a single day. With adequate help, it is not unthinkable to swap and trade letters from one copy to the next between flights. For the most part, however, this will keep you busier than a one-armed wallpaper hanger on a pogo stick. On the occasions when multiple flights are necessary within a short time

frame, e.g. for a football stadium crowd, this sort of juggling act truly becomes folly.

2. *Long term contracts* (i.e., the same copy on a repetitive basis such as weekly or monthly) will make additional equipment purchases necessary so that banner copies can be assembled and stored as a unit in themselves. Without the additional stock, it would become necessary to disassemble an existing banner to fulfill interim contracts, only to have to reconstruct the original long term copy at a future date. Such duplication of effort is annoying and clearly not worth the effort.

3. *The size of the banner equipment* - e.g. 3', 5', 7' - limits the length of the banner copy and thereby alters the inventory demand. This exists as an inverse relationship, i.e., fewer 7' characters would be needed than 3', etc., relative to the towing capacity of a particular aircraft (see Chapter 1).

Although the banner copy is physically limited only by the capability of the towing airplane, there is a practical upper limit to the number of characters any one copy can or should have. An "average" banner copy has around 25-30 characters. Depending upon the content of the message, more than 40 letters starts to become somewhat unwieldy to the eye (as well as to most tow planes). Advertisers should be discouraged from trying to cram too big a mouthful into a banner copy. Long messages are generally best communicated through other media.

4. *Aircraft towing capability* will clearly figure into any decision regarding equipment needs. An operator using a 300 Ag-Wagon would logically keep a larger inventory than his counterpart with a Cessna 150.

5. *Operational response time* should be given due consideration in selecting the assortment and quantity of your banner inventory. "Normally", one would expect some lead time between a request for a banner contract and the performance date, thus allowing time for ordering and receiving any additional equipment needed from the manufacturer. The real world, curiously enough, is chock full of impulsive types who demand (and pay for) immediate attention. This population seems to be skewed among the panic-prone subset of promoters, advertisers, and political special interest groups. The scenario goes something like this: A concert promoter sinks his advertising budget into high cost television advertisement during prime time to generate sales for an upcoming concert performance of a not-so-well-known rock group. Sales are below expectations, so he tosses in another couple of grand to dominate

the tube on the third day prior to curtain call. He is dismayed to learn that some international crisis has prompted all the networks to broadcast uninterrupted documentary coverage, thus preempting his advertising grand-slam forever. Furthermore, a more popular group will appear the following night and his surveys show that most of the teeny-boppers are saving their ticket money for the more fashionable group to follow. So he calls you up in a last ditch effort to salvage the concert (and his job, maybe) with banner advertising from sunrise to sunset for the next two days. If this sounds unrealistic or too fantastic to be true, just ask any experienced operator about his recollections of last minute requests. The author, for one, can testify that he is no less than $1,000 richer for a day's work *requested at 9:30 pm of the previous evening*. The point of all this rambling is that having an adequate inventory can make a significant difference in the accounts receivable department. The overnight capability of today's air freight system certainly narrows the distance between you and your banner equipment supplier but even this fantastic time service may prove just as good as never to the impulsive client. For a few impulsive-type copy openers:

> HAPPY ANNIVERSARY...
> HAPPY BIRTHDAY....
> CONGRATULATIONS....
> GRAND OPENING.....
> VOTE FOR.......

It's not a bad idea to keep these copies assembled and stored for immediate use, should the occasion arise. It will save time and reduce your response time to "same-day service" in exceptional cases. At the very least, you should be able to construct any of these messages from existing stock.

6. *Multiple aircraft* towing banners can obviously support a larger inventory, provided the business is there. If these aircraft are not based at the same airport, an even larger supply of banner stock may be justified to avoid shuttling equipment back and forth over long distances between launch sites.

7. *Excessive wear and tear* allowances should be considered if your particular pickup area is less than optimum. Rips, holes, and frays may appear in the equipment from launching over rough surfaces. This is best avoided by selecting a more appropriate pickup site, but in some cases this is not a practical or viable option. It is best, then, to pad the inventory as needed for such

nuisances for it would indeed be an expensive sin to become grounded for the lack of a single letter or numeral.

8. *Frequency of usage* of the alphabetic characters dictates the qualitative assortment of letters needed. It only makes sense to store more "E's" and "S's" on the rack than "J's" or "Y's." (I was recently asked to tow a banner copy reading: "YES-YES-YES-YES-YES-YES-YES-YES." Needless to say, the client wanted it "tomorrow." Since that request well exceeded my stock of "Y's," I quickly responded by asking for an alternative copy, such as "NO-NO-NO-NO-NO-NO-NO." This, for some reason, was unsatisfactory to the client. Happily, we came to an agreement on an alternative copy of a more conventional-type message.)

Numerals do not enjoy the frequency patterns seen among their alphabetical counterparts. Telephone numbers, addresses, dates, and prices account for the majority of requests for number symbols. As these tend to be random, little can be offered in the way of guidelines for inventory stock. Any special local circumstances, such as a telephone exchange of 555, etc. might be worth covering in the inventory. Above this, a stock of 2 each of the numerals would probably cover a fair majority of requests. There is always the option of spelling out a short number, e.g. CHANNEL TWO.

The letters "M" and "W" are fully interchangeable. Likewise, "6" inverts to read "9." The alphabetical and numerical "0" are identical. Duplication in the inventory is therefore an unnecessary expense.

The use of punctuation marks is not uncommon. In an attempt to curtail banner length or cram more information into the message, clients often make liberal use of abbreviations. It isn't always necessary to place a period after certain abbreviated words but it is probably worthwhile to stash a few periods on the shelf. An apostrophe is often an integral part of the construction of a name. At least one is recommended. Dashes help to interrupt and separate phrases within a copy and they are easier to use than successive word spacers. Commas, colons, quotation marks, and other such miscellany are best left for special order items as the need arises.

Special symbols such as the dollar sign, numeral sign, ampersand (&), and per cent symbol (%) are requested upon occasion. Again, they can be spelled out in many cases if banner length permits.

Table 10-1 has been composed to show, in a comparative way, the frequency of usage of alphabetical characters based on records

Table 10-1. Frequency of Letter Usage Based on Actual Assembled Banners (Columns 2 and 3). Figures Rounded to Nearest Percent and May not Add to Exactly 100.

LETTER	GASSER BANNERS, INC. (%)	DERRY AIR, INC. (%)	SCRABBLE[*] (%)
A	8	6	10
B	2	1	2
C	3	7	2
D	4	4	4
E	9	11	13
F	3	1	2
G	3	2	3
H	4	3	2
I	6	5	10
J	0	0	1
K	2	1	1
L	4	3	4
M	3	6	2
N	7	8	7
O	7	10	9
P	3	2	2
Q	0	0	1
R	6	6	7
S	6	7	4
T	8	5	7
U	3	3	4
V	2	1	2
W	3	4	2
X	1	0	1
Y	3	1	2
Z	0	0	1

of actual signs assembled. The data in the left hand column was kindly supplied by Gasser Banners, Inc. The center column lists the experience of the author over the preceding 18 months. Just for grins, the right hand column shows the letter distribution of the popular word game Scrabble®. (It would be an interesting exercise to formulate banner rates according to a letter-value system similar to that used in this game. Charge a base rate, say, $100/hour plus a graduated scale inversely proportional to the frequency of usage (e.g. $10/hour/Q, $3/hour/C, $1/hour/E, etc.) Score "double" for weekend jaunts and "triple" for special events, such as Labor Day weekend at the beaches, etc.)

From the preceding discussion, it becomes increasingly clear that the term "adequate inventory" is a relative one, at best. It seems reasonable, however, to extend some guidelines to the newcomer in the industry with the intentions of letting him adapt his own unique situation accordingly. For the support of a single plane operation, a character inventory of somewhere around 100 to

125 symbols (letters, numbers, punctuation marks, and special symbols e.g. "#," "$," etc.) would serve as a solid foundation from which to build. Additionally, 10 - 12 word spacer sets would be needed for convenience. (A harsh minimum would be five or six.) The minimum number of connector rods in any inventory should be compatable with the tow plane capability for the size equipment you intend to purchase (see Table 2-1). Double this number for a more comfortable margin to allow for breakage and multiple sign assembly. If the equipment is to be stored on racks, it would be a pleasant luxury to have one connector rod for each banner character in stock.

Ancillary hardware, including extra pickup assemblies, lead poles, tail assemblies, towlines, and grapple hooks are all nice-to-have items on the replacement shelf if and when needed. The demand for these items usually evolves with the business so that they might best be excluded from an initial equipment purchase.

Other items to always have on hand include the following:

1. *Safety links:* If there were a National Safety Link Association, their motto would be, "You never outgrow your need for safety links." A bargain at twice the price, these inexpensive components are best replaced according to the recommendations in Chapter 4 for maximum peace of mind. Order sufficient quantities to avoid frequent replenishment, shipping costs, and unwanted delays.

2. *Repair Tape:* This item is covered in some detail in Chapter 9. Suffice it to mention here that it is just plain wonderful stuff.

3. *Loose QD Clips:* As long as your supply of connector rods is comfortable, it probably isn't necessary to keep a large stock of spare clips around. Nonetheless, a few for leisure repair hours on rainy days won't throw a hole in your annual banner equipment budget.

Finally, an inventory is not a static entity by any means, but is constantly changing. Additional purchases add to the overall tally although not all changes are positive vectors. Time and use take a gentle toll on everything imaginable and ultimately the letters and such are destined to end up in the nylon graveyard. The section to follow deals with the problem of cataloging the inventory so that the current status is always at hand for reference.

CATALOGING

If for no other reason, the content of your inventory should be at your command when someone calls to inquire about a banner so that you know what the hell's going on. You should be able to determine at a glance whether you will need more equipment to fulfill a new contract or if it can be done with existing stock. Perhaps your memory is good enough to store this kind of information, but why waste the effort and risk costly effort when a systematic inventory sheet can provide you with accurate information when the fire lights are on? Imagine, if you will:

Ring. Ring.

"Good afternoon, Derry Air."

"Hi, this is Plick Spasmo. I'm with Triple A Advertising. Do you fly banners? How much are they? We want one for two hours over the rodeo and we want it to say 'MOM'S MOLDY MEAT MARKET—mmmmm GOOD.' Oh yeah, tomorrow okay?"

If this conversation is to be handled with any sort of grace, the person fielding the call must be able to return a plausible response based on his actual capacity to fulfill the request. Answers such as, "Let me check to see if I have enough 'M's' in my inventory and I'll call you back" will guarantee a never ending tailchase that always ends with, "I'm sorry, Mr. Spasmo is in a meeting right now. Can I take a message?" They are always in meetings. *Always*. Call him at home at 9:30 at night—his kid answers the phone, "I'm sorry. Daddy's in a meeting right now. Can I take a message?" Call him at 2:00 a.m. and his wife answers, "I'm sorry, Plick's in a meeting right now. Can I take a message?"

No—the burden of capturing this rare moment on the telephone with an advertising account executive is on *your* shoulders. He has, in one uninterrupted mouthful, asked several questions, omitted several others, and made numerous assumptions. He only wants one answer: "*Yes, we can.*" He is prepared to accept another possibility ("No, we can't") with the words "Thanks anyway" poised on his lips in a manner that borders on rudeness yet still clinging to the edge of being civil.

By the way, never say "No." You can say "Yes, but . . . " or "No, but . . . " but *never* leave the caller with an unqualified negative response. To draw an analogy, picture the jet-setter type at a relatively classy night club in hot pursuit of his new ladyfriend and eager to demonstrate his lust for tasteful music. He sidesteps his way through nearly a hundred feet of closely packed tables to announce his bid to the piano player: "Do you know *The Days of Wine and Roses*?" The keyboard artist gives him a stone-faced "No" and our friend slinks back to his table amongst the stares with "*moron*" written all over his face. A seasoned musician would have responded something like this: "No, but if you like Henri Mancini, I'll play *Moon River* for you." Now, feeling a bit more *avant-garde*, our hero struts back to his anticipating lady with great confidence, further armed with the name of another song *and* the composer.

Putting aside the issues of availability, scheduling conflicts, and the other jillions of factors that enter into delivering the finished product, Mr. Spasmo has asked you if your inventory will support tomorrow's performance. More to the point, do you have enough "M's" and "O's?" Table 10-2 offers a suggestion for storing and retrieving this data with minimal effort. In this particular example it is immediately clear that there is one less "M" than the customer wants—and one of them will have to be taken out of an already assembled banner. Having expended only the time it takes to count the "M's" in the proposed copy and reference the inventory sheet, you can respond with authority: "We can do it if you'll accept one less 'M' at the end of the copy, and they'll all be in caps: MOM'S MOLDY MEAT MARKET-MMMMGOOD."

Should this be unacceptable for some reason, you have the ammunition in front of you to fire alternatives in his face, such as deleting "mmmmmGOOD" and replacing it with a telephone number, address, special of the day, or whatever. Our client now feels that he is dealing with a pro who knows his business and he hasn't wasted much of his valuable time perforating an ulcer in the silence between the "um's" and "just-a-minute's."

Although this system is of proven value, there is an even better way—computers. Recent advances in this industry have slashed the cost of these electronic monsters to within reach of those not even able to *spell* C-O-M-P-U-T-E-R five years ago, much less afford one. They are, nonetheless, a substantial investment and one might consider substituting a few sheets of paper or a greaseboard for the time being if trimming pennies off of the investment budget is an issue.

STORAGE

Proper storage of banner equipment (here we are mainly concerned about the character symbols and *not* the hardware) simply means keeping it away from the things that harm the fabric or reduce its useful life, e.g. nails, dirt, sunlight, mildew, dogs & cats, kids, fire, rodents, and vandals, to name but a few.

Heat and sunlight, in time, will gradually exert a detrimental effect on fabric components. This equipment should not be stored in attics, haylofts, or in closed spaces with poor ventilation where

Table 10-2. Inventory Sheet. Keep a List of all Banner Characters in Stock and Note any that May be Down for Repairs. It is Also Handy to Keep Trace of the Letters Currently in Use as Elements of Existing Banners, Separate from Characters Stored Singly.

INVENTORY

Banners: #1 SOCK IT TO 'EM BRIAN   #2 VOTE FOR DEBBY   #3 CHRISTINE IS #1

	TOTAL	REPAIR	#1	#2	#3	READY		TOTAL	REPAIR	#1	#2	#3	READY
A	8	2	1			5	Y	3			1		2
B	4		1	11		1	Z	1					1
C	6		1		1	4		15		111	11	11	8
D	7			1		6	0	SEE ALPHA 'O'					
E	9		1	11	1	5	1	1					1
F	3			1		2	2	2					2
G	2					2	3	1					1
H	3				1	2	4	4	1				3
I	6	1	11		111	0	5	1					1
J	2					2	6	4					4
K	2		1			1	7	1					1
L	4					4	8	1					1
M	8		1			7	9	SEE 6					
N	6	1	1		1	3	.	2					2
O	7		11	11		3	-	3					3
P	3					3	'	1					1
Q	1					1	'	2		1			1
R	5		1	1	1	2	"	2					2
S	10	2	1		11	5	&	3					3
T	6		11	1	1	2	#	1				1	0
U	3					3	$	1					1
V	1			1		0	%	1					1
W	SEE 'M'						?	1					1
X	1					1	!	1					1

high temperatures are generated during the summer months, such as temporary buildings and automobile trunks. Protection from direct sunlight exposure should be no problem. Closed storage areas with thermal generating units, such as heaters, furnaces, and hot water pipes are best avoided for optimal protection.

Moisture, per se, is not destructive to nylon material but it *does* invite molds and mildews to flourish, especially in combination with closed container storage. This certainly attenuates the structural fibers, aside from the cosmetic disfigurement of the material.

175

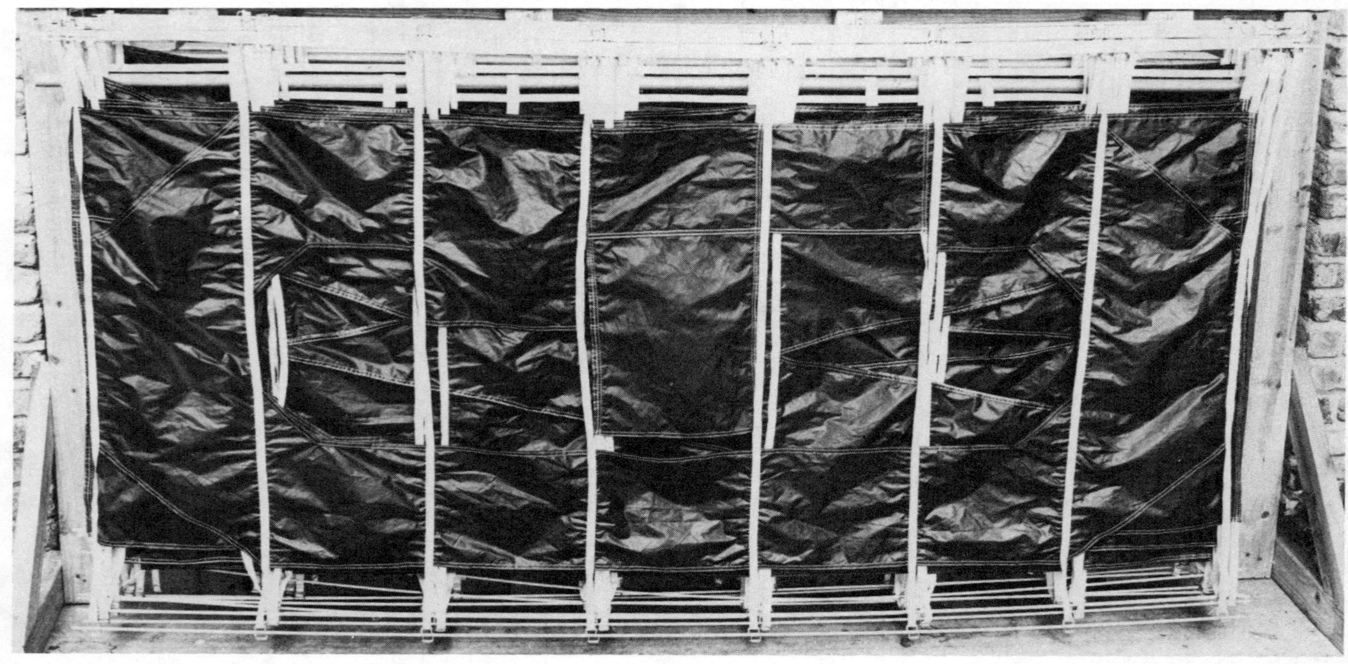

Fig. 10-1. Proper storage saves time and keeps the equipment in top condition. This rack can easily be constructed from common materials (see Fig. 10-2).

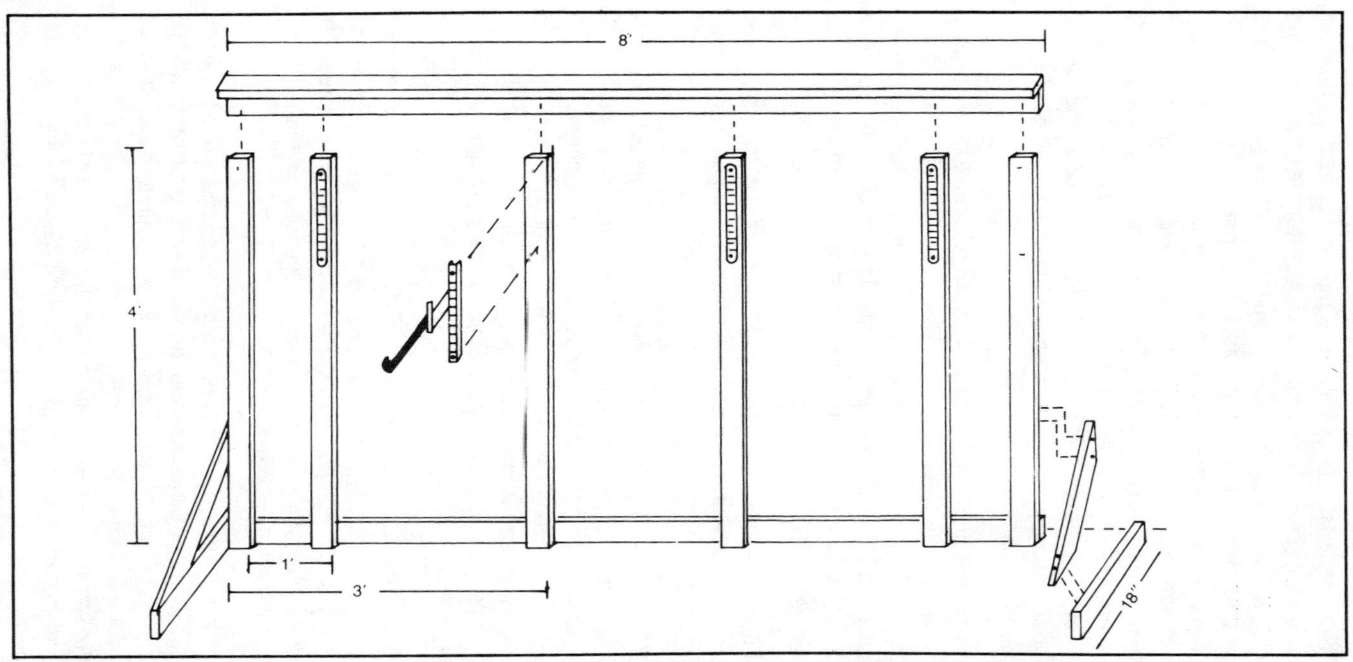

Fig. 10-2. It takes only seven 8-foot 2 x 4 studs to construct this rack. The job is completed by attaching four shelf brackets to support the connector rods holding each character. Rolled-up banners may be stored on top. This rack is lightweight and will stand flush against the wall.

With ample storage space, the simplest and best method for the safekeeping of characters and rods is the storage rack. Any kind of suitable rack will do. The adjustable slot-in-groove shelf supports that are commercially available at a reasonable price work quite well, though are not always practical to mount on some walls. The storage rack in Fig. 10-1 was designed to help overcome situations where supports cannot be directly affixed to walls, such as in metal T-hangars, etc. The letters are merely attached to connector rods at either leading or trailing edge and suspended on the horizontal supports. Pre-assembled banners can be rolled up and stored on top.

As a general rule, it is just as well not to disassemble a banner after you are finished with it. The next banner copy to be constructed is likely to have several identical letter sequences, perhaps even some of the same words. It is much easier to piece together completed segments of pre-existing signs than begin sign assembly *de novo*.

Figure 10-2 provides an exploded view of the banner storage rack, if the reader cares to entertain the idea of building something that won't fly.

If space is a problem, the letters can be sequestered in a relatively small space using boxes or crates. It is advisable to split up the inventory into bite-sized alphabetical segments, e.g. A through E, F through J, etc., in separate containers for ease in locating particular items. Rods and other hardware can then be stacked into a relatively small bundle for easy storage in a closet corner or shelf.

A pigeonhole arrangement has distinct advantages over crate storage for a slightly higher price paid for in space. Sawed-off cardboard milk containers are excellent for the purpose. These can be individually labeled externally for exceptional speed in selecting letters for banner assembly.

Nylon letters and symbols can be folded and stored in large manilla folders or envelopes and then filed alphabetically in a large box or even a filing cabinet. They can also be rolled up and wrapped with paper and tape. It is recommended that the adhesive on the tape *not* come into direct contact with the character fabric as with time, it becomes very difficult to remove and discolors the material.

A word of caution to the incorrigible obsessive-compulsive types: DO NOT IRON THE NYLON MATERIAL. To paraphrase a popular advertising motto, it melts on your iron - not in your hands.

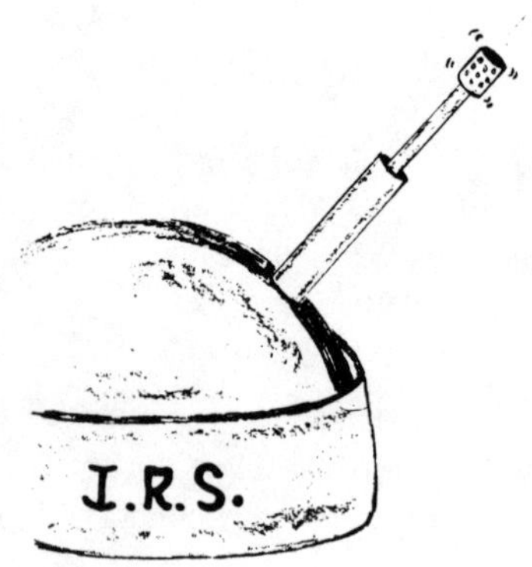

BU$INE$$
I.R.S.
A. TAYLOR

Chapter 11
Business

And so the final chapter has come. For most, it heralds the long-awaited respite from the drudgery of reading. For many, especially those with prior business experience, there is little point in continuing. For others, it is hoped, there will be some outline in the pages to follow around which to structure the genesis of a business as related to banner towing. In no way should it be construed that the forthcoming material is exhaustive, or for that matter, authoritative. It is rather a montage of subjects that deserve scrutiny by those souls seriously interested in marketing this specialized form of aerial advertising under the guise of a business. The text is largely based on experience and it should be clear that the author is devoid of formal training in the business world. Nonetheless, experience holds its own virtue, apart from the academic world, and may prove for some to bridge the partition between success and failure.

The topics presented in Chapters 1 and 2 serve well as prefacing remarks to usher in the subject at hand. The variables introduced therein ultimately provide the building blocks for the foundation of an individually structured business. With the assumption that such homework is already *fait accompli*, this discussion shall proceed to more detailed considerations.

GETTING STARTED

Putting a banner business together is, unfortunately, more complicated than sitting down and making a shopping list of things to accomplish and then going out and doing them. It involves the simultaneous focus of many loose ends into a viable format. Certain items, of course, have priority in a relative sort of way; ultimately, each subject must be aligned in the same reference plane. Some "first-step" ideas are presented here without regard to any order of importance.

Professional Assistance

Lots of folks are endowed with do-it-yourself talents from A to Z. Most of the rest of us fall prey at one time or another to the temptation to trespass into areas best left up to the pros. Accounting, legal, and medical problems are three such areas that come to mind, the latter hopefully not germane to this discussion.

> **Q**: How much for a lawyer?
> **A**: Lots.
> **Q**: How much for a C.P.A.?
> **A**: Lots.
> **Q**: Is it worth it?
> **A**: You bet!

It is no small joke to change accounting systems in midstream. Handing a shopping bag full of oil-stained receipts to an accountant for last minute salvage when tax time rolls around is probably considered poor form by most C.P.A.'s. Likewise, early consultation with a qualified attorney is sound advice. It certainly seems reasonable that, in the long run, professional service costs will be minimized by seeking guidance early. See Appendix B.

Business Format

Operating as a business necessitates the election of a specific business entity, e.g. sole proprietorship, partnership, corporation, etc. These organizational formats are treated differently for tax purposes, so it behooves the entrepreneur to choose the path that best suits his individual circumstances. The legal overtones aren't identical either, so limits of liability may play an important role in choosing a business organization. The I.R.S. publishes a guide (*Tax Guide for Small Business*, Publication 334) annually that is free for the asking and addresses the considerations presented here in some detail. Boning up on information of this sort will help to reduce time spent with professional consultation.

Business Name

Any name suitable for print will probably suffice. It is probably worthwhile to toss a few ideas back and forth in order to fine-tune a name that people will remember. This becomes increasingly important as the business matures, especially if there's more than one game in town. Depending on the organizational format, it may be necessary to establish an "Assumed Name" by formally registering within the city or county departments assigned to keep track of that sort of thing.

Tax Identity

It is necessary to be indexed according to a unique number for each business at the federal level for tax purposes. This may require applying for a federal tax identification number (employer identification number). Form SS-4 should be available at any local Internal Revenue Service office for this purpose.

Permits/Licenses

Launch a thorough investigation for required certification from governing authorities at the city, county, and state levels. Local ordinance restrictions of peripheral municipalities may require coughing up funds for outdoor advertising permits. Become informed about state and local tax collection responsibilities and secure the appropriate sales tax permits.

Banking

In most cases, a separate checking account should be established in the name of the business.

Flying Authorization

Determine and complete the necessary steps to become a qualified banner towing operator. At the very minimum, you must have access to a plane with a tow hitch that has been inspected and certificated for this purpose and a Certificate of Waiver for 91.18 of the FAR's. A visit to the nearest Flight Standards District Office (FSDO) or General Aviation District Office (GADO) should enable you to outline the required course of action.

Banner Equipment

Decide how much equipment will be needed and procure it. At the very minimum, you should get enough characters to assemble a sign for advertising your own business. This is not to suggest that you must spell out the name of the business—be simple and to the point. For example, "AIR ANTICS, INC. 762-9917" is less likely to get the message across than "BANNER TOWING 721-6290." In addition to the required characters and word spacers, the initial inventory list should include: tow hitch; grapple assembly; tow rope; mast assembly; connector rods; tail assembly; and pickup assembly (if pickup method of launch is to be used).

It is *highly recommended* that the following extras be added to the above list without hesitation: extra safety links (10); extra connector rods (three or four spares); repair tape ("Ripair") ; and extra QD clips (one dozen).

Operational Environment

Satisfactory arrangements must be clearly worked out for at least one launching area and drop zone. This may have to be coordinated with the FAA to comply with local restrictions as well as the airport manager if the operational arena is located on airport property. Keep in mind that the maintenance and/or grooming of the launch surface should be included as part of these initial arrangements.

Insurance

This is another area that may require professional assistance through a competent aviation insurance broker. Adequate aviation insurance coverage is a prerequisite *must* for every responsible operator in the banner industry. You will want substantial liability protection from a reputable company. Be certain to master a clear understanding of any restrictive endorsements, limitations, or exclusions that may be attached to the policy regarding banner towing operations.

Accounting System

Preferably on advice from your accountant, decide on whether you will be doing business on a cash or accrual basis of accounting. All other things being equal, the cash basis system of accounting is probably the simplest to work with in this business. The merits of the accrual basis, however, are worth investigating for comparison. A bookkeeping system is best established early and revised on an as-needed basis thereafter.

Office Arrangements

Without a doubt, any business must be conducted from some physical location; that is, an address. For many operators, the problem is already solved inasmuch as the addition of banner service is merely an extension of an already existing aviation service. Others, however, will find it necessary to establish a business locale. With total disregard for the economics of the matter, a formal business establishment would be "ideal." Acknowledging the fact that such a venture demands a substantial financial committment, one might consider the alternative of opening up under the roof of a small office warehouse. Currently, these are not totally off the wall price-wise and further provide equipment storage space. As an extension of this idea, a "T-Hangar" would serve the purpose well, should you be fortunate enough to occupy one. For many, the home office is a viable option, assuming that such an election does not violate local zoning restrictions or municipal ordinances. This needn't be terribly sophisticated, as a kitchen table corner will suffice at first. The point is to get the business situated so as to be able to receive mail. Actually, a post office box number is well suited for the purpose with the added benefit of not having to expose one's home address to the public. The U.S. Postal Service, on the other hand, heavily frowns on the storage of banner equipment in the post office boxes, so an additional location for that purpose must still be sought.

Telephone

Establish a commercial telephone number for the business, if you do not already have one. Any temptation to use a private home phone line should be dismissed. A preliminary investigation of the available telephone exchanges in your area combined with a little imagination might prove useful later on as an advertising gimmick. For example, 226-6377 corresponds to the alphabetical dial as "BAN-NERS." There should be little reason to rate this one any higher than "a cute idea" in the time-priorities department.

Telephone advertising will be mentioned under a different topic.

Business Forms and Stationery

Your banner towing business will be represented to many potential customers for the very first time by the information you send through the mail. For some, its appearance for all practical purposes will go unnoticed. For others, it may create a lasting impression—unfair as it may seem— of how well put together this act really is. Without further discussion, it is concluded that quality business cards and stationery are a basic priority for initiating a business.

Logo. It may be of value to design and use a logo to symbolize the business and further add an artistic flair to any printed material, thereby making it more attractive and professional in appearance. The logo need not incorporate a plane towing a banner, but the temptation to create such an emblem is undeniable. A reasonable facsimile of a banner in tow can be easily spread out across the width of a letterhead. Obvious problems erupt when attempting to compact the drawing to business card size. Therefore, be ever mindful of the business card dimensions if you choose to create a logo symbol. Graphic design assistance is commercially in abundance should you need to hire artistic talent.

Contracts. It is always appropriate to draw up a formal contractual agreement to cover each banner performance whenever possible. A written contract serves several purposes simultaneously.

1. A polished and professional setting is established by requiring the customer to commit himself on paper. It is well to leave the impression that this is a *real business* and not just some fly-by-day operation of hasty conception.

2. A contract delineates the obligations of both the customer and the operator. In this respect, there can be no ambiguity that might otherwise occur with a verbal agreement.

3. A contract provides some legal protection for the banner operator by securing a written document of the customer's request. This is particularly important to have around in the unlikely event that a question should arise about the banner copy or the date, time, or location of the performance. In the same spirit, a clause can be incorporated into the agreement that provides some legal remedies for failure of the client to cough up the bread as agreed.

Since a contract has roots in the legal world, it is best to consult an attorney before composing the document, or let him do it for you. Should you choose the latter, provide him with the substantive points, peculiar to the banner industry, that you wish to be highlighted in the agreement. At the very minimum, specify:

A. *Banner copy:* This should be an *exact* representation of how the completed banner will read. Unless special equipment is used, all letters will read in large case symbols (capital letters) with identical replication of spacing.

B. *Date(s) of performance*: Specify the exact date(s) or days requested by the customer.

C. *Rain date(s)*: The customer may want a banner dragged over the beach sometime during the weekend, but he may not particularly care if it is Saturday, Sunday or for that matter, the following weekend. A "rain date" is a nice tool to increase the percentage of contracts actually flown.

D. *Time of performance*: Be *very* specific about the time requirements set by the customer. This may vary from something as generous as "sometime during July" to a restrictive endorsement demanding a target time within one minute. Establish an "inside/outside" limit for the performance along with a "desired time" whenever possible to yield the maximum possible chance for completion of the contract. For example:

> Contract Time: 2 hours
> Desired Time of Performance: 2:00 p.m. - 4:00 p.m.
> Inside Limit: 12:00 noon (to begin performance)
> Outside Limit: 5:00 p.m. (to terminate performance)

This allows the pilot to get off early if the weather is forecast to crump during the preferential hours specified by the customer. Likewise, the agreement can still be fulfilled and the mission salvaged should a delay for whatever reason be encountered.

E. *Target area: Clearly* specify where the customer wants his banner to be displayed. This might be as specific as a street address or as general as the territory defined by the county limits.

F. *Equipment:* Record the height and color(s) of the banner equipment to be used. Any special equipment, e.g., logo panels, etc., should be fully described or drawn. If the customer is to pay for special equipment (and he should), state so. It may be advisable to include a clause regarding maintenance responsibilities for long term contracts.

G. *Rates*: Include *all* charges for banner services and equipment. Clearly state when the charges are in effect (e.g. " . . . from banner pick-up to drop").

H. *Payment terms*: The customer should be made aware of his financial compensation responsibilities to include deposits and/or pro-rated payments. Attach a clause stating the type of creek he will be up and his alloted number of paddles in the event of default.

I. *Special provisions*: Leave a space to write in any idiosyncratic requests or out-or-the-ordinary arrangements that would be pertinent to the terms of agreement.

J. *CYA*: Protect yourself in writing from the vicissitudes of the business. This should minimize misunderstandings and hopefully avoid potential litigation. For openers:

☐ Leave an option to contract the work out to another operator.

☐ Absolve yourself and the business from any guarantees of performance *for any reason*. Weather, maintenance, acts of regulatory authorities, etc. may be used as examples.

☐ Reserve the right to amend the contract as needed to compensate for unforeseen circumstances. The customer should be given the reciprocal right to cancel the agreement if the amendment is unsatisfactory.

☐ State the terms under which the customer may cancel or amend the contract, if any, along with the consequences of failing to comply with the contract.

☐ Include a statement to the effect that flying safety is the number one consideration of any responsible flying operation that overrides *any* provision of the contract.

The contract example shown in Appendix C is intended to serve as a guide only. You should have an attorney help you prepare this document locally.

Statements/Invoices. Your customers will expect to receive some written confirmation of services provided and payment accounting. This may be handled on a single statement that is best delivered or mailed as soon after each banner performance as practical. The format may be customized at the pleasure of the operator, but should include at least the following information:

1. Date
2. Performance time (actual)
3. Banner copy
4. Charges
5. Credits received
6. Balance due

Clients may request additional information for their records, such as pick-up and drop times and sites, airport of origin, type of banner equipment, etc. A sample statement might read as follows:

STATEMENT

Type	Date	Description	Time	Debit	Credit
—	06/30	Balance Forward	—	$200	
—	07/02	Payment Received - Check #204			$200
03	07/03	FIREWORKS TOMORROW NITE/CITY	2.0	$400	
03	07/04	FIREWORKS TONIGHT/CITY PARK	2.0	$400	
		Totals		$1,000	$200
		*Balance Due**		$800	

(03) 7'Black/5'Red Combination
(*) Terms Net 10 Days

DERRY AIR INC.

11005 GREENWILLOW DRIVE
HOUSTON, TEXAS 77035
(713) 721-6290

DATE____________

SOLD TO

CUSTOMER NAME
ADDRESS

YOUR ORDER NO. ________ DEPT. ________

DATE OF INVOICE	SHIPPED VIA	F.O.B.	TERMS	SALESMAN
	--	--	CASH	act

QUANTITY	DESCRIPTION	PRICE	AMOUNT
20 Hours	7-Foot Black Aerial Banner - Towing Services Copy: CITY NEWS WANT ADS GET RESULTS - 926-8237	$150/hr	$3,000.00

Fig. 11-1. Sample of typical invoice.

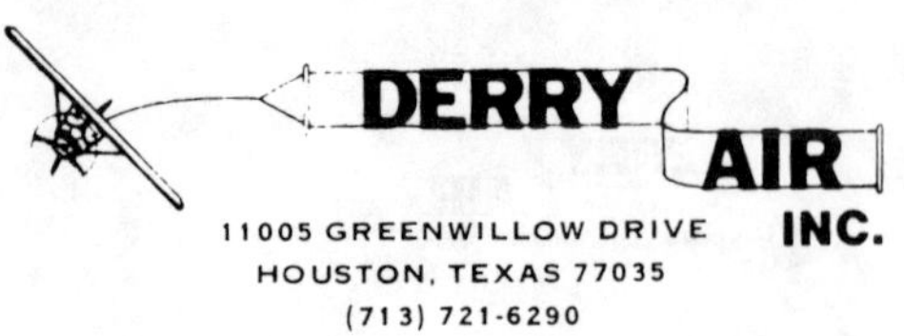

DATE ________________ ,19 ____

CUSTOMER NAME

ADDRESS

STATEMENT OF YOUR ACCOUNT · TERMS NET CASH

DATE	DESCRIPTION	✓	CHARGES		CREDITS		BALANCE	
8/2	Invoice Billing		3000	00				
8/2	Down Payment				1500	00	1500	00
8/15	Wx Cnx 2 hours *				300	00	1200	00
8/17	Payment Received				1200	00	00	00
	THANK YOU VERY MUCH							
	Total Advertising Time: 18.0 Hours							
	* Weather Cancel							

Fig. 11-2. Besides your invoices, it's also a good idea to mail out monthly statements.

Many customers will request an invoice for desired services. It will occasionally become necessary to submit such a form filled out in great detail for the customer, including such curiosities as purchase order numbers. The cost of blank forms is minimal. Imprinting your own logo adds to the attractiveness for a small increment in price. A sample of a typical invoice (Fig. 11-1) is shown here, along with a follow-up statement for billing purposes (Fig. 11-2).

Keep at least one copy for your own files in the event that the original is misplaced by the recipient or otherwise lost in the mail.

Rates

There must be some answer to the frequently asked question, "How much does it cost?" Converting all rates to a "per hour" basis as a common denominator, charges vary considerably nationwide. Individual operators have such variable expenses and peculiar circumstances that it is unlikely that any fee standards will be witnessed in the near future. The usual junk like fuel, oil, hangar/tie-down, insurance, maintenance, pilot expenses, and the like—*overhead*, for short—figure into everyone's price schedule. Rates are necessarily subject to adjustment according to what the market will bear and should be further commensurate with individual goals as discussed in the opening chapters of this book. Each operator may wish to consider any combination of the commonly used methods of charging for towing services:

Pick-up to Drop: This is an easy system to work with, especially if the launch/drop zones are located at the home aerodrome. Establish a separate fee for transit time if the banner is to be picked up at a remote site.

Over-the-Target Time: Customers may insist upon having a price quoted for time spent towing over the target. On a per hour basis, this is necessarily higher than charging from pick-up to drop because of the transit time involved. If your area of anticipated operation is large, it may be necessary to make individual quotations on this basis.

Minimum Time Over Target Fees: In some cases, clients will request services for a very short period of time, usually in association with a limited location. Birthday greetings are an example of this. Nobody wants to stand on his front porch watching a banner circle overhead for one entire hour. That predisposes the observer to vertigo in addition to the suffering of endless monotony. Such performances are best limited to about ten minutes to allow picture taking and rib poking. A flat rate for such requests is easy to administer. This type of rate system would enable a single operator to tug four or five signs over a stadium throughout the course of a single football game!

Linear Runs: Operators in the beach areas often charge according to the number of trips taken along the shoreline. This is especially useful if the towplane can operate out of airports at either end of the beach.

By-the-Letter: If your towplane can drag a lengthy load, you will soon learn that the extra work involved with banner assembly should have a price attached. You might start with a basic hourly rate that includes up to 25 characters and thereafter charge an additional fee for exceeding that limit. As a variant, establish a baseline fee for the pilot and plane and then charge by-the-letter on a per hour basis. Remember, it takes just as much work to assemble a word spacer as it does to add a letter or numeral.

Establish a minimum fee, *regardless* of your method of charging. Otherwise, some joker will ask for a 5 minute show and expect to get it for one-twelfth the hourly rate.

The cost and time for banner assembly necessitates a higher pricetag for at least the first performance. This can be assimilated over several hours or multiple performances so that the rates can be made more attractive for volume customers. Long term contracts enable the operator to assemble the banner and leave it in storage between towing performances, thus reducing time and energy for repeated assembly. Even so, some allowance for periodic repair must be considered, however small it may be. Some operators charge for banner assembly regardless of whether it gets flown or not. This scares a lot of folks away from using banners so the practice is to be discouraged.

It is not unreasonable to set an additional fee for the last minute callers. This is not so much to penalize the latecomers as it is to encourage timely orders for scheduling purposes. You may have to drop everything and even cancel personal plans to accommodate someone's afterthought, only to get weathered out on the following day. Here is an example of where a non-refundable fee is appropriate. In any case, it is best to require a week or more notice if you care to lead an otherwise smooth lifestyle.

Finally, it is always a good idea to get some "up-front" dollaroonies, especially from individuals and not-so-well-known businesses. It is fair to add a paragraph to the contract demanding forfeiture of all or part of a deposit for late cancellation.

It is especially important to keep a diary or other such log of business related activities during the formative period. Record names, dates, places, mileage, expenses, quotations, and other pertinent information that may likely need retrieval at a future time.

With the above squares completely filled in, a brand new aerial advertising business is on the threshold of birth. It should be obvious that one notable ingredient is missing—*customers*. They won't come knocking at the door, at least not until it is commonly known where the door is. You've got to go out and find them and tell them what you do and where to find it—there's no way around it.

ADVERTISING

As an aerial advertiser, you become a tiny gear in a monstrous machine. The number of bucks spent annually on advertising in this country is enough to make your socks roll up and down and turn your electrocardiogram flat. Just look at the advertising listings in the yellow pages of any major city directory. It's thick enough to make Leo Tolstoy's *War and Peace* look like a birthday card. And it's not just an alphabetical catalog of agencies. A prodigious array of miscellaneous services and products parade under this heading. If gathered under a single roof, the show would indeed be a spectacular circus. There seems to be no room for even a single addition to this already impressive roster of businesses.

But there *is*.

You will be telling your prospective clients that they should have a well-rounded advertising campaign by dividing funds among various media. It is probably wise to follow your own advice in designing a program to promote an aerial advertising service. The circumstances of each area of operation, combined with individual advertising budgets, will largely dictate the styles of promoting banner towing.

Nearly everyone will incorporate the following modalities at some level throughout the course of business.

Telephone

Aside from its necessity as a basic tool of business, the telephone has advertising potential.

Solicitation of business by telephone is a tedious chore that may produce immediate rewards. It does allow an informal survey of the potential market without a lot of footwork. It further introduces the business to the receiver of the call, providing a basis for either personal or mail follow-up.

The yellow page directory is probably a useful servant, but avoid putting too much emphasis here. The lead time necessary to get an ad in the directory may vary from six to eighteen months, depending upon the publication date. Be thoughtful of any decisions to purchase additional space. It might be better to spend those bucks to expand advertising coverage throughout the directory by cross-referencing. When people let their fingers do the walking, they may not know exactly where to look for banner towing services. It may be best, then, to make their fingers trip over a reference as much as possible. The categories to choose from are many. For example:

> ADVERTISING - AERIAL (This is a nice one to have since it will likely be the very first classification of the advertising headings.)
> ADVERTISING - OUTDOOR
> ADVERTISING - SPECIALTIES
> ADVERTISING - AIRPLANE
> AIR - ADVERTISING
> AIRCRAFT - ADVERTISING
> BANNERS AND FLAGS
> SIGNS - OUTDOOR
> BANNER TOWING
> AERIAL BANNER TOWING

Figure 11-3 shows telephone advertisement displays from successive years. The smaller ad is less than satisfactory because it does not state the words "BANNER TOWING" despite the implications of the logo. This is corrected in the larger ad, which also reflects some business expansion considered beyond the scope of this text.

Of course, incoming calls must be fielded, regardless of the availability of personnel. Telephone recording devices, call-forwarding features, and telephone answering services are generally cheap solutions. *Don't lose business because no one answered the phone.*

Mail Solicitation

In the author's experience, mail solicitation has been an abysmal failure, at the very best. It is most likely that, in spite of the profoundly interesting subject, such literature suffers an irreverant demise along with the material of lesser quality that we have all fondly grown to know as *junk mail*. On the other hand, follow-up contact through the mail is excellent adjunctive support for telephone visits and personal appearances. To be included in the stuffings:

☐ A letter, individually typed for customized service or a form letter that applies to the masses. Include information to acquaint the reader with the services available, emphasizing how cheap it is. (Appendix D)

☐ A rate card, with the specifics on *how* cheap it is.

☐ A photograph or similar depiction of what a banner looks like trailing behind its master. (Gasser has color fliers available at affordable prices.)

☐ A business card

And that's all you get for one stamp's worth.

Pull Your Own

If you are interested in tax-deductable flying time, there is a way: Advertise your own banner business with your own sign.

While you are learning the fine art of towing, it makes good sense to augment the training experience with business productivity. Later on, any time the urge strikes to go flying on a Sunday afternoon, take your banner along for the ride to remind the folks you're in town. As you might expect, this is one of the better forms of advertising.

Other Media

The remaining vehicles for advertising are limited only by the pocketbook: Newspaper ads, radio, television, magazines, posters, calendars, billboards, balloons, portable road signs, shopping mall displays, bumper stickers, T-shirts, pens, buttons, and so on *ad nauseum* - the list is endless.

Investigate the possibilities of swapping time with another advertising medium such as television or radio. In order to pull this one off, you have to get further up the ranks than, "Good afternoon, this is KXYZ, may we help you?"

The ultimate advertising stunt that may well override all of the above mentioned gimmicks is you and your personal time to visit prospective clients *in vivo*. Unless you have three heads and hair under your fingernails, there is no substitute for in-the-flesh appearances. There is no doubt that the increased rapport developed through personal contact will augment your chances of landing a client—and a contract.

The distinction between the terms *advertising* and *promotion* isn't crystal clear. Any debate over the issue shall be left to the academicians. For the purpose of this discussion, *advertising* generally refers to the solicitation of business from people or

Fig. 11-3. Yellow page advertisements. Lower ad omits phrase "banner towing," an omission that was remedied in the next edition (upper).

businesses you don't know. *Promotion* shall generally mean a more mature form of advertising that is directed at those with whom an acquaintance has already been made. The two terms quite often overlap, but so what?

Once you have established a reasonable rapport with a client or seemingly interested prospective customer, it is helpful to keep a high profile. You want to subtly remind him often that every day he doesn't use a banner, his life will be filled with anxiety, grief, hatred, and frustration . . . that his whole life has been wasted concentrating on the conventional media such as radio, television, and newspapers . . . that he may even suffer for eternity in Hell for such diabolic behavior if he doesn't shape up—soon!

Promotion doesn't have to cost fortunes to be effective. Simply passing out a few ball point pens (with the company name emblazoned on each) at your client's office will generate a lot of favorable feelings. People *like* to get stuff—*free* stuff. It doesn't have to be exorbitant or lavish—just *free*. And something for the kiddies at home? You scoundrel—right out of Chapter 1 of the Dirty Tricks Almanac. Balloons, combs, plastic rings, coloring books - it doesn't matter. Just give them some worthless piece of junk they can take to their precious little rug rats at home (who will subsequently destroy it in about 10 seconds or insert it into the dog's ear, or throw it on the floor and then eat it, or whatever else kids do for a living). You will be remembered for your thoughtfulness.

It is sometimes more appropriate to invest in a more substantial promotional gift, especially for the customers who have been faithful and regular. The range of prices and ideas need not be explored here; but as a suggestion, a handsomely framed picture of the client's banner has distinct advantages over a silver candle snuffer.

Not to be forgotten are the number of folks that participate, directly or indirectly, in the smooth operation of your business. The ground crew, the guys in the control tower, approach control, airport managers, etc. Although not necessary, it is a nice gesture to slip a box of candy or a bottle of booze under the table from time to time in gratitude for their assistance.

As this is a unique business, it has unique features that can be used to great advantage. How many times have you heard, "Oh, I've always wanted to fly, but I never had the chance.?" Well, dear prospective client, this is your chance. *Don't* go up and show them your dazzling aviator spirit by doing outside loops and inverted spins. Show them what their headquarters looks like from the air; show them the standstill traffic at rush hour; show them the packed beaches, stadiums, amusement parks, parks, and so forth. It's all tax-deductable flying time and it may land a contract.

HOMEWORK

As a professional, you need to be on top of what's going on in your area of operation for at least two reasons. Firstly, knowing the local current events will help in the process of scouting around for business. Secondly, it is important to be able to rattle off a list of events for your clients. This will increase their confidence in your ability to display their banner maximally.

Staying abreast of the local happenings isn't all that easy in a sizable community. It is well to stay in touch with the more conventional media (television, radio, billboards, etc.). Subscribe to the newspaper daily (now a tax-deductable expense) and scan it cover to cover to see who is doing the advertising as well as where people are likely to gather outdoors.

Get a detailed map of the area of operation for your banner towing business. Keep it within reach of the telephone for references to street addresses. Outline any areas where flying is likely to be restricted or prohibited. Get to know the territory well from the air.

BACKUP

Going it alone might be acceptable while the business is in the early stages of development. However, as things begin to roll along, it will become apparent that additional support in the operational department is needed. Finding pilots willing to fly is like finding pigs willing to eat. You can afford to be choosy in selecting pilots to help with the towing. Get any necessary clearances from the insurance company and secure an amendment to the Certificate of Waiver for banner towing to cover all pilots involved.

Another approach to the same problem has additional merit. If there are other banner operators in the area, it is well to establish a good working relationship with them. In addition to the backup help, you may find subcontracting to be a mutually beneficial tactic for excess orders. If operating with similar equipment, you may find it rather convenient to have another inventory nearby.

COST

Dollar assignments have been carefully avoided throughout this text. In part, this has been motivated by the inflationary trends of the present that would perhaps render this publication outdated in short order. Also, the spectrum of individual circumstances among the readership surely prohibits proposing a viable estimate of the required financial backing to begin banner towing. The summation of business costs, aircraft expenses, and banner equipment purchases should provide a clue to the magnitude of this endeavor. For those who dare, sufficient reserve must be available to absorb monthly expenses for at least a year.

PATIENCE

The shell of enthusiasm is very thin, especially when you've been in business for three or four months and the phone isn't ringing off the hook. Hopefully, patience and hustling will pay off in due time. It may take a year or even longer to start receiving regular business, though some may find good fortune much sooner. Don't become discouraged for the lack of instantaneous success.

There are more topics to pursue, more subjects to dissect, more anecdotes to share. But most of the big pieces are on the table, and further diversion would only amount to clutter. Admittedly, much has been left unwritten—but enough has been said.

Go *do* it.

Glossary

adapter pole—a banner component that permits the simultaneous use of different sized equipment, e.g. 7 foot followed by 5 foot characters.

banner—as a general reference, the fully assembled banner from the mast assembly to the tail assembly.

bridle harness—the ropes that attach to the lead pole.

bridle harness ring—the point of attachment for the tow line to the mast assembly.

character—any symbolic unit of the banner copy, e.g. alphabetical letters, numerals, punctuation marks, or word spacers.

connector rods—fiberglass rods used as supporting structures between each character in the banner copy. Each rod has eight evenly spaced clips (QD clips) to which the characters attach on the leading and trailing sides of the rod.

connector straps— the nylon loops that attach to the QD clips on each connector rod.

copy— the message of the banner.

drag chute—see **tail assembly**.

grapple assembly—the entire component system of the hook and cable that attaches to the tow hitch on the aircraft for the aerial pickup method of banner launch; includes the tow ring, short cable with nylon sheath, safety link, main cable and grapple hook. Its total length is about 30 feet.

grapple hook— the three-pronged hook that suspends from the grapple assembly.

lead pole— see **mast assembly**.

letter—an alphabetical character.

Logo panel— a custom prepared, heavy nylon fabric banner component upon which special emblems, insignias, designs, or logos can be painted or sewn in. This unit usually flies at or near the beginning of the banner copy for aerodynamic stability.

mast assembly—the components of the leading-most portion of the banner. It consists of the bridle ring and harness, the weighted metal pole, and an old-style connector rod to which the first character of the banner copy attaches. Also called **Lead Pole**.

mast collars— the metal guides for the bridle harness attachments on the mast assembly.

numeral— a character symbol of numerical value. Note that the numerical "0" (zero) is interchangeable with the alphabetical "O". Also, "6" and "9" are physically identical.

"0" ring connector— any of the welded steel rings as components of the tow line or cable assembly. One is always located at the leading end of the grapple assembly. Another is commonly used as the connector between the towline and the pickup loop. If the takeoff launch method is used, this steel ring will be attached to the cable that connects between the tow hitch and the safety link; also called **tow ring**.

pickup assembly—all of the components of any system that supports the pickup loop for the aerial launch technique. Classically, this consists of a pair of pickup bases and poles.

pickup loop— an endless loop of rope that forms the leading portion of the towline after an aerial pickup. The loop is draped over the pickup assembly as a target for the grapple hook.

punctuation—any of the interruptive character symbols, e.g. dash, comma, hyphen, parentheses, colon, etc.

quick disconnect clips ("QD")—the rigid metal clips that serve to attach any character to its respective connector rods. There are two different designs (new-style and old-style) depending upon the corresponding rods to which they attach.

"Ripair" tape—a commercial product of nylon fabric with sticky-back adhesive for repair of fabric defects.

safety link—a doubly-looped nylon strap with a break strength of approximately 500# that serves as the "weak link" in the towing system.

signal flag—any signal device operated by a ground assistant to provide visual cues to the tow pilot.

space symbol—see **word spacer**.

special equipment—custom equipment for special effects, such as logo panels, fraction symbols, custom color fabrics, adapter poles, etc.

tail assembly—the trailing edge unit of the banner, consisting of a connector rod attached to a fabric drogue chute.

tail chute—the fabric portion of the tail assembly that provides longitudinal tension and vertical stability for the banner in flight.

tail hitch—see **tow hitch**.

tail hook—the lever arm of the tow hitch that connects the tow ring to the aircraft.

tow cable—the cable portions of the grapple assembly; alternatively, the short cable that extends between the tow ring and the safety link.

tow hitch—the mechanical interlocking device that is installed on the aircraft frame. The tow ring is linked to the aircraft through this contraption.

tow hitch release handle—a lever installed in the cockpit or cabin that controls the release mechanism of the tow hitch.

tow hook—see **Grapple Hook**.

towline—the rope used for towing. Also called **tow rope**.

tow ring—see **"0" ring connector**.

tow rope—see **towline**.

uprights—the vertical support components of the pickup assembly.

word spacer—a basic banner unit of eight unadorned nylon straps.

. . . Those crazy skydivers are out every Saturday so I can't overemphasize the need to clear overhead at all times . . .

Appendix A
Checklist Guidelines

AERIAL PICKUP METHOD	TAKEOFF LAUNCH METHOD
Field Checklist	
1. Check approach clearance O.K.	1. N/A
2. Check departure clearance O.K.	2. Same
3. Check support bases (if used) for proper alignment	3. N/A
4. Check pickup loop position	4. N/A
5. Check pickup loop horizontal supports properly aligned	5. N/A
6. Tow rope inspection (including pickup loop)	6. Tow rope inspection (including safety link and cable connector to tow hitch)
7. Check tow rope connected to lead pole	7. Same
8. Mast assembly inspection	8. Same
9. Weighted end of lead pole properly aligned	9. Same
10. Check all connector rods and QD/strap connections (3X)	10. Same

11. Tail "TOP" properly oriented	11. Same
12. Banner copy properly spelled	12. Same

GROUND OPS

1. Inspect grapple assembly	1. N/A
2. Inspect or replace safety link	2. Same (more appro - priately included in field checklist)
3. Test tow hitch release mechanism	3. Same (may be delayed until just prior to takeoff)
4. Configure grapple assembly to tow hitch and secure cable	4. N/A
5. Stow grapple hook	5. N/A

TAXI/RUNUP

1. Recheck cable tension	1. N/A
2. Recheck cable stowage and confirm no fouling on cockpit items, shoulder harness, etc.	2. N/A
3. Recheck grapple stow- age secure	3. N/A
4. Check tow hitch re- lease handle in proper position	4. N/A
5. Connect tow rope (test tow hitch if not already accomplished)	5. N/A

Appendix B
Comments On Accounting

by Charles P. Taylor Certified Public Accountant

This section in relation to the subject of accounting might be compared to a few paragraphs on first aid as related to the subject of medicine. Yet the life-saving value of first aid treatment before regular medical aid can be obtained is held in high esteem.

Perhaps 90% of all accounting data can be referenced to *Cash receipts* and *Cash disbursements* (items 1 and 2 below). Fortunately it is almost impossible to fall behind in recording such information because the constant demand for up-to-date *Cash balances* (item 3 below) requires that the entries be current.

Whether the business is operated as a corporation, partnership or sole proprietorship, there is virtually no way to do without outside accounting and legal services. Regardless of the type of organization, items 1 through 6 below are equally applicable.

Accounting wants to be your internal control medium over the entire business cycle, on the production side—to receive an order for goods or services, ship the goods or perform the services, bill the customer, collect and deposit intact the amount received; conversely, on the procurement side—to give an order, receive the goods or services, be billed and pay for it.

The following records can be kept easily and well without bookkeeping or accounting training:

1. *Cash receipts* for which a book or columnar worksheet should provide columns to record the date of an item, name of payer, explanation in one or two words where possible, amount

received; and the supports for such transactions should be filed in an envelope in date sequence.

2. *Cash disbursements* for which a book or columnar worksheet should provide columns for date, check number, payee, explanation in one or two words where possible, and amount paid; and the supports would be filed alphabetically or alternatively by check number.

3. *Cash balances* should be kept currently. Checkbooks furnished by banks are usually kept so that the cash balance is brought down after every transaction or by the page; each month the cash balance per bank should be agreed, or reconciled with the cash balance per check book.

4. *Billing* to customers must be done on the same day the goods are shipped or the services performed. Credit extension is optional, fraught with some danger. Pre-numbered bills are a time saver and offer sequence control. A minimum of three copies should be prepared: - original for the customer, second and third copies for the numerical file and alphabetical file respectively.

5. *Payroll.* The services of others, if not independent contractors, should be recorded in a payroll record showing the payroll period, date paid, name of employee, gross pay, deductions for Social Security (also called FICA), Federal income tax withheld, state income tax withheld, and possible other deductions such as city tax, insurance—the net result being the amount of the check issued to the employee. Some banks print special payroll checks and there are many other sources of supply. The IRS and state authority should be contacted for identification number and current payroll tax guides.

6. *Depreciable property* can be listed from the foregoing records and files and depreciation can be computed. An investment tax credit can be computed on qualifying property which reduces income tax payable dollar for dollar.

A *Tax Guide For Small Business* is published by the Internal Revenue Service (Publication 334) and can be obtained from the IRS. The Nov. 1979 issue contains 192 pages well indexed in the following parts:

> I. The Business Organization
> II. Acquiring Business Assets
> III. Determination of Gross Profit
> IV. Determination of Net Income or Loss
> V. Disposing of Business Assets
> VI. Results of the Business Activity

VII. Credits, Self employment tax, etc.
VIII. Filled-in forms for:
    Sole Proprietorship
    Partnership
    Corporation
    Subchapter S
  Examination of Returns
  Tax Publications

Get one.

Appendix C
Aerial Advertising Contract Sample

DERRY AIR INC.   **DATE**
11005 GREENWILLOW DRIVE
HOUSTON, TEXAS 77035
(713) 721-6290

AERIAL ADVERTISING CONTRACT

DERRY AIR, INC., hereinafter known as the seller, agrees to display an aerial advertising banner for _________ __________________ , hereinafter known as the buyer, according to the terms and conditions set forth in this contract. It is understood that the seller may employ agents, contractors, subcontractors, and/or employees as necessary to fulfill the terms of this contract.

BANNER COPY:_______________________________________
DATE(S) OF PERFORMANCE(S):__________________________
"RAIN DATE(S)": ____________________________________
EQUIPMENT:______ 5-foot (RED)____ 7-foot (BLACK)_____
                Combination
                Description of special equipment, if any:
TIME(S) OF PERFORMANCE(S): ___(Desired)
                            ____ (Inside limit for launch)
                            ____ (Outside limit for launch)

TARGET AREA(S): _______________________________________

SPECIAL PROVISIONS: __________________________________

RATES: The buyer agrees to pay for____hours of performance time at a fixed rate of $_____per hour, according to the following schedule unless otherwise stated in the SPE-CIAL PROVISIONS section of this contract. Charges are in effect from banner launch to drop.

Total Contract Amount $ _________________
Less 50% Down Payment   _________________
Balance*   _________________

* The balance will be pro-rated and billed after each performance.

PAYMENT: Pro-rated payments must be received within ten (10) days of the billing date. In the event of default and/or non-payment by the buyer, the entire unpaid balance of this contract becomes due and payable. All attorney's fees, litigation costs, court costs, and/or collection fees will be added to the balance.

THE SELLER shall not be responsible for non-performance of flying due to weather, acts of regulatory authorities, or other unforeseen technical difficulties. Should the performance (s) fail to meet that contracted for, then payment shall be made only for the services actually performed at the rate stated above.

THE SELLER reserves the right to amend or change any terms, conditions, or rates stated in this contract by written notice to THE BUYER at least thirty (30) days prior to the time the amendment or change is to take place. THE BUYER shall then have the right to terminate this contract, without penalty, upon the date the adjustment becomes effective.

THE BUYER may request an amendment or change in the terms of this contract, provided he does so in writing at least fourteen (14) days prior to the performance date that the proposed change is desired. If equipment and scheduling permits, changes will be honored SOLELY AT THE DISCRETION OF THE SELLER. It is further understood that any alterations or amendments approved by THE SELLER may result in additional fees or charges and may also require an adjustment of the hourly rate.

DERRY AIR, INC. is committed to the highest standards of flying safety. Nothing in this contract, expressed or implied, will

supersede that commitment. The pilot-in-command of the aircraft
in use will be the sole determinant of appropriate flight routes and
altitudes and his judgement and decisions are final.
THE BUYER shall, upon signing this contract, hereby agree to all
terms and conditions as set forth above.

FOR THE BUYER: **FOR THE SELLER:**

_______________________   _______________________
            (Date)   *(Date)*

Appendix D
Sample Cover Letter

DERRY AIR INC.
11005 GREENWILLOW DRIVE
HOUSTON, TEXAS 77035
(713) 721-6290

It is indeed a pleasure to enter our third year of banner towing service in the metropolitan Houston area. Thanks to our many satisfied customers, our 1941 Stearman biplane has become a familiar sight in the Houston skies. We feel confident that it enhances our mission- TO GET YOUR MESSAGE ACROSS TO THE PUBLIC!

By conservative estimates, the "average person" is exposed to somewhere between 500 to 600 advertising messages *PER DAY*. It is no small wonder that he tunes most of them out. In this respect, aerial banner towing has a unique advantage over more conventional media. When a banner appears overhead, most people instinctively gaze with curiosity to find out just what the message says. (They don't change stations, or run to the kitchen for a snack.) Furthermore, those exposed to banners can usually recall the message *WORD-FOR-WORD*.

This service is available at the following general rates:

	7′ Letters	5′ Letters
1 Hour (minimum)	$195.00	$165.00
2 Hours	180.00/hr	150.00/hr
3 Hours	175.00/hr	145.00/hr

Rates include ALL costs of operation up to 24 characters (letters and spaces) per message. Additional characters and/or any specially manufactured equipment (such as logos, emblems, etc.) will be quoted on an individual basis. Charges are normally in effect from banner pickup to drop, except under special circumstances. Additional discounts are available for long term contracts.

We hope you will consider aerial advertising as a regular supplement to your current advertising campaign. PLEASE COMPARE YOUR ADVERTISING DOLLAR among the various media. We would welcome the opportunity to arrange a personal visit to discuss the benefits that our service can provide your organization, and tailor a special program to meet your specific needs. Your interest in DERRY AIR, INC. is warmly appreciated - please call today.

Very truly yours,

Appendix E
Tow-Hitch
Installations (AC 43.13-2A)

STRUCTURAL REQUIREMENTS. The structural integrity of a tow-hitch installation on aircraft is dependent upon its intended usage. Hitches which meet the glider tow criteria of this chapter are acceptable for banner tow usage. However, because the direction and magnitude of maximum dynamic banner towline loads occur within a more limited rearward cone of displacement than do glider towline loads, hitches which meet the banner tow criteria of this chapter may not be satisfactory for glider towing. Due to the basic aerodynamic difference between the two objects being towed, glider and banner tow-hitch installations are treated separately with regard to loading angles.

a. *Glider tow hitches.* Protection for the towplane is provided by requiring use of a towline assembly which will break prior to structural damage occurring to the towplane. The normal tow load of a glider rarely exceeds 80 percent of the weight of the glider. Therefore, the towline assembly design load for a 1,000-pound glider could be estimated at 800 pounds. By multiplying the estimated design load by 1.5 (to provide a safety margin), we arrive at a limit load value of 1,200 pounds. The 1,200-pound limit load value is used in static testing or analysis procedures per paragraph 127 of this handbook to prove the strength of the tow hook installation. When the hook and structure have been proven to withstand the limit load, then the *maximum* breaking strength of the towline assembly is established at the design load of 800

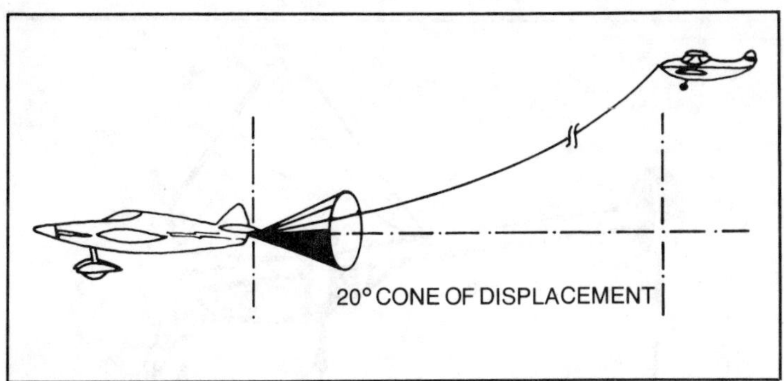

Fig. E-1. Glider tow angle.

pounds. Thus, the towline will break well before structural damage will occur to the towplane.

Another approach can be applied if the limit load carrying capabilities of a tow hook and fuselage are known. In this case, the known load value can be divided by 1.5 to arrive at the design load capabilities if the tow hook and fuselage limit loads are known to be 1,800 pounds. By dividing by 1.5 (1800 ÷ 1.5 = 1,200) we arrive at a design load value of 1,200 pounds. Thus, the maximum breaking strength of the towline assembly is established at 1,200 pounds and provides protection for the towplane.

Thus, in considering tow hook installations, one may establish maximum towline breaking strength by:

(1) Dividing the known limit load capabilities of the fuselage and two hook installation by 1.5; or

(2) Knowing the design load needs of the towline assembly and multiplying by 1.5 to arrive at a limit load. Then by analysis or static testing, determine that the hook and fuselage are capable of withstanding that limit load.

b. *Banner tow hitches.* Install the hitch to support a limit load equal to at least two times the operating weight of the banner.

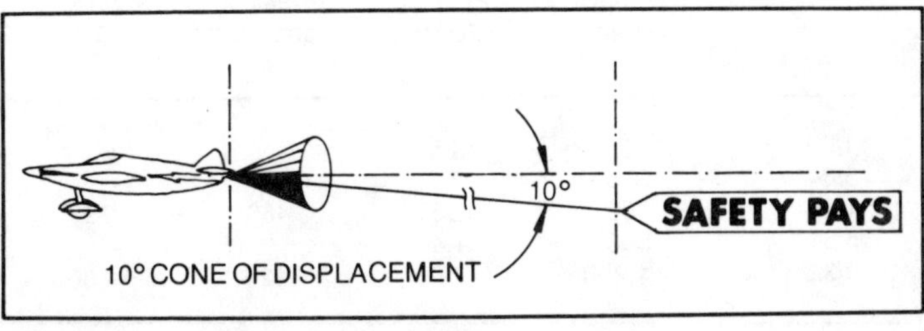

Fig. E-2. Banner tow angle.

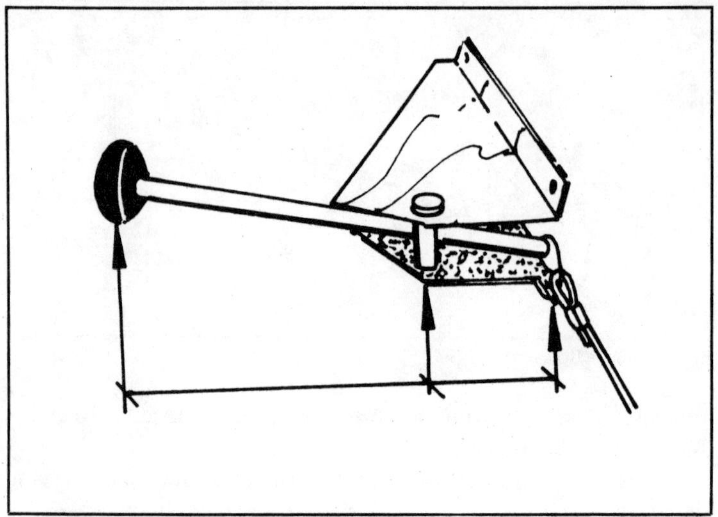

Fig. E-3. Typical tow-hitch release handle.

STRUCTURAL TESTING. Adequacy of the aircraft structure to withstand the required loads can be determined by either static test or structural analysis.

a. *Static testing.* When using static tests to verify structural strength, subject the tow hitch to the limit load (per paragraph 126 a or b) in a rearward direction within the appropriate cone of displacement per Fig. E-2. Testing to be done in accordance with the procedures of Chapter 1, paragraph 3, of this handbook.

b. *Structural analysis.* If the local fuselage structure is not substantiated by static test for the proposed tow load, using a

Diameter inches	Nonflexible carbon steel 1 × 7 and 1 × 19 (MIL-W-6904B)		Flexible Carbon Steel 7 × 7 and 7 × 19 (MIL-W-1511A and MIL-C-5424A)	
	Breaking strength (lbs.)	Pounds 100 ft.	Breaking strength (lbs.)	Pounds 100 ft.
1/32	185	.25	-	-
3/64	375	.55	-	-
1/16	500	.85	480	.75
5/64	800	1.40	-	-
3/32	1,200	2.00	920	1.60

Fig. E-4. Representative steel cable qualities.

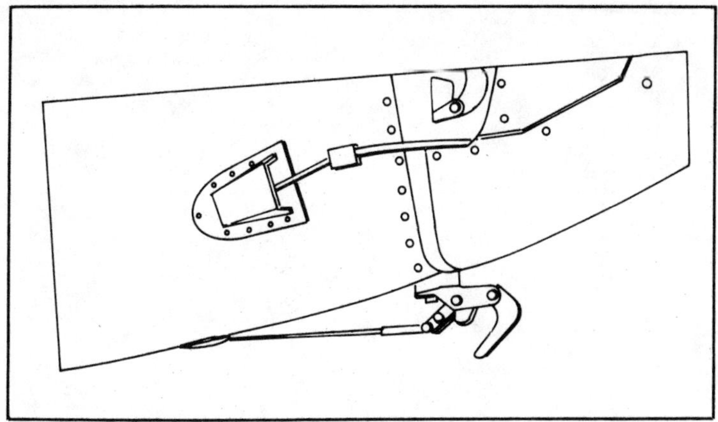

Fig. E-5. Tricycle gear aircraft.

method that experience has shown to be reliable, subject the
fuselage to engineering analysis to determine that the local
structure is adequate. Use a fitting factor of 1.15 or greater in the
loads for this analysis.

ATTACHMENT POINTS. Tow-hitch mechanisms are
characteristically attached to, or at, tiedown points or tailwheel
brackets on the airframe where the inherent load-bearing qualities

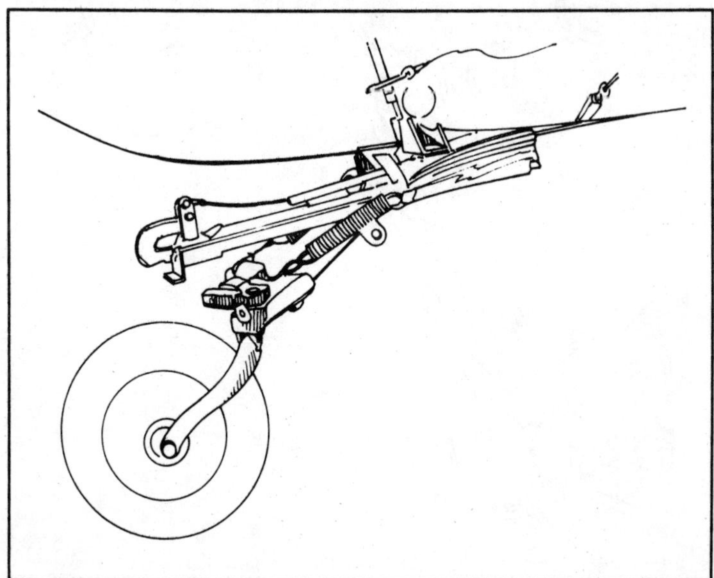

Fig. E-6. Conventional gear aircraft—leaf spring type tailwheel.

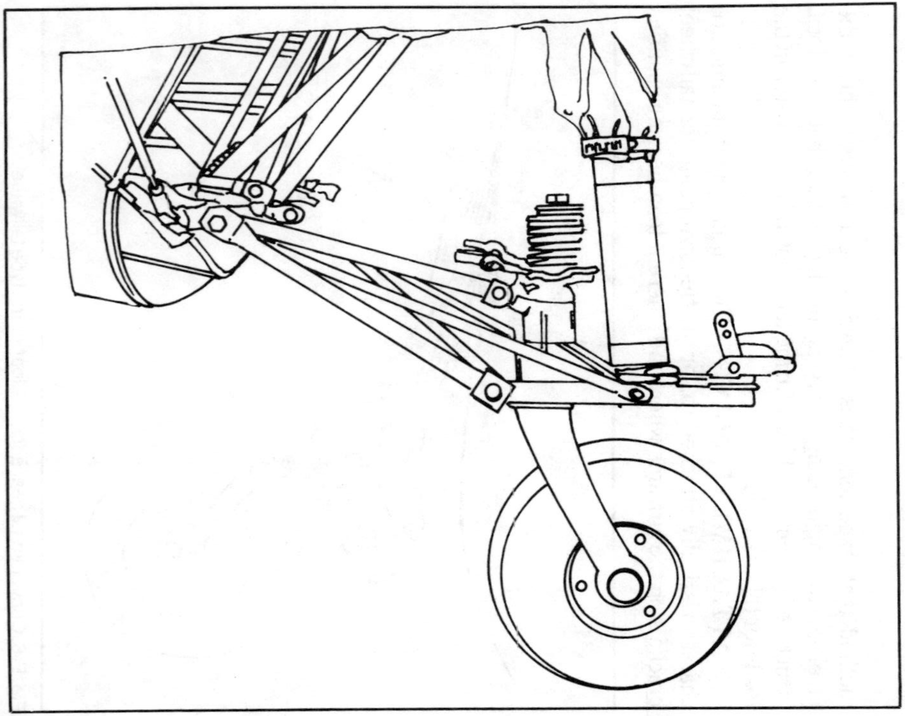

Fig. E-7. Conventional gear aircraft—shock strut type tailwheel.

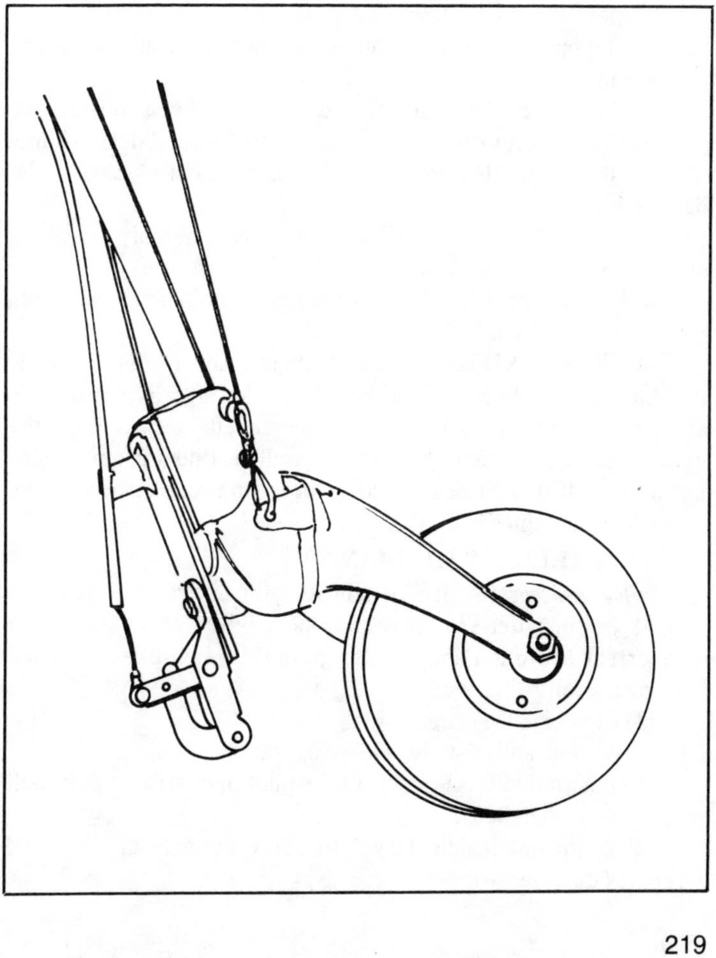

Fig. E-8. Conventional gear aircraft—tubular spring type tailwheel.

can be adapted to towing loads. Keep the length of the hitch-assembly arm from the airframe attachment point to the tow hook to a minimum as the loads on the attachment bolts are multiplied by increases in the moment arm.

ANGLES OF TOW. Test should be conducted on the system at various tow angles to insure that:

a. There is no interference with the tailwheel or adjacent structure.

b. The towline clears all fixed and movable surfaces at the maximum cone of displacement and full surface travel.

c. The mechanism does not significantly decrease the clearance from the tailwheel to the rudder.

d. The tow hitch does not swivel. Experience has shown swiveling could result in fouling both the release line and towline during operations by the towplane.

e. The opened jaw of the hitch does not strike any portion of the aircraft.

PLACARDS. A placard should be installed in a conspicuous place in the cockpit to notify the pilot of the structural design limits of the tow system. The following are examples of placards to be installed:

a. For glider tow—"Glider towline assembly breaking strength not to exceed ——— pounds."

b. For banner tow—"Tow hitch limited to banner maximum weight of ——— pounds."

WEIGHT AND BALANCE. In most cases, the weight of the tow-hitch assembly will affect the fully loaded aft c.g. location. To assure that the possibility of an adverse effect caused by the installation has not been ignored, enter all pertinent computations in the aircraft weight & balance records. (In accordance with the provisions contained in FAR 43.5(a) (4).)

TOW RELEASE MECHANISM

a. *Release lever.* A placard indicating the direction of operation should be installed to allay the possibility of confusion or inadvertent operation and the design of the release lever should provide the following:

(1) Convenience in operation.

(2) Smooth and positive release operation.

(3) Positioned so as to permit the pilot to exert a straight pull on the release handle.

(4) Sufficient handle travel to allow for normal slack and stretch of the release cable.

(5) A sufficient handle/lever ration to assure adequate release force when the towline is under high loads. (See Fig. E-3)

(6) Protection of cables from hazards such as:

(a) Wear and abrasion during normal operation.

(b) Binding where cables pass through fairleads, pulleys, etc.

(c) Accidental release.

(d) Interference by other aircraft components.

(e) Freezing and moisture accumulation when fixed or flexible tubing guides are used.

b. *Test of the release.* A test of the release and hook for proper operation through all angles of critical loading should be made using the design load for the glider or banner.

c. *Release cable.* Representative size and strength characteristics of steel release cable are as shown in Fig. E-4; however, it is recommended that all internally installed release cables be 1/16-inch or larger.

Index

THE
BARNSTORMERS

FLYING DAREDEVILS OF THE
ROARING TWENTIES

by

DON DWIGGINS

Printed in the United States of America

Reproduction or publication of the content in any manner, without express permission of the publisher, is prohibited. No liability is assumed with respect to the use of the information herein.

Copyright © 1981 by TAB BOOKS Inc.

Library of Congress Cataloging in Publication Data

Dwiggins, Don.
    The barnstormers.

    Includes index.
    1. Aeronautics—United States—History.
I. Title.
TL521.D87   629.13'0973   80-28414
ISBN 0-8306-2297-7

CONTENTS

To Bob Arabsky, of Winnipeg, Canada, a young man who is one of a new breed of airmen finding again the romance and adventure of barnstorming.

1/THEY RODE THE WILD WIND

BARNSTORMERS! It's a word to evoke a host of exciting images—of daring young men in bug-eyed goggles, caps set around backward, flying their flimsy machines through the air like erratic, colorful butterflies in high spirits, laughing at death!

Rooted in the carefree happy-go-lucky tradition of show business, the word was adopted by early aviators and made their very own . . . and so when you use it, do it affectionately, even with awe and respect, for that was the way people spoke it long before the first airliners flew, back in those dim years when man freed himself from the shackles of gravity and found a high road of adventure in the sky, that limitless arena for performing crazy stunts.

By long usage, the word now brings forth visions of wing walkers, parachutists and wild-eyed stunt fliers flashing low over county fairgrounds. scaring people half to death. True, that's the way it was in the Roaring Twenties, when the gypsy fliers spread their linen wings over the land to prove that flying was really safe after all, and not just for supermen.

But during the initial faltering steps toward the stars, when man first grew wings, when he dared to venture away from home and go chasing distant rivers, mountain ranges and far horizons, romance and adventure were part of the picture, too. At that time, barnstorming was only in swaddling clothes.

Curiously, it all began as a family affair. At first there were the brothers Lilienthal, the brothers Wright, and even the brothers Montgolfier, if one cares to go back to the first hot-air balloon days of two centuries ago. When the airplane was a *fait accompli,* a thing that really worked, the first to fly away from the nest were roving bands of barnstormers who called themselves by a different name—*exhibition fliers.*

Magician Harry Houdini bought Voisin biplane in Germany, took it to Australia, where he barnstormed in 1909.

In the year 1910, three exhibition teams held the spotlight both in the United States and abroad—the Wright Exhibition Company, the Curtiss Exhibition Company and the Moisant International Fliers. Because the story of the first two teams has been told enough times to become a part of the American legend, let's take a look at the Moisants, as daring and colorful a group of aerial wanderers as ever brought shrieks of terror from gaping crowds in the grandstands at county fairs.

John B. Moisant, leader of the clan, born in Chicago in 1868, was respectable enough when he studied architecture in a San Francisco university. It even appeared that he was headed for a humdrum life as a hotel operator when he and his brothers, George and Edward, went to Central America to seek their fortunes. Similarly, his sister, Matilde, seemed destined for a career as a school teacher.

Then, on a business trip to San Salvador, the brothers found themselves embroiled in a brewing revolution, and things began to happen fast. Their property was confiscated, and George and Edward were imprisoned. John set out for Nicaragua, alone and without funds. He didn't go to the American consulate and ask for help; instead, he angrily recruited a small army in the

Nicaraguan jungle and marched back to the San Salvador border at the head of a column of three hundred armed men.

To strengthen his hand, the President of Nicaragua supplied John Moisant with an armed gunboat, the *Monotombo,* plus a hundred hand-picked men from the Nicaraguan army and two hundred native Indians who were induced to "volunteer" for the reprisal expedition. At noon, June 12, 1907, the *Monotombo* chugged into the Bay of Acajutla and opened fire on the fortress. Within minutes a white flag appeared over a parapet, and before the sun went down, Moisant had carried the day.

With his brothers freed, John Moisant decided that business in Central America was too risky, so he sailed for France, the country of his forefathers where flying was now all the rage. Seeing a chance for new adventure, Moisant enrolled in Louis Blériot's flying school, and after three lessons he knew he had found where he belonged.

On August 24, 1910, at the controls of a brand new Blériot monoplane, John Moisant set out from Paris to establish a record by making the first international flight from one capital to another. His destination was London.

Wright Exhibition team flier Arch Hoxsey takes Teddy Roosevelt for plane ride over St. Louis in summer of 1910.

Sweeping out over the English Channel, Moisant battled unexpected winds that tossed his frail aircraft about like an autumn leaf. Finally, the white cliffs of Dover passed beneath his dragonfly wings, and as he headed inland, along the Thames, people ran into the streets to wave a welcome. Moisant happily waved back. Only six miles from his destination, however, his engine began to slow down and finally stopped. He glided down to land in a small field.

Not one to give up easily, Moisant tried again and made it on September 6, winning the London *Daily Mail* Trophy for being the first to cross "the Ditch" with a passenger, his mechanic, Albert Fileux. He actually had a second passenger on that historic hop—a striped cat that meowed happily for the photographers in London, glad no doubt to be back on the ground.

Air meets were now attracting large crowds in the United States, and so in October Moisant set sail for his homeland to participate in the Gordon Bennett Aviation Cup Race and International Meet at Belmont Park, New York, the first of a series of competitions he would enter under the auspices of the controlling Aero Club of America. Together with the Wright and Curtiss teams, Moisant's International Aviators held a corner on exhibition flying in those first years of barnstorming.

When Moisant returned to the United States from France, he brought with him the pick of the crop of European pilots, including such famed Gallic birdmen as René Barrier, René Simon and the great Roland Garros. Also in the Moisant troupe was an Italian pilot, Oresces Farrara, and two Americans who joined him in the United States, St. Croix Johnstone and John J. Frisbie.

With the coming of winter weather, the Moisant International Aviators headed south to barnstorm new territory. They were joined by another American pilot, Charles K. Hamilton, who had had a falling out with Glenn Curtiss. Hamilton owned a special plane, the *Hamiltonian,* powered with a 110-horsepower V-8 Christie engine. It was too hot to handle, and Hamilton had one bad smashup after another in it, frequently suffering a shower of scalding water from a broken radiator.

At Richmond, Memphis and Chattanooga, the Moisant troupe gave thousands of amazed spectators their first look at an airplane, and then they moved deeper into the South, putting on performances at Tupelo, Mississippi, and at New Orleans, Louisiana. There, as it did to many early birdmen, the way it also would to other subsequent barnstormers, sudden death came to John Moisant—wearing his flying boots.

John Moisant's death, however, did not eliminate the family name from the roster of active birdmen, or birdwomen. The following summer, his sister, Matilde, enrolled in a flying school John had started in Garden City, New York, along with a close friend, Harriet Quimby, a young and pretty drama critic of *Leslie's Weekly* who was fascinated by airplanes.

From the start, the two girls staged a friendly rivalry to see who would be the first to solo and the first female to win a pilot's license. A French in-

America's first great exhibition flier was Lincoln Beachey, shown here taking off on flight from San Francisco, March 14, 1915, which ended in fatal crash.

structor, André Houpert, took each girl in turn and explained all that he knew about airplanes. After thirty-three lessons adding up to four and a half hours air time, Harriet was the first to solo, on August 1, 1911.

Matilde wasn't too disappointed. On the 13th, she also ventured into the air alone. Defying superstition, Harriet promptly named her lovely Blériot monoplane *Lucky Thirteen,* for good reason: she had been born on a Friday the 13th; her first and last names began with the thirteenth letter; she applied for her flying license on the thirteenth day of the month and got it on the thirteenth of the following month!

On her twenty-fifth birthday, Matilde began her barnstorming career by following the sky trails blazed by her late brother, John. Joining the Moisant International Aviators group, she immediately set a world altitude record for women (1,500 feet!) and then traveled with the team to Mexico City to become (on November 13) the first birdgirl to fly there.

The laughing, dark-haired beauty caused gossips' tongues to wag. She was the first to wear a divided skirt, a necessity in the interest of modesty for riding in the open cockpits of her day. Harriet, not to be outdone, adopted an even more spectacular costume—trousers tucked into high-laced boots, a long-sleeved blouse with a turtleneck and a capuchin, all made of mauve-colored satin.

Back in the United States, Matilde aroused the wrath of the constabulary of Mineola, Long Island, by breaking the "blue laws" with an exhibition flight on a Sunday. When the sheriff sent men to arrest her, she spun her ship around, dusted them good and proper, and took off again, to land at nearby Moisant Field. There other officers hauled her into court, but a Hempstead justice ruled that flying close to heaven on a Sunday was in violation of no man-made law, or heavenly one, for that matter.

Matilde and Harriet decided to team up, and for a while they barnstormed the country together. Harriet, concerned over Matilde's apparent disregard for her life, admonished, "Honey, Number Thirteen may come up once too often for you, so look out!"

Harriet herself seemed driven to take bigger and bigger risks in flying. On April 16, 1912, she was poised on the Dover cliffs of England, the nose of her frail winged moth pointed toward France. She wanted to be the first woman to cross the English Channel by air. In her exclusive report in *Leslie's Weekly,* she wrote:

"I was hardly out of sight of the cheering crowd before I hit a fog bank and found my needle of invaluable assistance. I could not see above, below or ahead. I ascended to a height of six thousand feet, hoping to escape the mist that enveloped me. It was bitter cold—the kind of cold that chills to the bones. . . . A glance at my compass reassured me that I was on my course. Failing to strike clear air, I determined to descend again.

"It was then that I came near a mishap. The machine tilted to a steep angle, causing the gasoline to flood and my engine to misfire. I figured on pancaking down so as to strike the water with the plane in a floating position. But, greatly to my relief, the gasoline quickly burned out and my engine resumed an even purr. A glance at the watch on my wrist reminded me that I should be near the French coast. Soon a gleaming strip of white sand flashed by, green grass caught my eyes, and I knew I was within my goal."

Back home once more, Harriet Quimby was hailed as a heroine of the sky. Other young women eagerly read every word she wrote. In *Good Housekeeping,* she warned girls who might take up flying that "only a cautious person should fly; I never mount my machine until every wire and screw has been tested. I have never had an accident in the air."

As every pilot knows, it's bad luck to brag about your good fortune, and when you do, you're supposed to knock on wood. But where Matilde flaunted her superstitious fetish of Number Thirteen, Harriet never trusted to luck. It was true, nevertheless, that on each flight she did wear an antique necklace over her mauve blouse and that beneath it, close to the warmth of her body, there nestled a tiny, headless brass icon given to her by a French airman (she had accidentally decapitated it one day by slamming it to the floor when everything seemed to go wrong).

Early birdman De Lloyd Thompson, in Day biplane tractor, races speed king Barney Oldfield. The year is 1915.

But she thought it more important to leave the care of her Blériot monoplane in the hands of a competent mechanic and her barnstorming business in the care of a scrupulous manager, A. Leo Stevens. She was determined to make flying as safe as possible.

And on July 1, 1912, warming up her engine with special care at the edge of Harvard Field, Boston, there was no premonition of impending disaster. Riding with her was a passenger, William A. P. Willard, manager of the air meet, who had asked to fly with the popular girl barnstormer as a special favor.

Harriet was seen to smile back at him, finger her necklace for a moment, then open the throttle wide and swing her plane into the wind for takeoff.

Down the field she roared, lifting into the air as gracefully as a seagull, then spiraling upward, higher and higher, thousands of feet above the harbor. As the crowd below watched, Harriet's tiny Blériot machine suddenly nosed into a vertical dive, and an object was seen to plummet from it—the body of a man! Neither occupant wore safety belts, and whatever happened—no one is certain to this day—the sudden dive catapulted Willard from the plane to plunge through the sky, clawing at the air until he struck the water and died.

Many of the screaming, shocked spectators following Willard's body with their eyes failed to see a second form hurtle from the diving monoplane, but those who did cried out: "There goes Harriet!"

11

The impossible had happened; the unbelievable now was to be believed. Harriet Quimby, the darling of the skies, was mortal, after all, and flying was still the most dangerous game. Shocked most of all, of course, was Matilde Moisant, who had been in the air at the time and had seen the whole thing happen. Stealing her nerves, she immediately landed and hurried over to where rescuers had taken the bodies of the victims.

Reporters crowded around Matilde, asking for some comment. Wiping away tears, she told them, "No accident except that to my brother Jack affected me so much . . . when I think how she was always scolding me for my carelessness, and here I am after all my accidents, while she had to die in her first mishap. Well, after all, flying's like that—just a game of poker. . . ."

Harriet's death was a severe blow to Matilde, but pluckily she kept on flying her frail little Blériot, *Lucky Thirteen,* until a series of near-fatal accidents prompted her family to beg her to give up flying. At Shreveport, Louisiana, her plane's undercarriage caught on a hummock during a landing and *Lucky Thirteen* did a complete somersault, landing on top of its pilot. Still, she escaped unscathed. A few days later, at Wichita Falls, Texas, flying again proved to be "a game of poker" to Matilde when on a landing approach she saw a swarm of people suddenly sprint out onto the field, directly in her path. Reacting instinctively, she advanced the spark lever and the idling engine caught hold, but not quickly enough. She pulled up into a stall, then came crashing back to earth, the fuel tank ripping off and bursting into flames.

Those nearest to *Lucky Thirteen* grabbed a wingtip and tried to haul the blazing wreckage away, but the girl pilot was nowhere to be seen.

"Hurry!" a woman cried frantically. "She'll be cremated!"

Then, as if by a miracle, Matilde Moisant wriggled free of the pyre. Her hair was singed, but otherwise she was unharmed. It was obvious, at last, that death was hovering too close for comfort. Heeding her parents' wishes, Matilde finally hung up her helmet and goggles for good.

Ruth Bancroft Law, who had learned flying from a spectacular exhibition flyer, Lincoln Beachey, made her first public flight over Boston Harbor immediately after the death of Harriet Quimby. Until she retired from active flying in the early 1920's, her barnstorming career followed the same pattern as Matilde Moisant's, that is, as a family affair.

Ruth's brother, Rodman Law, was first to become airminded when, on February 2, 1912, he took up parachuting—by leaping from the upraised arm of the Statue of Liberty. Living through that, Law, a steeplejack by profession, made more headline jumps—from the Brooklyn Bridge, from the roof of a New York skyscraper, and finally, on April 13, 1912, from a Burgess-Wright hydroplane over Marblehead, Massachusetts.

As the second man in the world to leap from an airplane (following Captain Albert Berry, a St. Louis balloonist and parachutist) Law proved a big

Ruth Law barnstormed the countryside in Curtiss pusher after learning to fly under Lincoln Beachey.

attraction at county fairs. Until World War I erupted, he traveled about the countryside, leaping from planes, as one of aviation's first true barnstormers.

Ruth once, in August, 1914, flew her daredevil brother up to jump altitude over Salem, New Hampshire; and on December 17, 1915, anniversary of the first flight of the Wright biplane at Kitty Hawk, North Carolina, she became the first woman to loop the loop. When the country went to war, Ruth began flying Liberty Bond promotion tours; and when the Statue of Liberty was first illuminated, she circled it at night with magnesium flares blazing from her wingtips.

In November, 1916, Ruth set out from Chicago in a Curtiss pusher to attempt a nonstop flight to New York City. She ran out of gas over Hornell, New York, and landed "deadstick" at a racetrack infield, having covered 590 miles in 5 hours 45 minutes to break the former nonstop flight record of 462 miles, set earlier that year by Victor Carlstrom.

During the war, Ruth and her husband, Charles Oliver, went overseas to assess the state of flying in France, and on her return she attempted to enlist as a combat pilot. Her offer refused, she joined the American Red Cross to help out in the first Liberty Loan drive. She once buzzed the length of Penn-

Rodman Law gave up exhibition jumping to join Signal Corps as a balloonist in World War I.

NATIONAL AIR & SPACE MUSEUM

sylvania Avenue in Washington, D.C., her wingtips grazing trees lining that thoroughfare.

The war over, Ruth and her husband sailed for the Philippines where she personally assisted in the establishment of a pioneer air route on the island of Luzon. She returned to barnstorm America for a few more years, then quit, as Matilde Moisant had done, when death seemed to be hovering too close. On October 4, 1921, at Long Branch, New Jersey, flying a ship with a rope ladder dangling over a speeding car, Ruth had been shocked to see a stunt girl, Madeline Davis, leap for the ladder and miss. Madeline struck the ground at better than 60 mph, and was killed.

Retiring to Beverly Hills, California, Ruth Law told an interviewer why she gave up flying: "Things are so proper now. . . . A pilot has so many rules and regulations to follow. . . . I couldn't skim over rooftops today or land in the streets or on a race track. The good old crazy days of flying are gone."

In the "good old crazy days of flying," perhaps the best known family of barnstormers was the Stinsons—Katherine, Marjorie, Eddie and John—who hailed from Jackson, Mississippi. Katherine was barely seventeen when she went up for her first plane ride at Kinloch Field, St. Louis, with Tony Jannus, instructor at Tom Benoist's aviation school. Later, the Stinsons moved to Chicago, where Katherine acquired her pilot license on July 24, 1912, to become the world's youngest female flyer.

Not until the next spring, at Cicero Field, Chicago, did Katherine begin her career as an exhibition flier, in a brand-new Wright Model B pusher. It was the start of a barnstorming tour that would take her to Coney Island, Arkansas, Montana, Louisiana, Texas, North Dakota, Michigan and Missouri.

Her older brother, Eddie, served as Katherine's traveling companion and mechanic. In repayment she taught him to fly. Wintering at Fort Sam Houston, San Antonio, all four of the Stinsons became expert pilots and opened a flying school of their own.

When in San Francisco, in 1915, the noted stunt pilot Lincoln Beachey was killed, pulling the wings off a swift little monoplane, the *Beachey Special,* Katherine bought the wreckage, salvaged the rotary engine and had a skilled plane designer, Matty Laird, install it in a special exhibition tractor ship. With it she thrilled crowds by looping the loop and, before Ruth Law, by flying at night with flares on her wingtips.

In 1917, Katherine sailed for the Orient, barnstorming through Japan and China, then returned when the United States and Germany went to war. Like Ruth Law, she tried to enlist as a combat pilot, but her services were declined. She and Marge then settled for training pilots for the Royal Canadian Air Force. With the school running smoothly, she helped the American Red Cross over the top in its Liberty Loan drive by a spectacular cross-country flight in a Jenny from Albany, New York, to Washington, D.C.

On the first leg of that flight, Katherine raced and beat the crack Empire State Express passenger train into New York City, then followed the "iron beam" into Philadelphia with a railroad timetable for a map. She finally made it to the national capital, landing beside the Washington Monument. A cheering throng of five thousand people hailed her as a real heroine when she handed a check for $2,000,000, the proceeds from her spectacular benefit aerial performances, to Secretary of the Treasury William Gibbs McAdoo.

Katherine gave up flying when she married at the age of twenty-five, while Eddie went on to become the "dean" of aviation, the first to log more than fifteen thousand hours in the air. As a plane manufacturer, Eddie Stinson built the transatlantic ship flown by Ruth Elder and George Haldeman and designed a popular line of private aircraft that is still flying.

Selling airplanes meant traveling around the country, and in that sense Eddie Stinson joined the ranks of postwar barnstormers whose air paths frequently crossed in a number of small Midwestern towns. For a time he was joined by another traveling airplane salesman, Carl Squier, who would later become the president of Lockheed Aircraft Company.

2/THE HERO

I WAS SIX YEARS OLD when I first met a real life birdman, gypsy flier, or barnstormer, call him what you will. I don't remember his name, but I will never forget what he looked like.

It was a warm summer day in 1919 when he buzzed our summer home at Canada Lake, New York, and with a catch in my throat I watched him finally zoom off, waggling his wings, and head for the community of Wheelerville, some five miles away.

"Daddy!" I cried. "Let's go see the airplane!"

By the time our Model T Ford was bouncing across the clover field toward the beautiful winged machine, the pilot, a tall, lanky fellow in leather coat, whipcord breeches and shining puttees, already was busy pouring gasoline from a five-gallon tin into the wing tank through a chamois skin inside a funnel.

I jumped down, raced to the airplane, and stared at it in wonder. Dad said it was a Jenny, but I thought it should have a more godlike name, for certainly the man who was up there on the wing was some kind of god.

His lean face was oil-streaked, and a trace of a smile flickered beneath a precise waxed moustache. The corners of his eyes crinkled. Goggles on his forehead made another pair of eyes against the leather helmet he wore, the ear flaps turned up.

"Here, kid!" he called, tossing me the empty gas can.

I caught it and set it down, feeling proud that I had been noticed.

"Want to go for a ride?"

My heart stopped beating. "Sure!" I cried. "Can I, Dad?"

The pilot turned to my father. "Five bucks for five minutes. And your sister can go too," he said to me.

Phoebe's face broke into a smile. Dwig and Betsy, our understanding parents, looked at each other, and Mother bit her lip. "Let's let them go," she finally said. "It will be a wonderful experience!"

Moments later, strapped side by side into the open front cockpit of the wonderful biplane, we were bouncing over the field and then suddenly flying! With a great leap the Jenny hurtled a fence, shot across the Wheelerville sawmill pond and began an easy, graceful climb toward Canada Lake.

Half out of the seat, I peered over the side, the wind bringing tears to my eyes and the sweet smell of burning castor oil to my nostrils. We were flying! There was nothing holding us up! Nothing but thin air, a miracle!

Phoebe's elbow nudged me. She pointed down to a bend in the shoreline. There it was—our home!

Above the hammering of the OX-5 engine I faintly heard the pilot's voice: "Sit down! You wanna fall out?"

I reluctantly slid back onto the seat cushion and then gasped as he dropped the left wing. We were going to tip over! But we didn't. Down we flew through the spiral glide, around and around as our home, the "Dwigwam," appeared pinned off the end of the wingtip, like a butterfly on a specimen board.

At once my thoughts were down there, where I liked to lie on the boathouse roof, look up at the clouds and wonder what it would be like, being a bird.

Now here I was, seeing the world from a new point of view, and marveling. I knew then what I wanted to be when I grew up—a pilot! Free to wander through the clouds and look down on the greens and browns of the countryside and follow wheeling hawks that hunted along the swampy shores of a hidden lake, the secret place of my boy's world.

Suddenly we were streaking back toward Wheelerville, the five minutes running out all too soon. I felt a sadness as we came back over the millpond and the pilot banked around, diving down toward the field where Mom and Dad stood waving beside the car.

With a jarring bump the wheels struck a clump of grass, sending the airplane careening to the left. The pilot struggled with the controls to straighten it out, but around we went in an ever-tightening spiral, until the right wing began dragging the ground.

There were no brakes on the Jenny, but if there had been, it would have been too late. We had already ripped the cloth off the wingtip. We'd ground-looped.

The Model T Ford came bouncing across the field toward us, Mother waving frantically.

"Is everything all right?" she cried as the pilot jumped down and walked toward the wingtip.

"Sure!" he replied. "I always land this way!"

He went back to his cockpit and got out a black satchel, selected a piece

The Aviator—The Superman of Now

The world has its eyes on the flying man. Flying is the greatest sport of red-blooded, virile manhood.

Make your vacation the greatest you ever had by joining the Wright Flying School. Live in the open—in the aviators' tent city. Convenient hotels for the fastidious.

A short course at the Wright camp will fit you to fly any type of machine. Expert instruction in flying, assembly, upkeep, motor-overhaul, etc. Dual controls. Pupil flies the first lesson. The school is located on Hempstead Plains—the greatest aerodrome in America.

Send for New Booklet

WRIGHT FLYING FIELD, Inc.
60 Broadway, New York

By 1916 the Wright brothers had added a cockpit to their pusher planes and opened a flying school on Long Island, New York.

of canvas and a can of dope. In less than a minute he then expertly slapped the fabric over the wingtip and glued it down with the sharp-smelling stuff.

"Now," he grinned at my folks, "anybody else want to take a chance?"

He was one of an army of ex-military pilots who roamed the land with their war surplus crates, bought for $600 all boxed up, brand-new. (That's why airplanes came to be called crates, one theory has it.)

They were the gypsy fliers, out to spread the gospel of aviation to the grass roots country and to work a miracle by getting America into the air, first to wonder, then to enjoy flying, and finally to travel as a matter of course.

Airports were few and far between, but these small two-seater training planes, the JN-4D Jennies and Standards, didn't exactly need airports. Any quarter section would do, and you could tell by how green the grass was whether the field was too wet to land on, and all you needed for a wind sock was a cow's tail. Every pilot knows as well as a farmer that cows eat facing upwind.

The gypsy fliers knew it was better to land in a field near a road because then you would not have to walk so far to get gas, and if there was a barn handy, it was better to stake your airplane down behind it overnight, to keep it out of the wind.

Thus, gypsy fliers became barnstormers.

Each band of barnstormers consisted of individual heroes, for anybody who flew was obviously superior to simple ground-bound folk. They dressed the part, they lived the part—and if they swaggered a bit, they were forgiven. A country must have its heroes, and while it was fine to give special notice to generals and admirals who fought great battles, the barnstormers were of a different breed.

Facing the threat of a storm alone . . . flying into the teeth of a gale . . . winging off to remote wonderlands beyond the horizon . . . this was a new kind of stimulus, something one could understand and imagine emulating. Today one would say these were heroes "to identify with."

In the year 1919, when I met my first barnstormer and thrilled to the new world of aviation, World War I was over. The big Standard Aeroplane Factory in my home town of Plainfield, New Jersey, bulged with training planes nobody needed.

Elsewhere across the country, from Buffalo, New York, to Dayton, Ohio, and to Sy Christofferson's little Jenny factory at Palo Alto, California, thousands of war surplus biplanes posed a threat to the aircraft industry that had burgeoned.

America's air strength had been slow in growing. In the first eight years of the Army Aviation's existence, from 1909 to 1916, only 142 aircraft had been built and delivered. But the following June, Congress, facing the threat of war, made the largest single appropriation ever—$640,000,000—for a skyful of airplanes.

Flying schools mushroomed to train combat pilots for the planes we would eventually build—too late. Nearly 15,000 cadets received air training in this country and another 1,800 in Europe. By March, 1918, as the war drew toward a close, Army Aviation's strength had zoomed to 11,000 officers and 120,000 enlisted men.

Not all of these trainees were pilots, of course. At the time of the armistice, November 11, 1918, the country had at the front only 757 pilots and 481 observers, with 740 planes and 77 balloons, while another 1,402 combat-ready pilots, 769 airplanes and 252 balloon observers had entered the zone of advance.

Most of the combat planes were of foreign make, but U.S. war plants were beginning to hum at top speed, turning out copies of British designed De Havillands and Handley-Pages fitted with the American Liberty motor, our greatest technological contribution to the war in the air.

There were plenty of war aviator heroes, though. An elite of 63 Yankee flyers was credited in World War I with destroying 462 enemy planes out of the American total of 491.

LOS ANGELES COUNTY MUSEUM

World War I aviators, out of a job when hostilities ended, turned to barnstorming for a living.

These aerial killers made names for themselves that would live in history—
Eddie Rickenbacker, Kiffin Yates Rockwell, Frank Luke, Didier Masson,
Gervais Raoul Lufbery. They would be wined, dined and feted over and
over and showered with confetti, all drawing admiring glances from pretty
girls and some of them gaining fortunes.

But there were other unsung heroes who came home with only a few
dollars in their pockets, a dedication to flying and no job to return to. Many
were young men who had gone off to war from college and so found themselves with no career to follow. Others were country lads about whom a song
had been written that asked, "How you gonna keep 'em down on the farm,
after they've seen Paree?"

You couldn't. For aviation had opened a brand new world of adventure,
and the taste of flying behind stinking, oil-throwing engines was still strong
in their mouths. These were the men who would become the gypsy fliers.

People were optimistic at first that the Post Office Department, the Army,
and the Navy could absorb the hundreds of thousands of surplus warplanes
in storage, and a glowing future was seen for sport aviation by simple conversion of the trainers.

At the beginning of 1919, one authority, Archibald Black, a Navy aeronautical engineer, wrote that "in the case of the sporting field, practically all
existing types except the largest are already fairly well adapted to commercial
use. The smaller and lower-powered machines, like the Curtiss JN and the
Standard J, are particularly suited. The extent of this market is, of course,
problematical and will depend greatly upon the large number of men of
independent means who are now in the service: flying boats and seaplanes
may be expected to be very popular among these men. . . ."

True, there were plenty of rich men's sons in uniform, and already one
aviation enthusiast with money, Rodman Wanamaker, was organizing an
Aviation Section of the New York Police Department. Wanamaker, a police
commissioner, had contracted Glenn Curtiss just before the war to build a
trimotor seaplane that could compete for a London *Daily Mail* $50,000 prize
offered for the first transatlantic nonstop flight. Now, with the war ended, it
seemed to some as though aviation would become a rich man's game.

That same year in Hollywood, California, a movie studio owner named
Thomas H. Ince posted another $50,000 prize for the first transpacific flight,
which was to start from his Ince Aviation Field at Venice and finish either
in Australia, Japan, the Philippines or Asia.

Others in aviation took a dim view of the future of civilian flying. W. T.
Thomas, president of the Thomas-Morse Aircraft Corporation, observed, "I
believe there is a market for low-priced aircraft for civilian fliers, but here
again, a great deal of press propaganda is needed to educate the public sufficiently to insure a sale of enough machines to allow their manufacture at a
low price."

Commented Ottorino Pomilio, another airplane builder, "Unfortunately, a large number of people, and generally not through their own fault but essentially due to censorship which during the war only permitted the publication of incomplete or uninteresting news, have not yet formed an idea of what the airplane is today . . . ."

Pomilio was right. Aside from a few thrilling war dispatches about the exploits of combat airmen, to the general public an airplane was still a crazy contraption that had no practical use.

A few tentative steps were being taken, however, in this first year after the war. Albany, New York, established the nation's first municipal aerodrome, and the Post Office Department moved its New York "aerial mail" terminal from the Belmont Park racetrack to Newark, New Jersey, easily accessible via the Hudson tube.

There was big talk about inaugurating passenger airlines and spanning the country with air mail and cargo routes, but it was just that—big talk. The nation was caught unprepared; dreams of filling the skies with wealthy flying sportsmen and linking distant cities with scheduled airlines remained just dreams.

The frantic war economy had ground to a sudden halt. Plane factories, shipyards and munitions plants no longer swarmed with war workers. The painful aftermath of the great conflict had begun.

From the signing of the armistice to March, 1919, government aircraft contracts totaling $469,000,000 had been canceled. Searching about for some way to stimulate postwar flying, the federal ban on civilian flying, in effect through the war years, was lifted.

Flying schools rolled open their rusty corrugated metal doors for business, hopeful that the returning veterans might see a future in the sky. Among the first was the Curtiss Aviation School at Garden City, Long Island, New York, reopened under the direction of a Curtiss test pilot, Roland Rohlfs. At Tarrytown on the Hudson, the Castle School for Girls offered a course in airplane mechanics.

These were bold steps. The rich man's son, the women's rights enthusiasts, the sporting set, all found flying a fascinating new pastime, more thrilling and less expensive than yachting. But something else was needed to get people into the air, Pied Pipers who could lead the way to America's new destiny in the sky.

They finally came along, discharged Army Air Service pilots who had won a few dollars in a shipboard crap game on their way home from Europe, their worldly possessions in a duffel bag slung over their shoulders as they strode down the gangplank singing "It's a Long Way to Tipperary." They were looking forward to some kind of a career, but there just weren't any careers around. That didn't bother them, however, when they heard you could buy a brand-new flying machine for a few hundred dollars!

Almost overnight, barnstorming was born.

In tiny fields across the nation they set up shop. They didn't need to hang out shingles; everybody in town knew when they'd arrived, looping, rolling and diving overhead to draw customers out to the edge of town to buy rides. Lester Gardner, the worried editor of *Aviation* magazine, observed in October, 1919: "One of the most interesting phases of present aviation activities is the great number of small companies engaged in exhibition flights

Jennies, Standards, Canucks, De Havillands hit the barnstorming trail when the war ended.

and in passenger flights of short duration. Such work has not yet reached the dignity of aerial transportation work, but nevertheless both activities have considerable educational value for the general public."

Educational? What was going on was a wild, uninhibited aviation explosion that would reshape America! The public loved it, tearing through the sky over the old folks' farm, doing loops and spirals, and buzzing girl friends' houses. It was a time for release, a time for rebirth. It was the springtime of civil aviation, and the saps were rising.

Poor Lester Gardner's voice was lost in the whirlwind when he warned, "Stunting should be avoided. Even if the passenger does ask for stunt flying, he will not enjoy it. He will come down congratulating himself on being a brave man but with the feeling he has had a very serious experience. The passenger should come down feeling that he has had a perfectly safe and normal experience which he would like to repeat . . . ."

From coast to coast they flew, these wild men of the sky, war pilots free of military discipline and enjoying immensely the role of hero.

And the public loved it. For a dollar a minute they could buy a lifetime of thrills and have something to talk about for months and years.

But there were fatalities. Through the natural course of events, the less skilled pilots killed themselves off (along with their unlucky passengers) until, by 1920, a toughened band of gypsy fliers, many of whose names are remembered today, simply took over postwar aviation.

Remember Jack Knight? Clyde Pangborn? Eddie Stinson? Frank Hawks? Art Goebel? Roscoe Turner? Carl Squier? Didier Masson? Ormer Locklear?

The list was a grease-stained honor roll of men who flew dawn to dusk, day after day, in sweaty uniforms behind overheated inline engines.

Dale Seitz . . . Frank Clarke . . . Martin Jensen . . . Boots LeBoutillier . . . these were the wood-and-wire heroes of the Roaring Twenties.

And there was Arrigo Balboni, the world's first flying junkman, who opened an airplane bone yard on Riverside Drive in Los Angeles and kept a famous Gold Book in his oilstreaked office, where aviation greats were glad to sign their names.

Jimmy Doolittle, Charles Lindbergh and Wiley Post were among those who stopped by Balboni's yard to pick up a used engine cowling or a wing strut, and the latest gossip.

But first you had to sign the Gold Book.

"Your name will be worth money some day," Arrigo would say glumly. "When you kill yourself."

3/FROM HAWKS TO DOVES

THEY WERE LEGENDARY FIGURES who lived legendary lives, but now the war was over and their days of glory were past, as dead as yesterday's newspaper. What would become of them, those thousands of stalwart war heroes, the gladiators of the sky? And what of the planes they flew?

After all, the war to end all wars was only a ghastly memory, and who needed warplanes with chattering machine guns? For that matter, who needed men trained to kill, men who knew no other profession than flying into the face of death?

Charles Herbert Veil, at twenty-two a veteran of the French Foreign Legion and the legendary Lafayette Flying Corps, Spad 150, *Group de Combat 16,* pocketed his *Croix de Guerre,* his *Medaille Militaire,* and his dog tags, climbed into his Spad and roared off toward Paris, anger in his heart.

He adjusted his goggles, wiped a slick of oil from his windscreen and began a long, shallow dive toward the City of Light. It was in just such a dive that he had ripped through a flight of nine Fokkers, scattering them like birds. He shot five down to win his ace rating, then hedgehopped back over the front lines, his plane riddled by bullets.

That was yesterday. Now, dead ahead, crowds filled the Champs-Élysées, cheering, crying men and women watching marching soldiers, returning heroes who had ended war forever. Veil dove down low, zipped across the Seine and threw the Spad at the columns, fingers tight on the controls. Men broke and ran. Was this fool crazy?

Barnstormers got publicity by posing pretty girls with their surplus ships, like this Nieuport painted with famed death insignia of French ace Charles Nungesser.

The Arc de Triomphe, gleaming white in the morning sun, loomed dead ahead in the crowded Place de l'Etoile. Veil made for it, his wheels inches above the crowd. At the last second he snapped over into a vertical bank and knifed through the arch, then zoomed up and slow-rolled away, grinning.

27

They couldn't court-martial him now! He was free to go where he wished, do as he pleased. But just in case there were any MP's around, Veil landed the Spad on a back road outside Paris, hitched a ride to the city, and boarded a train at the Gare du Nord, his destination Poland. As one of the few Americans who chose to remain abroad after the fighting, Charlie Veil barnstormed from the Bosporus to the Baltic, and in 1920 settled down at last to organize the first air line between Paris and Warsaw.

John J. Niles recalled a flying buddy named Hawkins who was bitter because a Frenchman had landed a Caudron right on the roof of the Galeries Lafayette, the big department store in the heart of Paris' shopping district. He was arrested, but he didn't mind; he had won a large bet.

Hawkins fell in love with a bridge, a lovely stone-arch bridge spanning the Seine at Choisy-le-Roi. Every afternoon for a week he flew out and looked at his bridge, swooping low to estimate the width of the span. It would be great to fly beneath it! One evening, Hawkins and Niles drove out to look it over more closely. Yes, there was enough width for his Camel, but only two-and-a-half feet of headroom to spare between the top of the arch and the water.

The next day, following a heavy rain, the skies cleared and Hawkins swung the Camel out toward Choisy-le-Roi, where Niles and other squadron mates were waiting on the bridge to see if he could make it. As Niles recalled it, the stunt ended this way:

"At the last possible moment, everything went wrong! We could almost see the whites in Hawkins' eyes as he pulled his controls back into his stomach and zoomed his plane off the river in a furious attempt to clear the bridge. His motor was going like a million. There was no reason for what had happened, as far as we could tell, except pure last-minute loss of nerve.

"As the plane shot up into the air, we had a fleeting glance at its oil-bespattered underneath. The circles of red, white and blue stood out before us like huge bull's-eyes. We had never had a chance to see the bottom of a Camel airplane up so closely—and most of us didn't care to see another under exactly the same conditions. . . . With the zoom Hawkins had pulled, considering his forward speed, the plane was thrown into an almost vertical climb. When he reached the top of the climb he began to settle a bit. Realizing this, he flattened out for an instant and tried to zoom again.

"He had cleared the bridge! But the settling movement which followed his first zoom landed him directly in front of a latticed metal telephone pole and a cross-arm strung with wires. Could he possibly raise his plane over this unforeseen obstruction or would it tip him over and land him in the river just beyond? There was a ripping sound. The metal telephone pole rocked. Jangling wires fell all about us. We dared not look up any longer. We had seen enough.

"To our great surprise, Hawkins' motor continued running. We knew he couldn't hang up there on the telephone pole forever. When we looked again, he was heading for our field. He had withstood the shock!

"Not one of us had moved. . . . Finally, we cleared out, and back at the camp we found that Hawkins had used rare judgment in landing in a seldom-used part of the field. He had brought back a unique souvenier—a dozen strands of wire tangled about his landing gear.

"After mess, Hawkins told us what had happened. As he approached the bridge he saw that there was not enough space for him to get through. We denied this—we had measured the distance the evening before! But later in the day we found our mistake—the rain had raised the stage of the river; in fact, the Seine was two feet higher than it had been the night before."

These were the ex-heroes of the First World War who turned to barnstorming . . . men seeking an outlet for pent-up emotions, trying hard to fit into a peaceful way of life. The way to do it was to buy a surplus ship and go into business, save a few dollars, and then do whatever seemed best.

Paul Baer, officially credited with destruction of nine enemy planes and unofficially with eight more, came home a hero and found the country overrun with others just like him. His Distinguished Service Cross and a nickel would buy him a cup of coffee. Nobody was interested in hearing how he had finally been shot down while trying to save a buddy or how he spent the last year of the war as a POW. Heroes were a dime a dozen.

Taking off for South America, Baer worked for a while as an airmail pilot, then drifted off to China, where he found similar employment, guiding a war surplus Curtiss amphibian in and out of Shanghai to establish a flying mail route there. For several years Baer enjoyed the life of a barnstormer in the Orient, until on December 9, 1930, he collided with a Chinese junk on the crowded Yangtze River, clipped off its mast and dove headlong back into the water. He was killed instantly.

Luckier was Lieutenant Robert S. Fogg, who had fought the war at Love Field, Dallas, Texas, as an advanced acrobatic flight instructor. Fogg spent the first three postwar years barnstorming through Texas, Oklahoma, Missouri, and Kansas, then settled down in New Hampshire to operate a seaplane service on Lake Winnipesaukee. More than thirty thousand passengers flew with Fogg without mishap, and in 1924 he was awarded the first star route R.F.D. airmail contract. Fogg became something of a legend in 1927 when the Post Office Department enlisted his aid in flying mail and emergency supplies into seriously flooded regions. For two months he flew six hundred miles a day, landing his ski-equipped plane in open fields and on mountainsides to keep the marooned inhabitants in touch with the outer world.

Earl S. Daugherty was already a veteran pilot when World War I broke out, having soloed in a Curtiss pusher at Los Angeles in 1911. Because of his experience, Daugherty was assigned as a flight instructor at North Island, San Diego, and later at March Field, near Riverside, California. After the war he joined the horde of barnstormers beating the bushes for passengers willing to take a chance on flying, five minutes for five dollars.

In July, 1921, Daugherty and another well-known barnstormer, Frank Hawks, made some kind of history with the help of an agile wing-walker named Wesley May by bringing off, the hard way, the first mid-air refueling. It was a stunt, pure and simple, dreamed up to bring out the crowds to Daugherty's small field in Long Beach, California, where he was trying to make a go of it as a fixed-base operator. By the mid-twenties, audiences already were becoming bored with routine wing-walking and parachute jumping. Some wild new stunt was needed to attract the crowds and then sell them rides. Wes May had the answer. Hawks described what happened this way:

"When the time came to carry out this hazardous and daring plan, a five-gallon can of gasoline was strapped on Wesley's back and I took off with him crouched on the top wing of my Standard. Daugherty already was up in his Jenny and after maneuvering long enough to build up the right amount of suspense among those watching— or rather, to give our barker on the ground a chance to build it up—we flew alongside the other ship.

"May stood up and Daugherty jockeyed his Jenny in until his lower wing was overlapping our upper one. Then the daring wing-walker, who would have been courting death under any circumstances during such a mid-air transfer from plane to plane without a parachute, but was doubly handicapped now with his awkward burden of fuel, reached up and grabbed the wing skid of Daugherty's ship, lifting himself calmly onto the other plane as I dropped down to give them leeway. Then, in full sight of the crowd and to the accompaniment of mad cheering, which none of us could hear, he walked down the length of the wing to the fuselage and poured his can of gasoline into the Jenny's regular gas tank. The first successful refueling flight in the world had been completed!"

They're all dead now—Hawks, Daugherty and May—-each giving his life while following the meteoric sky trail of the barnstormer. Daugherty, who won considerable fame as a movie stunt flier, was married in an airplane, lived a full career in an airplane, and died in one on December 8, 1928, when he pulled off a wing thousands of feet over Long Beach. Hawks, who is remembered as the "Meteor Man" for his series of record-smashing speed flights above America and Europe, was killed just ten years later, when his ship flew into high-tension wires and burned. And poor Wesley May, one of the great wing-walkers and parachute jumpers, died after landing in a tree in a cemetery. He slipped from his perch and fell, fatally fracturing his head on a tombstone.

Comedy was a part of barnstorming from the start; folks not only liked to be scared to death, they enjoyed a great belly laugh, usually at somebody else's expense, boondock fliers were quick to learn. One of the favorite aerial

Earl Daugherty and his spectacular wing-walker stunts lured customers to airport for rides.

EV HOSKING COLLECTION

circus clowns was William (Wild Bill) Kopia, of Newark, who began his routine by appearing at an air show when it was already half over. Dressed as a female "opera star," he would approach the box office and buy a ticket for a passenger hop. Of course, the presence of the notable performer was announced over the loudspeaker, and a ripple of applause would greet "her" as she gingerly climbed into the passenger seat of a Jenny parked in front of the grandstand with engine idling. At the last moment, the pilot would suddenly climb out and dash for the hangar to get something he had "forgotten."

The opera star, waving to the crowd, would accidentally hit the throttle and the Jenny would lurch forward, spin around, charge for the grandstand and then race off downfield in a cloud of dust. While women screamed and men yelled in excitement, the obviously petrified passenger would wave her arms helplessly. As the Jenny neared a fence, it would suddenly leap into the air and stagger back. From there on, Wild Bill Kopia put on a beautiful demonstration of skilled aerobatic flying—using a second set of controls in the front cockpit. The crowd loved it.

Not many years ago, an ex-marine named Walter O. Geary was up in a Piper Cub for his final check ride before becoming a civilian pilot, his instructor, Roland Maheu, in the front seat. As a marine, Geary naturally had lived an adventurous life, and, as far as he was concerned, Maheu was just another pilot who couldn't possibly know what real danger and excitement were. The engine suddenly stopped, and there they were, 2,500 feet up over a sea of green forest near Auburn, Maine, with no place to land.

"What'll we do?" Geary yelled.

"Get out and crank it!" Maheu grinned back at the Marine.

"I'm serious!" Geary retorted. His flight test forgotten, he was visualizing what would happen when the Cub came down in the midst of a clump of pines.

"Well," Maheu replied, "so am I!"

As the marine watched in amazement, Maheu unfastened his safety belt, flipped the door open and swung out, balancing on the landing gear and holding onto the wing strut with one hand. Then with the other he reached forward and gripped the propeller blade from behind.

"Contact!" he yelled.

Geary turned on the ignition switch, Maheu snapped down on the prop and in a moment it was purring in a steady rhythm. Maheu climbed back inside, and after admonishing his student pilot to be sure to use carburetor heat next time he started a glide on a humid day, to prevent icing, they flew back to the airport and landed.

As shocking as the stunt appeared, it was old stuff to Maheu, who had been doing it for years as a barnstorming act. Only once did he find himself in a tight spot, and that was when the propeller came to rest in a vertical position, out of reach. By turning over and reaching his leg out as far as he

Roland Maheu was capable of incredible aeronautic feats. Once when his engine stopped in midflight, he stepped out onto wing strut and pulled prop through while his copilot made contact.

could, he caught the top blade with his foot and pulled it through until the engine came to life.

The Post Office Department was out of its mind, pilots said back in 1918, when the beginnings of a transcontinental airmail service were proposed to include a night run over the dangerous Allegheny Mountains from New York to Cleveland. Called the "Hell Stretch," it was a region of wild turbulence, blinding fogs and hard-core clouds that concealed mountain peaks at their centers.

The ink was barely dry on the armistice agreement when the forward-looking Otto Praeger, a postal official, picked the worst day of the year, December 12, to attempt to inaugurate mail flights over the dreaded run. Not a single plane got through, nor did they make it on the second attempt on January 2, 1919. Snow squalls blinded the pilots of the Standard JR1B's, and their Hisso engines were unequal to the task of battling the elements.

Angered that so much was expected of them, the airmail pilots staged history's first airmen's strike.

When some sort of scheduled mail service between New York and Chicago began later in the spring of 1919, the Post Office Department eagerly turned its eyes westward, dreaming of the day when transcontinental airmail would revolutionize the service. They had the planes—surplus Jennies, Standards, and De Havillands—but from the start they were handicapped by a lack of competent pilots. The Army was using the airmail runs to train its own birdmen.

What they needed was a special breed of airman—rugged, weather-toughened, versatile, adaptable, and with plenty of guts—someone who could find his way through a blinding snowstorm and land in a hayfield in the middle of the night with only a bonfire to guide him. Looking about, they discovered their man in the barnstormer.

Back from the wars, unemployed, beating the back country to eke out an existence hopping passengers at county fairs, these aviators fitted in with the Post Office Department credo, "The mail must go through!"

In order to get an airmail system started at a time when there were no beacons, no radio aids to navigation and few landing fields, they needed a man with more courage than brains. A smart pilot wouldn't fly under those conditions for a million dollars! But the barnstormer not only needed work, he had proven himself capable merely by staying alive at his profession and not starving to death.

With unbelievable audacity, during the closing days of President Woodrow Wilson's administration, airmail officials tried a desperate experiment to keep their service alive—a night mail run over the transcontinental route. What they needed next were men to hurdle the mountains!

Amazingly enough, the nerve-center of such an operation, an accurate terminal weather forecast system, simply didn't exist. In 1919, true, the Air Service had offered gypsy fliers what weather aid they could with thirty-hour forecasts of questionable value; these were issued "without assuming any responsibility for accuracy." In fact, they pleaded: "Aviators making a flight following a prediction of the meteorological officer are requested to advise him as soon as possible of the accuracy of the report"! If they were right with their guesses, fine. If not, well, it was back to the weather maps. . . .

The roster of the early airmail barnstormers is filled with heroes, the best known, of course, being Charles A. (Daredevil) Lindbergh. There were Gil Budwig, Christopher V. Pickup, Walt Shaffer and Jimmy James, to name a few. And there was Jack Knight.

In the cockpit of his De Havilland mail plane, Knight carried a notebook he'd put together during a brief barnstorming career that took him into practically every small town landing field east of the Mississippi. In the notebook were hand-drawn sketches showing the best places to land and the telephone numbers of farmers along the routes he usually followed. By phoning ahead before taking off on a cross-country flight, he had a better picture of the weather situation than he could get from official Air Service forecasts.

Flying without the aid of blind-flying instruments, it was simply a matter of luck to go up on top of the weather and hope to get back down in one piece. A couple of times Lindbergh decided the best way back was by parachute, but such a system didn't inspire much public confidence in the airmail service. Knight preferred to sneak along under the weather, following a winding road through a canyon or buzzing along the "iron beam" of a railroad track to get to where he was going. A Rand McNally road map was helpful, but in a pinch there was nothing like familiar landmarks, and Knight knew the rooftops of eastern America like the back of his hand.

On one run from Chicago to Bellefonte, Pennsylvania, Knight managed to cruise along the highways under a low-lying fogbank, then follow a powerline east until he hit the Allegheny foothills. There was nothing to do but land, or else climb on top and hope for a hole to sneak through up ahead. After a couple of hours of skimming his wheels over the cloud tops, Knight had to make a decision—jump or try to land his ship, almost out of gas.

Some of the early barnstormers flew the mail with a cigar as their only instrument of navigation: when they'd used up two inches it was time to land. But Knight was more scientific: he simply set his throttle for an easy letdown and took hands and feet off the controls, letting the ship descend on its own. It worked well unless he encountered rough air, which was likely to throw him into a "graveyard spiral"—and that's what happened over Bellefonte.

Groping his way down through the white mists, he knew that mountain peaks jutted up more than a thousand feet higher than the airport he sought. And in the swirling air currents, his DH, given its head, began to wander around in circles. Knight's compass spun crazily, and the wind whistling through the flying wires began to moan like keening mourners. Gingerly he pulled back on the stick, but that only made the ship dive faster. He was in a tight spiral dive! Fighting disorientation, which made everything seem upside down, he closed the throttle and jiggled the stick some more. He suddenly broke out beneath the cloud deck, plunging earthward in a sickening dive. He righted the ship just above a narrow road winding through a canyon, the mountains rising high into the clouds on either side. In relief he followed this path into Bellefonte.

Knight figured the gods were with him on that trip; in fact, he had written his last will and testament on the flyleaf of his notebook before coming down. It was this same Jack Knight who would make aviation history by carrying the night mails east from North Platte to Omaha, Nebraska, and on to Chicago on February 22, 1921, to help inaugurate the first transcontinental airmail service. Following bonfires for beacons across the lonely prairie country through the black night, he was one of a relay of pilots who performed the feat in the tradition of the old pony express. Four pilots had started off in the morning, two from New York, two from San Francisco. Both westbound planes were forced down, and the pilot of one of the eastbound ships crashed and died in Nevada. But Jack Knight, who had learned his lessons barnstorming, flew both his run and that of the Omaha pilot who was grounded in Chicago. Two other pilots completed the trip from Chicago to New York, setting a new record for transcontinental mail—33 hours 20 minutes. Through barnstorming, men trained to be hawks had proved worthy of the task of flying like doves.

4/THE GYPSY FLIERS

ONE OF AVIATION'S most exclusive organizations today is a group of old-time pilots who call themselves Quiet Birdmen, a fraternity that had its beginnings early in 1919 when some five hundred ex-World War I pilots banded together in New York with the express purpose of keeping flying alive.

Airplane factory shutdowns and deactivation of scores of military training fields had threatened to bring the aviation industry to a standstill just when it should have been launching itself on a great era of expansion.

At the time, inasmuch as it was generally conceded that wealthy sportsmen and returning war heroes would comprise the bulk of America's pilot fraternity, the QB's were the first to organize on that concept.

It got under way originally as the American Flying Club, and it was born in France on Armistice Day, with five hundred combat pilots for charter members. Its initial clubhouse was an old mansion off Fifth Avenue on 38th Street in New York City.

Opening in a blaze of publicity, the club tossed a gigantic party well attended by those listed in New York's social register, and a champagne dinner for members of the 94th Aero Squadron, which included such aces as Eddie Rickenbacker, Reed Chambers, Doug Campbell, Thorn Taylor, Weir Cook and Jimmie Meissner.

Handling publicity for the club was Captain Harry Bruno, who represented the Manufacturers Aircraft Association, a business organization dedicated to getting people talking about flying and overcoming their fear of getting off the ground.

The club closed its doors not long after its opening, but after that some of the members began meeting once a week at a Washington Place Italian

Shades of yesterday! Rebuilt Jenny flies over Los Angeles freeway.

restaurant called Marta's. There a few believers met who held to the illusion that aviation was destined for a bright civilian future—Jimmy Doolittle, Fiorello La Guardia, Clyde Pangborn, Cy Caldwell, "Pop" Cleveland, Jimmy Haizlip and others.

Bruno arranged for Casey Jones to give a plane ride to the editor of *Ace High* magazine, Harold Hersey, who had never been off the ground. Impressed, Hersey dedicated the next issue of his magazine to "the brave quiet birdmen who are patiently working in these pioneer days of aviation toward the definite goal of commercial flying."

Today, the QB's are something of a power in aviation, their secret meetings restricted to pilots only. But in 1919, they were preoccupied by the nagging question of what to do about the warehouses full of surplus warplanes that threatened to glut the market.

Bruno, working desperately for both the QB's and the Manufacturers Aircraft Association, surveyed the problem and was shocked. Nearly a billion dollars had been appropriated to darken the skies with American-built warplanes, and yet only 196 had gotten to the front and none into combat. Where had all the money gone?

There were plenty of fine aeronautical engineers and designers in the United States, and yet about the only innovation we had come up with was the Liberty engine, installed in foreign-designed craft like the De Havilland-4. A Congressional investigation was ordered, but proved little.

The upshot of this situation was that the warehouses were glutted not with combat ships but with trainers, trainers, and more trainers, mostly JN-4D Jennies, Standards and Thomas-Morse Scouts. A worried Congress, beset by aviation industry lobbyists, finally passed a law ordering the Army and Navy to unload some of their surplus planes as salvage. To the dismay of returned war pilots, these craft were ordered smashed and sold as scrap, by the pound, to keep them off the market.

In his biography, *Wings Over America*, Captain Bruno tells of being present at an army air base in 1920, "when a soldier, acting on orders, crashed a sledge hammer into the vitals of a Hispano-Suiza engine in a wartime DH-4. The sight saddened me beyond words."

A major standing beside him, William B. Robertson, also could not stand

Clyde Pangborn (left), a founder of Gates Flying Circus, and Colonel Roscoe Turner were charter members of QB (Quiet Birdmen) Club.

to see the lovely ships disposed of in so callous a fashion and decided to buy up the wrecks and salvage what he could. Thus was born the Robertson Aircraft Corporation in St. Louis, where Major Robertson managed to assemble from the junk pile a fleet of 225 Standard J-1's, 165 Hispano-Suiza and Curtiss-motored Jennies and 75 Liberty-motored DH-4's. With these ships he opened a flying school, and for chief pilot and instructor he hired a slender, tow-headed youth, Charles Augustus Lindbergh.

Another concerned party, alarmed at the prospect of war trainers flooding the airplane market, was the Curtiss Aeroplane & Motor Company. Opening negotiations to buy back its own products from the government, a Curtiss spokesman explained, "To sell them without guarantee among unknown buyers would be to reap a harvest of accidents and retard the development of flying."

What he meant to say was that the Curtiss factory would have to shut down and scrap all its plans to enter commercial production of sport and mail planes, and so, backed against the wall, there appeared to be only one way out. Curtiss negotiated a deal to buy 2,176 Jennies and 4,608 OX-5 motors for $2,700,000. Reconditioned, these were placed on the market for what they would bring, sometimes as low as $300 each.

Down in Houston, Texas, early in 1922, a young man named Benjamin Odell "Benny" Howard heard about the Curtiss deal and applied at the company's warehouse there for a job assembling ships. Fascinated with the beautiful wood and wire wonders, Benny paid ten dollars down on a Standard biplane, looked up a man who claimed to be a pilot, and after three trips into the sky made his first solo hop.

The wind in his face and the smell of castor oil in his nostrils were a heady combination to Benny, but getting around the sky and back to earth in one piece was more a matter of luck than skill; his instructor also had only just recently soloed for the first time, and left it up to Benny to figure things out for himself.

Experimenting, Benny found that when he pulled the nose up, the singing of the wires changed pitch, and that meant it was time to put the nose down. The technique was rudimentary, but it seemed to work fine, until one day, to pay for a needed tank of gas, he took up his first passenger, who wanted some stunts.

With his unsuspecting passenger in the front cockpit, Benny staggered up into the sky as high as he could get, which was about 1,500 feet, and there attempted his first "stunt"—a steep turn. Around and around they went, until the wires began screaming in protest. To correct things, he simply hauled back on the stick to make the nose come back up where it belonged and stop that banshee scream. Of course, all that that did was to tighten the turn and force the DH-4 into a sickening spiral.

The deadly dive continued, Benny pulling the stick back for dear life, exactly the wrong thing to do. The green earth was a blur, until finally they smacked into it with a splintering crash. Benny went to the hospital and his passenger to the morgue, but some time later Benny was up and around, walking with a limp that proved to be permanent.

Instead of being discouraged, he bought another surplus ship, sold it, and went barnstorming with the customer who bought it from him. They made it together all the way to Roanoke, Virginia, and there Benny decided he had found his career.

Having learned how to get out of a spin, Benny turned to designing airplanes and whipped up a backyard wonder that got up into the air and hit 100 mph with a 90-horsepower OX-5, a good thirty per cent faster than the Jennies and Standards flew. At eighteen, Benny, now a pilot, designer and builder of a plane that flew well, discovered that money could be made with his ideas. He sold the home-built model for $1,000, twice as much as surplus crates were bringing, then contracted to put together a kind of cargo plane for a Texas bootlegger to run whiskey across the border from Mexico.

Benny's first home-built was called the DGA-1, and the second, DGA-2. When asked what the initials meant, he replied, "Damned Good Airplane, what else?"

After a stint with the Alexander Aircraft Company in Denver as a $200-a-month "research engineer," Benny joined Major Robertson's airmail line in St. Louis, flew Ford Trimotors on the St. Louis to Chicago run before one day walking across the field and joining Transcontinental Air Transport, which had offered him more money.

After that Benny Howard found his niche building racing planes. DGA-3 was a hot little job that streaked across the sky at close to 200 mph with only a 90-horsepower Wright Gypsy engine. Called "Pete," it swept the field at the 1930 National Air Races, winning five "firsts" and two "thirds" under Benny's now-skillful piloting. DGA-4 and DGA-5 were twin racers called "Mike" and "Ike," which he flew barefooted with such breathtaking skill in closed circuit races that he was ever after known as Benny the Pylon Polisher.

With his next design, DGA-6, Benny built an amazing, four-place, high-wing monoplane with a wing-loading double that was accepted for land planes in 1933. Called *Mister Mulligan,* it was clocked at an amazing 287 mph at sea level under full 830-horsepower.

Entered in the 1935 Bendix Air Race, *Mister Mulligan* was a sleeper. The favorite to win was a souped-up Wedell-Williams Racer flown by Colonel Roscoe Turner, another barnstormer turned speed pilot, but when the two ships flashed across the finish line at Cleveland, the timers checked and double-checked their stopwatches. Vincent Bendix, the race sponsor, finally went to the microphone and announced, "The winner, by twenty-three seconds, is Benny Howard!"

DGA-6 eventually went to war as a production model designated DGA-8, her clean lines adapted to a whole line of liaison transports, secondary trainers and observation ships. Benny himself went to war in 1939 as director of flight tests for Douglas Aircraft Company's DC-4, a 240-mph cargo and troop carrier.

Another colorful QB who joined the ranks of the gypsy fliers, Captain Basil Lee Rowe, enlisted with the U.S. Army Air Service in World War I as an engine mechanic with the 871st Aero Squadron. Following his discharge, Captain Rowe bought a surplus Jenny and went barnstorming along the eastern seaboard, blazing sky trails that would be followed by scores of other itinerant pilots.

A typical back country flier who loved his work, Captain Rowe, a native of Shadaken, New York, dressed the part of the daring aviator and sported a thick mustache that gave him a look of dashing carelessness. It was only a pose; Captain Rowe was a careful, studious fellow who had big ideas about the future of flying.

In his book, *Under My Wings,* he wrote of this early period, "Flying haphazardly around the country, I developed a system of picking the profitable place to land. To test a town for its interest in flying, I would buzz it a couple of times. If the people continued about their business, I did the same. But if the animals and fowl took off for the woods and the kids tried to follow me, it indicated virgin territory. In that case I looked for a farmer's field from which to operate and, when I found one, buzzed the town to get the whole population following me out to the field like the children of Hamelin following the Pied Piper. The farmer usually let me use his field for a free ride for himself and his family.

"I would rope off my loading and unloading area to keep the people out of the propeller. I'd pick a couple of helpers from the crowd, offering them a free ride for their services. To refuel, I would walk into town with my cans and get them filled up at a gas station. We used the same fuel as automobiles. I carried my personal belongings in an old suitcase up in the front seat of the plane along with spare parts. I slept under the wing of the plane or sometimes in the farmer's barn."

After barnstorming by land from 1920 to 1926, Captain Rowe began exploring the West Indies, and for the next three years kept busy establishing a mail and passenger service that became the West Indies Aerial Express. Later on he joined Pan American Airways in Miami as chief pilot.

As an island barnstormer, Rowe lived through many unforgettable experiences. One of the most harrowing occurred during a night spent on the shore of Lake Enriquillo, after an emergency landing with his Waco on a flight from Puerto Plata to Port-au-Prince, Haiti, in 1930. He described the incident this way:

"I was over the middle of the lake when my engine began to labor and slow down. . . . I cut off the throttle and began to search the lake shore for a suitable place to set down. Logs littered the shoreline. . . . I was down to about a thousand feet when the engine started to hammer and bang."

42

Rowe continued on down, but just before landing he saw one of the logs suddenly scamper off; the logs were crocodiles! Rolling to a stop, Rowe leaped from the cockpit and lay on the wing long enough to make sure there were no crocodiles under him, then got busy and removed the engine's crankcase. A bearing had burned out.

He had forgotten to take along any emergency rations or water, and under the hot sun began to feel pangs of hunger and thirst. He set about gathering leaves to make a fire and boil some of the foul lake water, but then he remembered he didn't have any matches with him. Resourcefully he thought of the engine's magneto and stripped it off. Turning it by hand, he managed to create a spark that lit the leaves. He soon had a can of water boiling.

Night was falling, and he heard in the distance the frantic barking of the wild dogs of Haiti, vicious beasts descended from animals brought over from Spain by slave-runners. He rigged a tarpaulin shelter, climbed back into the cockpit with a stout stick and settled down to wait for dawn, wondering what he should do then. Something bumped his airplane. In alarm, he grabbed the stick and sat up.

"A big fight started below me in the night—great-jawed, savage crocodiles versus fanged, hungry dogs. . . . With every lurch I almost stopped breathing. With only a linen cloth separating me from what was down there, I also did a little serious praying."

The gutteral roar of the bull crocodiles and the hysterical barking of the wild dogs gave Rowe a sleepless night, but finally dawn broke, and as it did, there was a loud explosion. Rowe jumped up, then groaned. One of the tires, slashed by sharp crocodile teeth, had blown.

Wondering what to do about the bearing, Rowe stared at his feet, kicking the sand. An idea came to him: why not use the heel of his shoe? He got busy with his jackknife fashioning a bearing liner, reassembled the engine and removed the good tire to keep the Waco from ground-looping on takeoff.

Within minutes, after a precarious takeoff, he was flying again and heading for his home field at Barabona. When he arrived, his partner, Bill Wade, looked at his heelless shoe and inquired, "What happened to you?"

"I ate it," Rowe grinned.

From his barnstorming days, Captain Rowe picked up many tricks of flying that can only come from experience. In testing a new Pan Am Sikorsky S-40 Clipper on a flight from Kingston, Jamaica, to Cristóbal, Canal Zone, he used a unique application of Breguet's Law, a flight formula that says an airplane's range depends upon its flight efficiency and the ratio of fuel weight to gross weight at takeoff. This meant the lower the takeoff weight, the farther a plane could fly.

"I stripped the ship of every available ounce so I could carry extra fuel," Captain Rowe recalled. "Each reduction of six pounds meant an additional gallon, which was equivalent to two more miles. I stripped the cabin of everything, including the floor. I even sawed off the handle of my toothbrush, cut my shaving stick in half and made strip charts of my maps to save the weight of paper."

Government regulation of barnstorming was non-existent until the Air Commerce Act of 1926 placed licensing of pilots and aircraft under federal control. In 1919, the only way to stop a foolhardy gypsy flier from endangering people on the ground was to let him go kill himself first.

On January 1, 1919, Associate Editor Ladislas D'Arcy of *Aviation and Aeronautical Engineering* magazine warned that "the attitude of the Government during the war has prevented all commercial flying. The issuance of individual licenses should be commenced at once and the freest encouragement given to all who desire to enter the aeronautical field as sportsmen or to those seeking an industrial opportunity."

As it was, anyone with a few dollars in his pockets could go buy a surplus Jenny, climb in and fly it off to operate from some cow pasture. True, barnstorming was financially a risky venture to begin with, and in the words of one ex-military aviator, Dick Depew, "the most dangerous thing about flying is the risk of starving to death!"

Even so, hordes of daring young men were lured by the prospect of owning their own airplanes and going into business for themselves, a sort of rebellion against society's regimentation, a rebellion dating back to their years in Army uniform. Wasn't it better to fly the open sky in weather-beaten ships, hopping from town to town, following the county fair circuits than to fly a desk in some dull office?

There were plenty of customers in the early days, men and women who would step forward from the crowd and heroically hand over ten dollars for ten minutes in the sky. Unforgettable was the thrill of being strapped into the front seat of a moth-like biplane, stomach knotting at the expectation of what was to come, staring at the frightening panel of dials and gadgets, hearing the ear-splitting thunder of the engine roaring to life and feeling the hot blast of exhaust fumes on your face.

There was the prideful wave of assurance to a friend waiting to watch you take off—and probably crash—a weak grin of bravado, and then the sensation of gathering speed as the frail biplane bounced along over the grassy field, leaped a fence and was *flying!*

A look back at the pilot's face, hidden beneath a tight leather helmet and goggles, brought a wave of confidence in this superman in the back seat who held your life in his hands. Then came the gradual release of tensions as you dared to look over the side and see the beauty of the world spread out below—patterns of farms, rock fences and orchards, and over there, the schoolhouse! You forgot your fears and in the supreme enjoyment of the new experience you were getting far more than your ten dollars' worth. You were introduced to a whole new world, the world of the barnstormer.

Back on earth, you shook hands with your pilot, who casually lit a ciga-

Canuck was Canadian-built version of the famed Jenny.

rette and looked over the crowd for more customers. He was a rakish sort of devil, one who seemed always to be leaning into the wind, squinting at far horizons, smelling the air for stormy weather. He was a hero.

Then there were the wing-walkers, the parachute jumpers, the aerial stuntmen who teamed up with barnstorming pilots to help gather crowds by cheating death while clambering over the outside of the fabric planes as they swooped low over towns. One such youth was described by a small-town social worker, Olga Edith Gunkle, in an article she wrote for *Scribner's Magazine* in 1929. He had come to her office to discuss giving a benefit performance to help him get his nerve back after watching a pal fall to his death from an airplane's wing:

I had finished a list of names for him, but I still kept my pencil on it. I was rather reluctant to let this blue-eyed youth go. He was such a contrast to the battered and bescarred bits of human wreckage who came asking meal-tickets and lodgings. He made me think of limitless expanses of sky—flying clouds—wind-swept spaces—youth indomitable.

"Ever see anyone hurt except your pal?"

"Lots of them. One guy stumbled and fell and got the top of his head chopped by the propeller. An' then one kid—aw, you don't like to hear about

45

it, do you? But you see you don't mind it quite so bad after a while, except when it's your pal."

"Ever get hurt yourself?"

"Once in a while."

"Badly?"

"Last time I got hurt, something went wrong with the plane and they had to land before I could climb back up the rope. I got dragged. Was unconscious for three days."

"My gracious, I'm glad I don't have to watch anything like that. I'd hate to see anyone hurt."

"Oh, you'd get used to it," he gravely remarked.

"Don't the crowds that watch you make you nervous?"

"No, ma'am. You never even think of them. You see, it's like fighting a hundred-mile gale up there, with the wind blowing and the plane moving along. You've got to spend all your time hanging on, and it sure takes every muscle in your body."

As he spoke, he gripped in imagination a wiry bit of rope and I saw the muscles in his hands and neck tense and swell. His whole body was fighting the hundred-mile gale. The musty volumes in my office faded away and I was one of the spectators staring upward with bated breath, while far up in the air a tiny figure swayed and twisted and clung to a bit of flying rope; only I was one of the spectators who knew just how young and strong he was, with his curly hair and gallant blue eyes. And I was fearful—horribly fearful —lest he too lose his grip and come crashing, crashing, downward. It would be such a pity! Something fine and splendid would be gone from out of the world.

Wichita, Kansas, a city destined to become famous as the "Air Capital of the World," was the launching place for Sidney Q. Noel's barnstorming career, one day in June, 1919, when he walked down the street and spied a gorgeous Curtiss JN-4 Jenny displayed in an automobile parking lot next to the Arnold Brothers Motor Company.

Noel and a buddy, James A. Ellison, just returned home from duty overseas, fell in love with the Jenny at first sight and decided then and there to buy her. That was fine with Mr. Arnold, who had purchased a pair of ships from the Curtiss factory as an investment.

"How much for that airplane?" Noel asked, kicking the tires.

Arnold sized up his customer and knew, from the light in his eyes, he had a pigeon. It was quite obvious that Noel had already made up his mind.

"Three thousand dollars and she's all yours."

"Three thousand—?" Noel cried.

When Arnold turned and started back to his office, he felt the expected tug at his sleeve. "Okay, it's a deal," Noel said. He had to have that Jenny!

Noel was no novice; he had learned to fly in the Army and completed one hundred missions over enemy lines with the 148th, a scouting squadron, and later trained on French Spads before the armistice ended his military aviation career. By then he was sold on flying.

So it was that Noel and Ellison climbed into the two cockpits one bright morning and headed south on their barnstorming adventure through Kansas, Texas and Oklahoma. Noel recalls, "We would circle a town, and by the time we landed, mostly in stubble fields, the whole town would be there. We didn't have to ask them. They came up and asked us for rides."

There was no need for wild stunt flying. The towns Noel and Ellison flew into had never seen an airplane before, with the exception of one or two when early Curtiss and Wright pushers visited back in the prewar Early Bird days.

By Christmas, 1919, the barnstormers had made enough to recoup the $3000 they had invested, and Noel bought out his partner. The next spring Noel teamed up with two other gypsy fliers to form the Salina (Kansas) Airplane Company, one of the first "flying circus" teams to barnstorm the Midwest.

While the exploits of the Salina Airplane Company were not particularly death-defying, they were representative of the growing wave of gypsy fliers

Fred Kelly (right) opened up Cuba to flying in 1920 with a JN-4D Jenny. Later he flew for Western Air Express with Jimmy James (left). Man in middle is Herbert Hoover Jr.

who were spreading the gospel of aviation across the land, getting people used to flying in airplanes instead of looking upon them as instruments of death and destruction, an image that dated back to prewar years when birdmen like Linc Beachey, Arch Hoxsey and Hubert Latham created the legend that it takes a superman to fly.

There was, of course, another kind of gypsy flier who seemed to fly with suicidal purpose, men like Captain Frank T. Dunn, ballyhooed as "the only flier who has successfully looped the bridge of a navigable stream." Basil Rowe went Dunn one better by having a screaming blonde beauty in the front seat while diving down beneath the Philadelphia-Camden Bridge and pulling up and around in a complete loop.

The girls, naturally, went for the gypsy fliers in a big way, and the fliers in turn tried hard to live up to the heroic image of the returned war veteran who had killed enough enemy pilots to rate as an ace. While few of the early barnstormers had actually earned that status, it was only necessary to dress the part. Many held a superstitious attachment to a favorite piece of clothing. Clyde Pangborn refused to fly without his soft chamois vest, and the great Casey Jones insisted on wearing a loud green jacket. Others made sure some kind of icon dangled in their cockpits to get them safely back to earth.

If most of these early fliers placed a large amount of faith in sheer luck, there were a few who went about their work with what they believed was a "scientific" approach. One Texas barnstormer known as Crazy John flew regularly between El Paso and Dallas, and to keep from getting lost he stuck two arrows on his compass, one reading, "El Paso This Way" and the other, "Dallas That Way." In spite of this "science," he never got lost.

American barnstormers introduced aviation to Cuba after the war, although the United States Army Signal Corps had operated observation balloons at San Juan Hill during the Spanish American War. Perhaps the happiest barnstormer to find a paradise in that country was a former military flying instructor named Fred Warren Kelly, who left his job at Gerstner Field, near Lake Charles, Louisiana, in 1919 to seek his fortune in Cuba.

Shipping a war surplus Jenny to the island by boat, he and a Cuban friend, Rafael de Zaldo, established Cia. Airea Cubana, that country's first airline, so to speak, based in Havana. When a windstorm wrapped their only aircraft around a tree, Kelly went to the United States to bring back another JN-4D and a Curtiss Oriole, in which he taught Zaldo to fly.

For a while Kelly enjoyed himself stunting over town, shooting off night fireworks and hopping passengers in search of aerial thrills. When the novelty wore off, he and Zaldo flew their planes to the hinterlands in search of new customers who had never seen a plane before.

"We attracted a lot of attention, and the Cubans were eager to use our services," Kelly recalls. "We landed at Cardenas, Cienfuegos, Santa Clara,

These five barnstormers started up Western Air Express: (left to right) Fred Kelly, Jimmy James, Al DeGarmo, Maurice Graham, C. C. Moseley. Graham froze to death in a snowstorm after a crash during a mail flight over Utah.

Ciego de Avila, Camaguey, Holguin and Guantanamo."

In December, 1920, at the edge of Santiago de Cuba, Kelly put the Jenny down in a small clearing before he was able to realize it was too small for a takeoff. The natives pitched in at his urging and by the next day had chopped down enough trees to make a decent runway. In return, he gave the mayor's pretty daughter a thrill ride.

Among Kelly's new-found friends in Cuba was another beautiful girl, Lolita Bacardi, daughter of the owner of the Bacardi rum distillery, who decided that the American *piloto,* along with his pal, Zaldo, were some kind of heroes who needed a medal. Lolita threw a huge party, complete with suckling pig and plenty of rum, then kissed both fliers on their cheeks and hung on their necks medallions on which was inscribed, *"Kelly y Zaldo— primer vuelo Havana-Santiago de Cuba,* December 31, 1920."

To repay her kindness, the barnstormers one morning flew over the Bacardi estate, looping and rolling and buzzing low to get everybody out of bed. When he saw Lolita in the patio, waving her handkerchief at him, Kelly banked over and threw out a bouquet of flowers.

49

Kelly ended his barnstorming in 1925 and returned to Los Angeles, where he became one of the first airmail pilots to fly the treacherous mountain route to Salt Lake City for the old Western Air Express. Not only was it tough flying, it could be rugged on the ground, too: A fellow pilot, Maury Graham, who had become a war hero by finding the Lost Battalion of the 77th Division, crashed in a snowstorm on that run in 1930 and froze to death trying to walk out.

5/SAGA OF THE BIG FISH

AS FAR AS SHREWD old Orville Wright was concerned, the only thing wrong with flying was getting back onto the ground. Alone up there in the sky, a man had only the birds to contend with, but if his engine quit and the irresistable clutch of gravity grabbed him, look out!

At the beginning of 1919, with America entering her first peacetime year since the armistice, the airplane engine that wouldn't quit on a whim hadn't been invented yet—Orville knew. The mass-produced Liberties, the Curtiss OX series, the Hall-Scotts, the Duesenbergs, the Packards, all were great powerplants when they kept turning, but no wise pilot would risk his neck flying cross-country over difficult terrain where he couldn't glide down "dead stick" for an emergency landing when his engine quit, leaving him in a most uncomfortable silence, sitting there counting the rivets on wooden propeller blades that suddenly windmilled to a stop.

"To make flying perfectly safe," cautioned Orville Wright on January 1, 1919, "good landing places must be provided every ten to twelve miles." Orville figured that a pilot flying one mile high then would have a chance of gliding for six to eight miles and setting down on a good field.

Quite obviously, Orville had his head in the clouds on that idea. To build the hundreds of thousands of airports he envisioned scattered across the country was simply out of the question economically. Until the day came when engine reliability could guarantee that a plane would stay up for a reasonable length of time, pilots were going to have to trust to luck and their own skill in making forced landings.

A scattering of sport planes developed in America during the war years paid serious attention to the problem of short-field landings. Aircraft Engi-

America Trans Oceanic Co. opened offices in West Palm Beach, Florida, in 1919. For certain passengers it sometimes needed bigger. boats than this Curtiss Seagull.

neering Corporation's hot little Ace biplane could land in fifty feet, its makers claimed. Their slogan was, "The Country Road Your Airdrome."

Among the first to buy an Ace was barnstormer Eddie Stinson, but at $2,500, while "ideal for the ranch owner, the pilot of the aero mail, the sportsman, and the explorer," it couldn't compete with the war surplus market of Standards, JN-4Ds and Thomas-Morse Scouts that could be picked up for a few hundred dollars each.

Among the thousands of leftover military planes that would find their way into the ranks of the barnstormers, several dozen flying boats were put on sale by the government, a fact few people remember. Here were aircraft that needed no landing fields. Lakes, rivers, bays, and the oceans themselves provided all the room to roam a pilot could want, without worrying about a place to set down when trouble developed.

And of all the surplus flying boats purchased for civilian use, none had a more remarkable career than the *Big Fish*. Randolph Baldwin, the man who flew as her copilot and mechanic, grows ecstatic when he recalls the beauty of her sleek lines, her rugged stability, and her unique paint job. And he remembers her balky Liberty engines, this with a shrug of amusement that seems to say: "It was all great fun, but I wouldn't want to trust my life to them again!" More than once Randy Baldwin and his chief pilot, George

52

A. Page, Jr., found themselves with one engine out whistling down through the sky to land in heavy sea swells somewhere between Florida and the Bahamas, where the pioneer "overseas" airline they flew for, America Trans Oceanic Company (A. T. O.), operated right after the war.

The *Big Fish* was a Curtiss H-16C, a twin-engine flying boat that had been purchased from the Navy by David H. McCullough, who, on February 24, 1920, began scheduled passenger operations between Miami, Florida, and Bimini, some forty-five miles east of the Florida coastline, in the Bahamas.

McCullough, in fact, was an old hand at making water landings with flying boats. On May 17, 1919, as one of the pilots of the Navy transatlantic flying boat NC-3, attempting an ocean crossing from America to Europe, he had become lost in the blinding whiteness of a deep fog bank, and was forced to land as the only way to find his bearings. Unable to take off again because of high seas, McCullough and his crew performed a miracle by sailing the flying boat through a savage storm for 205 miles into the port that had been her destination—Ponta Delgada, in the Azores.

McCullough typified the postwar barnstormer—courageous, energetic, daring and adaptive. Bringing the NC boat through a fierce Atlantic storm that threatened to batter her to pieces with each crashing wave was an epic

In 1914, Glenn Curtiss built flying boat America *for wealthy sportsman Rodman Wana-maker to fly the Atlantic. Craft was underpowered and crossing was never attempted.*

53

feat in itself, but it took real guts to refuse an offer from the skipper of a passenger liner, the *Harding,* to tow them into port.

"We came this far; we're going the rest of the way!" McCullough signaled. Chugging into port on two damaged engines, he rated a 21-gun salute from a shore battery. The NC-3 was every bit as big a hero as the NC-4, which alone made it across the Pond while they sat drifting.

The Curtiss H-16 *Big Fish* herself had an interesting history, being a development of the original *America* flying boat that Glenn Curtiss had built in 1914 for Rodman Wanamaker to fly the Atlantic. The *America,* christened on June 22, 1914, was launched on Lake Keuka at Hammondsport, New York, where Curtiss' airplane factory was located.

From the start, it was apparent that the *America's* two Curtiss OX-2 engines, of 100-horsepower each, were not strong enough to give her the speed and range necessary for the ocean hop, so a third engine was mounted atop the upper wing. Outbreak of war canceled all plans for the venture; her pilots, Navy Commander John H. Towers and Lieutenant John C. Porte of the British Royal Naval Air Force, went on active duty, and the craft was placed in storage behind the Curtiss flying boat hangar at Port Washington, New York.

Six miles from the Curtiss plant at Garden City, New York, hummed another wartime plane factory, the A. S. Heinrich Corporation, where Page and Baldwin were employed, and where they formed a lifetime friendship. "There never was another pilot like him," Baldwin says of Page today, his hero worship unconcealed.

Wartime responsibilities separated these two eager young men, Page joining Curtiss in his booming Navy flying boat business, and Baldwin moving to the Naval Aircraft Factory near Philadelphia, where he set up the metal and machine departments for production of Model MF flying boats.

Thus, following the armistice, Page and Baldwin were well versed in the lore of flying boats, Page as an engineering pilot and Baldwin as a mechanic, and they were immediately hired by McCullough to man an H-16 war surplus craft that, with a piscatorial paint job, would become the *Big Fish.* A tough, seaworthy craft, she had a ninety-five-foot wingspan and originally carried a crew of four—pilot, copilot, gunner-bombardier and a wireless-operator-gunner in a compartment behind the pilot's cockpit.

Converted for barnstorming, the craft carried two passengers up front and nine in the hull, in addition to pilot and copilot-mechanic. Her two Liberty engines were rated at 350-horsepower, but Baldwin managed with special domed pistons somehow to get 400-horsepower from each of them.

"We switched from castor oil to regular automotive oil and used the best grade of aviation gas we could get," he recalls. The boat's 235 gallons were enough to keep her in the air at 75 mph for 5½ hours nonstop—if she happened to fly that long without a breakdown. The Liberty engine, Page and Baldwin discovered, had a bad habit of chewing up her distributor shaft gears.

Big Fish *rests on water off Bimini.*

The *Big Fish* got her name from a giant, forty-ton whale-like fish landed by a prominent angler, Captain Charles Thompson, who later became president of the Bimini Bay Rod & Gun Club, a resort whose guests often flew the American Trans Oceanic (ATO) boat from Miami on its triweekly run. The attractions of Bimini were more than just fishing and shooting; the Eighteenth Amendment had gone into effect June 30, 1919, and Bimini was a favorite spa for the thirsty.

Rumrunners, in fact, operated on a schedule almost as regular as that of the *Big Fish,* a fact that once worked in Randy Baldwin's favor during a hairy adventure that followed a routine forced landing, when the flying boat on a flight from Nassau had to land on the sea after one of the engines' distributor gears stripped. The landing itself was uneventful. Page, who was doing the flying, was skilled at water landings and set the *Big Fish* down between two swells not far from Andros Island, biggest of the seven hundred islands in the Bahamas chain.

"There were two native sponge boats nearby and one came over to us," Baldwin remembers. "It was a small boat, but we off-loaded our passengers and put them ashore."

All that afternoon they taxied the *Big Fish* on one engine until they found a sheltered cove, and the next day Page did a masterful job of threading the flying boat along a narrow waterway passage into Low Sound. For the next eleven days Baldwin camped out on Andros Island while Page searched the United States for parts, which he finally located at Fort Myer, Virginia. Randy quickly became disenchanted with the swarms of gnats that buzzed around him at night and finally moved aboard the *Big Fish* to sleep. When Page didn't return during the second week, Baldwin, worried, hitched a ride with a passing rum boat piloted by a bootlegger named Bruce. Baldwin and Page finally found each other, returned to repair the Liberty engine, and flew off once more to Florida.

In Miami, they made their headquarters at the Halcyon Hotel—there were no others of any consequence—but spent most of their time on the flying boat or on a small island owned by a squatter named Chris, who helped them haul fifty-gallon gasoline drums out from the mainland in his rowboat.

Navigating the open seas with only a wildly swinging compass was something of an art, for unlike their fellow barnstormers who operated over land, Randy and George had few landmarks—only an occasional cay or a distant island to help them get their bearings.

"We had no drift indicators, so we took along a ball of cotton and tossed some out from time to time. By looking back," Baldwin explained, "we could see if we were drifting to the right or the left of the cotton."

The only other flight instrument on board was a spirit level attached to the side of the hull, which served as a pitch indicator that told them when they were diving or climbing in bad weather, and there was plenty of that. On one trip they returned to the Florida mainland only to find the Miami shoreline hidden beneath a black squall line stretching from horizon to horizon. Over the inlet at Palm Beach, where they detoured, they were battered by a fierce gale that threw the flying boat all over the sky. Two women passengers up front thought Page was giving them some free acrobatics and yelled in glee, but Baldwin was saying his prayers.

"My God!" he yelled. "Get this thing back out over the ocean!"

Page circled low over the palm trees, which looked like so many agitated windmills under the force of the storm, then eased around to land inside the bay, the spray drenching everybody to the skin.

Another time, carrying eight passengers on a run to Nassau, George and Randy saw the sky ahead blacken with towering nimbus clouds. Page decided to set down the *Big Fish* to ride out the storm on what looked like smooth water behind a wind-lashed cay. At the last second before touchdown Page saw that the waves in all directions were breaking over jagged coral rocks.

He yelled, "Pull up! Pull up!" Both men hauled back on the controls. They lifted the *Big Fish* clear just in time to prevent her hull from being ripped wide open.

While the *Big Fish's* runs between the Florida coast and the various islands of the Bahamas were essentially barnstorming ventures, they served another historic purpose; in thrice-weekly flights between Miami and Bimini, they pioneered scheduled airline transportation in America many months before a competitive line, Aeromarine West Indies Airways, Inc., began hauling passengers from Key West to Havana, Cuba.

Aeromarine, which had flown that route since November 1, 1919, carrying airmail by single-engine flying boats, has long been mistakenly credited with starting the first "overseas passenger operation." Actually, their passenger runs did not get going on a regular schedule until October, 1920. Their ships were modified Curtiss F-5L Navy "flying cruisers," colorfully named the *Santa Maria, Nina, Pinta, Columbus, Balboa* and *Ponce de Leon*. But by the time they were in full swing, ATO and the *Big Fish* already had airlifted several hundred passengers from the United States to the British territory of Bimini on something like scheduled flights.

At a "special rate" of $25 each way, passengers could dash over to Bimini in a little more than half an hour. The *Big Fish*, "piloted by one of the best-known fliers in the world and assisted by an expert mechanic," advertised that it would leave Miami for the forty-five-mile run at two P.M. sharp on Wednesdays, Fridays and Sundays, returning at the same hour on Thursdays, Saturdays and Mondays.

Thus, while other owners of surplus Curtiss F-5L twin engine boats flew occasional charters to the Bahamas, the *Big Fish*, with Captain Page as pilot and Baldwin as copilot, was first to advertise regular runs to the British spas. It may be considered ironical, but the Volstead Act did provide a vital stimulus to the early start of scheduled airline service, for as Prohibition went into effect, all possible liquor was hastily shipped out of the country and piled on the beaches in the Bahamas in row after row of cases. Naturally, those who liked a drink followed, for runmrunning had not yet become organized. On North Bimini Island, a Miami group erected a 105-room clubhouse which became not only an important fishing, gambling and drinking resort, but also served as a terminal office for ATO.

By February 23, 1921, the *Big Fish* had spread her fins further south, carrying tourists to Havana. One traveling VIP was Margaret Alphonso de Bourbon, a cousin of the King of Spain.

After two full seasons on the "Bimini booze run," as the scheduled Bahamas flights came to be called, the *Big Fish* left the lucrative island trade to her competition, forsaking the beauty of the Bahamas and the Antilles green-water hops. In the spring of 1922, she pointed herself northward, like a homesick duck, and joined the flights of migratory birds winging toward Canada to enjoy the long days of summer.

At Lake George, New York, she came to rest, after a brief stop at Port Washington, where Page turned her over to a new pilot named Johnny. Randy and Johnny operated her the rest of that summer, hopping passengers throughout upstate New York. Then, when autumn leaves began falling, it was time to return to Port Washington.

Pure of heart to the end, the *Big Fish* abstained, as she had in Florida, from joining in the lucrative bootleg business along the Canadian border, leaving that field to other barnstorming Jennies and Standards that slipped off overhead in the dark of night with their bellies painted black. Her short and honorable career came to a sad end in October, 1922, when Randy and Johnny winged down the Hudson River on her last flight from up north.

The lights of Poughkeepsie slipped by under her left wing as she pierced the darkness, flying lower and lower to stay beneath a fog bank that obscured the Tappan Zee from view. Baldwin flipped on a flashlight and held it on the compass for Johnny, then looked up just in time to see the Albany night boat looming dead ahead.

Swerving violently, they missed the steamer by inches. In leveling out to attempt an emergency landing, Johnny banged the hull on the water with bone-shaking force. Waves ripped off the false step on the bottom, leaving a gaping hole in the hull.

"We came to a stop, engines running, sitting there slowly sinking," Baldwin recalls. "I suddenly remembered we had passengers in back—a woman and two small children. I looked around. They already were knee-deep in water!"

RANDOLPH BALDWIN COLLECTION

Up in flames goes the Big Fish, *her owners having sadly put the torch to her broken hull.*

Although furious with Johnny for wrecking the *Big Fish,* Baldwin was grateful to him for saving his life just at that moment. "I'd jumped up and started around to get the kids, forgetting about the propellers. Johnny grabbed me by the arm and pulled me back, just in time to keep me from getting my fool head cut off!"

The *Big Fish* finally came to rest floating on her wings. Randy dove overboard, braving the rough chop and swimming for shore two miles away. He encountered a passing motor launch and yelled for help. The boat stopped, reluctantly picked him up, and returned to tow the *Big Fish* ignominiously to shore. She was taken to Port Washington on a railroad flat car.

Later, Baldwin and Fred Golder, ATO's airport manager, returned with a can of gasoline and put the torch to her. Randy sadly watching her go up in flames, snapped a picture to remember her by. After that, ATO's operations went into a decline. The outfit flew a few more years with smaller boats, but already competition was strong as other gypsy fliers with a Navy background moved down to Florida to help skim off the cream of the tourist market. West Palm Beach was a favorite hangout for flying boat barnstormers, but few, in the mid-1920's, were able to duplicate the success of the *Big Fish.*

6/DAREDEVIL LINDBERGH

A SLEEPY, Midwestern university town, Lincoln, Nebraska, had wide avenues lined with shade trees, and was dominated by the four-hundred-foot tower of the Nebraska State Capitol building. One day in June, 1922, as the town basked in the languor of a summer sun, the throaty Hisso engine of a slowly climbing Lincoln Standard biplane droned on far above in the sky. The craft circled with the sweeping grace of a hawk riding a thermal, its eyes scanning the grain fields below for a sign of movement that would reveal a scampering rabbit or squirrel. In the Standard's rear cockpit crouched a pilot named Erold Bahl, a slender man with his cap on backward and wearing a business suit in the manner of the late, great Lincoln Beachey. In the front cockpit was another slender young man wearing helmet and goggles. His name was Charles Augustus Lindbergh, and at twenty he was about to make his first parachute jump, the hard way.

Leveling off at jump altitude, Bahl was suddenly active, businesslike. He rocked the control stick, signaling Lindbergh that it was time. The helmeted youth up front waved, then unbuckled his safety belt, gripped the center wing strut and stepped out of the cockpit onto the narrow catwalk along the fuselage. He smiled at Bahl, then, remembering his instructions, made his way carefully along the wing spar, keeping his balance by holding onto the maze of flying and landing wires that criss-crossed from top wing to bottom.

He paused to steady himself, feeling the slipstream tearing at his clothing and filling his mouth with ram air. A misstep now could mean death, for his parachute lay out there near the outer bay strut, clipped to the flying wires, waiting for him to reach it and clip it to himself.

Far below the Standard, a small group of people stood at the edge of Lincoln Field: Ray Page, the president of Nebraska Aircraft Corporation, Bud Gurney, a close friend, Charley Harden, the man who had rigged his parachutes for him. Yes, there were two parachutes, and Lindbergh intended to use them both.

He had watched Harden make a dangerous double drop only a few days before, and it had become a compulsion to do the same. He would later write, in *The Spirit of St. Louis:* ". . . When I decided that I too must pass through the experience of a parachute jump, life rose to a higher level, to a sort of exhilarated calmness. The thought of crawling out onto the wing, through a hurricane of wind, clinging on to struts and wires hundreds of feet above the earth . . . left in me a feeling of anticipation mixed with dread, of confidence restrained by caution, of courage salted through with fear."

Finally clipping on the double chute and sitting down, staring at the patterns of fields between his feet, Lindbergh committed himself by swinging down to hang beneath the bright yellow wing. Below was "nothing but space . . . terrible . . . beautiful."

In a moment he pulled the release knot and fell away from the Standard, until at last the parachute canopy filled out and he swung gently, noticing the redness of the sunset, the position of the landing field below, all fear gone. A second pull on the knife-rope cut him free again, to plunge on earthward, waiting for the tightening of the harness when the second canopy blossomed. There was a too-long wait and the ground rushed up, far too fast. Then, a jerk, and the final glorious ride to earth.

Harden was white when he ran up to see if Lindbergh was all right. The second chute had streamed and nearly did not open at all!

After that, Lindbergh was inwardly amused to find his stature had grown considerably higher than the six feet three inches he stood in his socks, and when visitors came to the field, they pointed to him and said to each other, "That's the parachute jumper!"

It had not yet been six months since young Slim, as he was known, got close enough to his first airplane to touch it, inside the plane factory at Lincoln. He had already decided to quit the University of Wisconsin and become an aviator, instead of going into politics like his father, a congressman from the Sixth Minnesota District, or becoming a farmer.

It was at the factory that he fell in love with the acrid smell of nitrate dope, the bright yellow wings stacked along the side wall, the big Hispano-Suiza engines on their test blocks, the slim, trim fuselages of the Standard JR1's that could do 86 mph in level flight and climb to 10,000 feet in only 20 minutes.

On the bright spring morning of April 9, 1922, young Lindbergh and a buddy from the sandy prairie country of Nebraska, Bud Gurney, eagerly climbed into the front cockpit of a brand new Standard and belted themselves down, side by side, for their first airplane ride. Their pilot was a man named

Otto Timm, a Minnesotan who looked not unlike Lindbergh and who had been flying homebuilt airplanes since 1910.

The two boys held their breaths as Timm, all business, waved to the mechanic in front and yelled, "Contact!" A former barnstormer himself, Timm was now "chief engineer" at the Lincoln Standard plant, and his voice rang with authority.

Slim and Bud watched the mechanic throw the weight of his body into a strong pull on the varnished propeller blade, at the same time stepping backward with the grace of a ballet dancer. As the engine coughed and settled down to a rhythmic idling, the mechanic pulled the chocks from in front of the wheels and Timm gunned the ship forward, careful not to blow dust into the hangar where two other mechanics were painting wing fabric with the heady dope.

Gurney and Lindbergh nudged each other and grinned. This was the life! They looked back into Timm's face, half covered with big oval goggles and a tight cloth helmet. Timm ignored them, busily studying the wind, the clouds, the drift of shadows, the thousand and one things that are a secret part of the airman's world, the things he instinctively senses with a skill that is called airmanship. Lindbergh studied his face and saw all this, and was determined to be such a man.

DAVID D. HATFIELD COLLECTION

Lincoln Standard biplane, a barnstormers' favorite, was the ship in which Charles A. Lindbergh learned to fly.

The flight was over too soon. They had climbed high and swooped low over distant hills, chased wheeling hawks, and headed toward a drifting white cumulus cloud. The boys winced as Timm flew right through it, half expecting to feel an impact of fabric wings on the cottony substance. Toward the end of the ride, Timm banked the Standard over steeply so that the flying wires were parallel to the horizon, and when he pulled back on the oak joystick, they felt a tightening in their stomachs and a tugging that opened their jaws.

Ray Page, the company president, took Slim's personal check, and he was enrolled as the firm's one and only flight student. His instructor, Ira Biffle, a dark-featured, embittered veteran instructor left over from the war, did not share his student's poetic eye. Flying was simply a dirty, stinking, underpaid job, the more so since a close friend of his, Turk Gardner, had died in a crash.

Lindbergh had eight hours of dual instruction time in his logbook when he learned that the factory's only trainer was being sold to Erold Bahl. Furthermore, Page refused to let him solo without putting up a bond in case he cracked up on landing. He simply didn't have the money. He was at a dead end. It was then that Slim appealed to Bahl to take him along when he went barnstorming, and when Bahl reluctantly agreed, Lindbergh was launched as a barnstormer. Not as a pilot, at first; his jobs were to wipe down the ship at the end of the day, pull the prop through to start the engine, and coax the small-town crowds to fork over a few dollars for a few minutes in the sky.

One day Lindbergh volunteered to climb out onto the wingtip when they flew over town, to attract more curious customers.

"Go ahead," Bahl shrugged. "But don't put your foot through the wing fabric!"

Launched as a wing-walker and parachute jumper, Slim now rated equal billing with the pilot on the brightly colored handbills they tossed out of the cockpit while flying over Midwestern towns on a Saturday afternoon.

DAREDEVIL LINDBERGH! the handbills said. Slim then went on his second barnstorming tour, this time with a pilot named H. J. (Cupid) Lynch, a short, stocky man who flew a plane to Slim's liking. He decided early in the game that he could live longer by avoiding the crew of hard-drinking, swearing, grease-stained "airport bums" whose misadventures in the sky too often ended in the graveyard.

In the year 1923 alone, reported the authoritative *Aircraft Year Book,* gypsy fliers were responsible for 179 serious plane accidents in which 85 people died bloody deaths and another 162 were injured. Listed as "probable causes" were such chilling reasons as *stunting at low altitude; plane taken up with only a pint of gas; bad landing on bad field; plane plowed into crowd; stunt flyer failed to come out of barrel roll and tailspin* and *control stick broke*. By the following year, barnstormers were held accountable for two out of three fatal crashes.

Slim Lindbergh's second fling at barnstorming was an unforgettable experience, one that instilled in his heart a greater love for America than he had ever known. Winging low across western Kansas and eastern Colorado, he smelled the rich, warm pungency of the prairies in harvest, and he squinted at the vast expanses of bronzed land in a way that would line the corners of his blue eyes with tiny crow's feet.

The silver-hulled Standard seemed to belong in that part of the sky, winding its way jauntily between towering cloud pillars and dropping down to follow the muddy Platte and Powder Rivers, which smelled of damp algae and catfish, or skirting around the proud escarpment of the Big Horns, so full of color in the glory of October.

Lindbergh's eyes saw all this and loved it, for the grand panoramas of his country formed a giant relief map that supported the sky, and so flying over it, hedgehopping or higher up in the clouds, he was still a part of the land. The tug of gravity was ever-present, and when he tumbled down from the sky in free fall at the start of a parachute drop, it was like coming home. The flying season ended in Billings, Montana. Lindbergh packed up his parachute and shipped it back to Lincoln, then set out on a river adventure in a two-dollar boat on the Yellowstone.

During the winter months, Slim visited his father in Minneapolis, and there told him of his decision to buy an airplane and go barnstorming on his own the following spring. The elder Lindbergh saw the stubbornness in his son's eyes and smiled; he even signed Slim's note for $900, borrowed from a bank in Shakopee.

Lindbergh's first airplane was a beautiful, brand new JN-4D Jenny, painted olive drab and complete with two extra wooden propellers and a book of instructions, which he read over carefully after handing the salesman $500 at the Souther Field warehouse near Americus, Georgia. He memorized the specifications the way other youths memorized their girl friends' faces.

Heart of the Jenny, of course, was her V-8 engine, the Curtiss OX-5, an amazing powerplant developed by an engineer named Charles Kirkham at Glenn Curtiss' Hammondsport (New York) factory, where thousands had been mass-produced during the war. Developing 92-horsepower at 1300 rpms, the OX-5 weighed 434 pounds, while the gross weight of the whole ship was 1920 pounds, 50 pounds lighter than the Standard.

Flying the Jenny would be a new experience, Slim realized. The Standard's Hisso engine, rated at 150-horsepower, gave the JR1 a sea level speed of 86 mph, compared to about 70 for the Jenny. And as neither engine was supercharged, the higher you flew the slower you went.

Lindbergh hadn't been in the air since October and felt rusty, and what's more, he had never soloed, having only handled the controls of the Standard a few times when Bahl wanted to catnap. Hence, his first solo hop was nothing he would ever brag about, for on landing, he bounced back into the air and had all he could do not to groundloop.

Taxiing back to the hangar, he was embarrassed to find a stranger standing there, watching him. But the stranger didn't laugh; he was there to help him check out. After a few more circuits and bounces he was advised to wait until the air grew still at dusk. At last, going up alone as the sun sank over the Chattahoochee, Lindbergh began to feel at home in the Jenny. He flew in lazy circles over the Georgia countryside until the shades of night rendered the landscape flat and shapeless. Then he reluctantly idled the engine and glided down the sky to land. A watchman, alone on the field, came up grinning widely and congratulated him for "flyin' higher than any other fellers did!" Lindbergh accepted the compliment happily, then began to think of the future.

For the next year, Slim Lindbergh would follow the barnstormer's trail as he had planned, through the deep South and back up north to the Nebraska and Minnesota country he loved. Wisely, he waited until he had accumulated some five hours of flight experience in his Jenny at Souther Field before venturing off to join the gypsies of the airways. At that, in Meridian, Mississippi, he came close to killing himself and his first passenger.

His fare loved it, though. A huge man, he had eased himself into the front cockpit, and Slim had had to remove the control stick to make room. "Go ahead and do any nipups you like, son," he yelled. "I flew with the Lafayette Escadrilly!"

With close to three hundred pounds extra weight, Slim felt the Jenny wallowing like a bull walrus over the soft cow pasture he'd chosen to operate from. He picked up the tail and opened the throttle wide. With a great leap the Jenny got off the ground just in time to hurdle a split-rail fence. He mushed along, barely above a stall, unable to rise higher until crossing the crest of a ridge. Then he dove down into a small valley, gathered speed and managed to struggle back around and land.

"That's the stuff for me!" the big passenger boomed as he stepped out. "I shore like that there flyin' right off the ground! Scared hell outa the enemy that way, I did!"

Lindbergh swallowed and tried to smile, wiping sweat from his face. It was obvious his fare hadn't realized how close to disaster they'd been.

Early the next morning Lindbergh was off again for greener pastures. His goal was a little town 125 miles west of Meridian, but somehow he ended up 125 miles to the north, following a new compass he hadn't yet installed. Worse, a severe storm front was rolling toward him, and there was nothing to do but set down in the nearest field. It looked nice and green from the air, but he rued his choice, learning a lesson the hard way. Green grass means wet, soft earth, hidden stumps and ditches. He hit a ditch and nosed up on the propeller, which splintered to bits.

When a spare propeller arrived, Slim installed it, made a short test hop and then had time to earn a few dollars hopping passengers before moving on. On his last hop he took up a farm hand whose fare had been paid by two friends who thought it would be a fine joke if Daredevil Lindbergh scared the wits out of him. Slim agreed, although he'd never done acrobatic flying before. He tried a loop, but didn't have enough speed to get over the top. The Jenny fell off on one wing and wallowed back down the sky, much to his embarrassment and disgust.

Learning as he flew, Slim Lindbergh slowly added new maneuvers to his repertoire and practiced them daily at the end of a cross-country hop to a new town. It served both to sharpen his own flying and to bring out the customers.

Flying up into Arkansas, the Jenny pointed its nose west over Fort Wayne and headed into the Panhandle country of Oklahoma and Texas, then swung up into the Kansas wheat country. At Alma he survived his second crash landing, which occurred when his left wing hooked on a rock hidden by the tall grass in the field he had picked out for an airport.

Sleeping beneath the wing, Lindbergh rose early, patched the torn wingtip with fabric and dope he carried with him and moved on toward Lincoln, Nebraska, to say hello to his old friends there and show off his new plane. There Bud Gurney joined Slim and together they barnstormed west through Nebraska and into Minnesota, where for the third time his Jenny came to grief. Heading for Shakopee to see his father, he ran into a heavy rainstorm that soaked the OX-5 and drowned out three cylinders. His engine sputtering sadly, Slim headed down for the nearest field and plopped the Jenny into swampy terrain. This time the Jenny nosed over completely. He wired to Americus, Georgia, for his second spare propeller.

Finally reaching his home town, Lindbergh offered to fly his father around the state on a political campaign tour. It was a short, sad experience that ended within twenty-four hours when something went wrong mechanically with the temperamental Jenny and the Lindberghs crashed to earth at Glencoe. The local newspaper quoted C. A. Lindbergh, Jr., as telling their reporter that "the mechanism had been tampered with."

Slim's father escaped with a pair of broken glasses and a bloody nose, which came from striking the instrument panel, and went on to complete his campaign by automobile, but he was defeated at the polls, not that Slim didn't do his best to make up for what happened. Whenever an opponent stumped a town to make speeches, Slim was right there with the Jenny, hopping passengers and making so much noise that the politician couldn't be heard.

When October came, Lindbergh flew off to St. Louis to attend the International Air Races at Lambert Field. There he ran into Bud Gurney again. Bud was making double parachute jumps himself now, the way Slim had done, but when he went up with Lindbergh, the Jenny couldn't lift the heavy load of the two youths and the two parachutes more than 1,700 feet. Bud jumped anyway and on landing broke his left shoulder.

To Slim Lindbergh, it was quite an experience rubbing elbows with the great pilots of the day. Lieutenant Alford J. Williams of the Navy, a crack acrobatic pilot, thrilled him by flashing over a speed course around three pylons at an average of 243.67 miles an hour. There was a minor aviation boom on, he discovered, and the price of Jennies had rocketed. He sold his ship to a man from Iowa, taught him to fly it, then in December picked up a bargain in an OX-5 Canuck, the Canadian version of the Jenny. The man who sold it to him had won it in a raffle, through a 75-cent ticket held by his young son.

After barnstorming around Illinois during the early winter months, Slim teamed with an auto salesman in St. Louis, Leon Klink, and together they headed south to seek warmer weather. Lindbergh had taken his entrance examination for the Army Air Service at Chanute Field, near Rantoul, Illinois, and had until March, 1924, to finish barnstorming before being inducted at Brooks Field, Texas, as a flying cadet. Swinging down through Kentucky, Tennessee, Mississippi, Alabama, Florida, Louisiana and Texas, Slim cracked up only two more times—not bad for a gypsy flier.

One smashup was attributable to engine trouble, which led to a forced landing. and which in turn led to a shattered propeller. The other happened when Slim, who had made an emergency landing in a tiny pasture, tried to take off from the main street of Camp Wood, Texas, his forty-four foot wings barely squeezing between two rows of telephone poles forty-six feet apart. He missed. One wingtip hung up on a pole when he hit a bump in the road and around he went, the Canuck's nose ramming through the wall of a hardware store. When he offered to pay for the damage, the proprietor wouldn't hear of it. "Look at all the free advertising I got!" he declared.

Near Pumpville, Texas, Lindbergh and Klink ran into more misfortune when they landed near a railroad siding and in the morning had to hack a runway through hundreds of feet of sagebrush, cactus and mesquite. When they tried to take off, a Spanish dagger cactus impaled the plane's wing like a butterfly on a pin, ripping away the bottom fabric. Down they came again, blowing a tire.

Klink decided he had had all he wanted of flying and decided to hop a freight train for California. It was March, and Lindbergh was due at Brooks Field anyway. They shook hands and parted, and soon Slim was bouncing the battered Canuck down the sand strip, through the sagebrush, and into the air, one tire gone, the fabric on the lower wing shredded and a wing-spar cracked and held together with rope.

When he landed the Canuck later in the day at Brooks, mechanics and pilots gathered around it, shaking their heads in utter disbelief that it could even fly. Slim took their ribbing good-naturedly, but when the Commanding

Officer showed up, it was another matter. Turning beet-red, the C.O. ranted, "Get that damned thing out of my sight!"

Lindbergh stammered an apology and got back in his Canuck, flying it over to a commercial airport. Four days later he was formally inducted as a cadet in the United States Army. On his first day he soloed—in a Jenny.

Shy, quiet, withdrawn, Charles Lindbergh was just another unknown gypsy flier until he astounded the whole world with his thrilling solo flight on May 20–21, 1927, from Roosevelt Field, Long Island, to Paris' Le Bourget Airport. At twenty-five, Slim Lindbergh, ex-barnstormer, became the most admired man in the history of aviation and an American phenomenon.

He had done equally dangerous things many times, as a wing-walker, parachute jumper and then as an airmail pilot. He had become "King of the Caterpillars" by saving his life four times in emergency jumps, more than any other member of the exclusive Caterpillar Club, and rated only scant attention. But his Atlantic flight threw a spotlight on him, and even the cynical press realized that a new hero was born.

For some sadistic reason, newspapermen delighted in making life hell for this introverted airman who saw visions too big for them to grasp and horizons too distant for them to comprehend. There was the time he was hired to fly film plates through murderous weather to give the Chicago *Tribune* a scoop over other papers on the story of the tragic rescue attempt of Floyd Collins, who was trapped in a Kentucky cave. At Chicago's Checkerboard Airport a figure appeared and identified himself as a *Tribune* man. In innocence, Lindbergh handed over the films—to the representative of a rival paper, the *American*.

Despite such treatment, America at large took Slim Lindbergh to its heart and worshipped him. Whereas he had been denied the right to take up a plane alone in Lincoln, when he wanted to make his first solo, the Navy Department ordered all stations to make available to him any plane he wanted to fly. And on December 10, 1927, Colonel Charles A. Lindbergh was decorated with the nation's highest award for valor—the Congressional Medal of Honor.

In truth, Lindbergh, the boy who barnstormed the wheat fields of Kansas and the hills of Wyoming, who crashed his way across the country on his first glorious trip with his own Jenny, became a part of the American dream. In an era of bootleg booze, gangsters, and moral decadence, he did what no other living person could have done. He gave his country something good, clean and honorable to admire and to love. And he started the biggest flying boom in history.

7/HAWAII OR BUST

MARTIN JENSEN, king of the gypsy fliers, began barnstorming out of San Diego in 1924 and became the first pilot to cover the United States, coast to coast. With Peg, his pretty, red-haired wife, Marty crammed more adventure into a decade than most airmen would encounter in a lifetime.

On an endurance flight over New York, while Peg flew the ship, Marty amazed press photographers in another ship by crawling out with a set of wrenches and working on the engine two thousand feet over the city. His aerial wanderings were first done in a clipped-wing Jenny, then in a Thomas-Morse Scout and in a few planes he built himself.

In 1927, Jensen was barnstorming the Hawaiian Islands when Lindbergh hopped the Atlantic, and so when the Hawaiian pineapple tycoon, James D. Dole, posted $35,000 in prize money for a race from the mainland to the Islands, Marty hurried back to San Diego to get a ship he felt could compete in the race.

The Dole Race became one of aviation's black marks, the most tragic event of its kind ever held. But it was a test of nerves and skill for America's leading barnstormers, and in the mad flying craze that began with Lindbergh's great flight, the race did have its hour of triumph. Here is the story of that ordeal as related to the author by one of only two pilots to reach Hawaii—Martin Jensen:

San Francisco lay a thousand miles behind us, and somewhere ahead lay a tiny pinpoint in the vast Pacific—Wheeler Field, Hawaii. The time was 10 P.M. Honolulu time, on that black night of August 16, 1927. Except for my navigator, Paul Schluter, I was alone over the Pacific graveyard, in bad trouble and frankly scared.

Our little fabric monoplane, the *Aloha,* pierced the inky night with only

COURTESY MARTIN JENSEN

Martin and Peg Jensen pose with Earl Daugherty's Jenny in Long Beach. Despite good luck billikin insignia, Daugherty pulled off a wing and died in plane crash.

my instincts as a barnstormer to guide her. Sure, I had a ball-bank gadget on my panel, but until now, I'd never learned to use it on a night flight through a suffocating fog blanket.

My throat felt as dry as sand as I peered left and right from the *Aloha's* cockpit—and saw nothing. And then I heard the whine of my Wright J-5 engine suddenly increase in pitch. Under my tense fingers, the control stick became rigid with the grip of rushing air. A feeling in the seat of my pants (the barnstormers' friend!) told me death was rushing up at better than two miles a minute. We were in a screaming spiral dive, that awful, sickening plunge that only gets worse when you try to pull out. I glanced quickly at my altimeter. We were still four thousand feet above the wind-lashed waves of the Pacific.

If we were in a spin, I thought wildly, I could recover. I had made a living as a county fair stunt flier and knew how to get out of a spin. An idea suddenly occurred to me; it seemed crazy, but it was the only way. I chopped the throttle, eased the stick back into my stomach until the *Aloha* shuddered, then slammed on full right rudder. The ship seemed to hang there in the black

70

night, until with a sickening plunge the nose dropped and we were whipping wildly around and around, heading straight for hell.

I let her spin for maybe a thousand feet. Then I dumped the stick forward, got off the rudder and pulled out of the spin into level flight once more. My disorientation subsided and I exhaled slowly, wiped the sweat from my face and settled down to the toughest flying job of my life. I had to learn to fly on instruments, there in mid-Pacific, or die trying.

I felt a sudden jab in my back that made me almost jump out of the cockpit. I remembered Schluter, back in the cabin, strapped in among the extra gas tanks that cluttered up the space. In my excitement I'd forgotten about him. Paul was poking me with a broom handle, our only means of communication other than a clothesline.

I reached down and felt the line running over the pulley, and a note came forward to the cockpit, clipped on with a clothespin.

"For God's sake!" Schluter had scrawled hurriedly. "No more stunts until we get to the Islands!"

I had to laugh, despite the narrow squeeze we'd come through. I'd taken Paul on only three days before the great Dole Race from Oakland to Honolulu had begun. He had answered an ad I'd placed in the San Francisco papers.

Aloha, *ship built by Vance Breese, was flown in Dole Race by Martin Jensen and navigator Paul Schluter.*

At first I'd been skeptical because Paul, an "old man" of thirty-seven, was the oldest entrant in the race. And besides, he had never flown before. He had been a ship's navigator on a coastal vessel, the *City of Nome*. We were already on our way before I fully realized what that meant. Schluter's navigation tables were good only on the surface of the ocean—he had no way of correcting for altitude. In order to get a position fix, I'd have to drop down to within a hundred feet of those whitecaps while he shot the sun or the stars.

Our second navigation problem was simply that a great fog blanket covered most of the Pacific on that unforgettable night, and I was having trouble getting on top. Each time I would try to climb up through the overcast, the *Aloha* would dive off into a screaming spiral. I had no alternative but to grit my teeth and fight it out.

The engine coughed. I felt my mouth full of cotton again. That one tiny engine would have to keep running steadily for a full day—twenty-four hours —or else. But it caught hold again and I relaxed enough to go back to battling those crazy instruments. I was thinking about how, only a few months before, another J-5 engine had taken "Lucky Lindy" across the Atlantic, from New York to Paris, in 33½ hours. In fact, it was because of the great Lindbergh flight that I was now out over the Pacific, bucking rising headwinds, groping through the black night and flying by the seat of my pants.

James D. Dole got the idea for the great Pacific air race while returning to Hawaii from the mainland on a Matson Line steamer, when the ship's wireless crackled with the news that Lindbergh had done it—flown the Atlantic solo. Dole well knew that Hawaii was a potential tourist mecca and he was not blind to the world-wide publicity another ocean flight, this one over the Pacific, could generate.

Already, of course, the route had been spanned by air; two years earlier, Commander John Rodgers almost made it in a flying boat, but navigated the last three hundred miles by water after a forced landing. In July, one month before the Dole Race was scheduled, Ernest L. Smith and Emory Bronte barely made it to the island of Molokai from Oakland before running out of gas.

From the start, Dole's "Pineapple Derby" was a disaster, plagued with unbelievably bad luck. The prize money—$25,000 for the winner and $10,000 for the next ship in—had worked like a magnet. Forty hopeful barnstormers dug down in their pockets and paid the entrance fee.

Unlike Lindbergh, these were not the thorough, precision airmen one would expect to find competing in so hazardous an event, a wild publicity stunt from the beginning. They came from the ranks of the barnstormers, men dazzled by the blinding picture of sudden wealth and fame, spurred on by the Lindbergh hero phenomenon. If Lindy had made it over the 3,600-mile route in 33½ hours, it should be a snap to fly 2,400 miles to hula-hula land nonstop. But there was a difference—Pacific headwinds.

By the time August rolled around the field of forty starters had narrowed to fifteen. Worried government officials, fearing a tragic end for pilots who

Jensen finished tragic Dole Race after Art Goebel. They were the only two pilots to make it. Here Paul Schluter (left) and Jensen grin happily with Martin's wife, Peg.

had no over-water flying experience, insisted on rigid proficiency examinations. One pilot was flatly disqualified because his ship couldn't even carry enough gas to cover the distance. Two others had second thoughts as the countdown continued and wisely withdrew. Another three wrecked their ships on trial flights.

Ben Wyatt, a Navy lieutenant who was handling the pilot checks, made me promise to do exactly what Schluter would tell me to do. I agreed, reluctantly. I'd fly and Schluter would navigate, but first I wanted to find out if he was subject to airsickness. After all, it would be a rugged flight. I didn't want a greenfaced navigator on my hands.

I met Paul for the first time at the little tent city that had sprung up at the end of the 7,000-foot sandy runway on Oakland's Bay Farm Island, where the other Dole entrants worked feverishly getting their ships into zero-hour readiness for the race.

"Let's take a little ride," I said to Paul. We climbed into the *Aloha* and were quickly bumping down the takeoff strip. I thought then that we'd be lucky even to get out of there with a full load!

I climbed the *Aloha* out over San Francisco Bay, circled back near the Oakland Airport, and then suddenly shoved the stick forward. I heard Paul yell something, but held the ship in a dive until the airspeed climbed to a screaming pitch. Then I hauled back hard on the stick, until it seemed I'd sink through the seat. The little monoplane gracefully pulled up into a vertical climb, then arched across the sky and down again through a beautiful loop.

I spent the next ten minutes putting the *Aloha* through her paces and finally cut the throttle and glided back to earth. On the ground, Schluter managed a weak smile.

"You okay?" I asked.

"Fine!" Schluter said, then turned a light green. I could see he was going to be sick. He had spirit, though, and I knew I had a courageous navigator. Now, all he had to do was find Honolulu for me.

We were supposed to go on August 12, because on that night a full moon would rise at sunset, giving us some light to fly by all night long. We would need it. In 1927, instrument flying was still in its infancy. Smith and Bronte were experienced pilots and had good radio equipment, as did Army Lieutenants Lester J. Maitland and Albert F. Hegenberger, the only other aviators ever to make the Hawaiian Islands. The Dole Race pilots, on the other hand, were barnstormers by and large—men used to following iron beams and making sure there were plenty of cow pastures to land in if their engines quit.

Of the eight final starters, there was one exception—Art Goebel. A veteran Hollywood stunt pilot and a member of the famous Thirteen Black Cats aerobatic team, Goebel *was* an experienced instrument flier, and his ship, the Beech monoplane *Woolaroc,* did carry a two-way radio.

Goebel had been on a Universal Pictures studio location job when Lindbergh hopped the Atlantic. He decided immediately to have a go at the Pacific race when Dole posted his big prize money. Art had done all there was to do with an airplane, carrying wing-walkers like Gladys Ingle and Ivan Unger on thrilling plane changes and performing hair-raising newsreel stunts such as diving under the Pasadena Colorado Street Bridge with two girls standing on his top wing. But the Dole Race was an entirely different kind of stunt, calling for special new skills.

The moon rose full and bright on the night of August 12, but nobody was ready to start. We all signed an agreement to delay the race four days, and in that decision lay death.

At noon, the first ship took off. It was the *Oklahoma,* with Ben Griffin and Al Henly as crew. A crowd of ten thousand people cheered lustily as the *Oklahoma,* overloaded with gas, staggered off the end of the runway and headed for the Golden Gate.

The next two starters didn't even get off the ground. And from the way those ships, the *El Encanto* and the *Pabco Flier,* floundered in the sand, I could see trouble ahead for me. The fourth off was the *Golden Eagle,* a sleek Lockheed Vega piloted by Jack Frost and sponsored by the San Francisco *Examiner.* Frost lifted off gracefully and swung into the west as the next ship, *Miss Doran,* thundered off with three people aboard—her pilot, Augie Pedlar, her navigator, Lieutenant V. R. Knope, and pretty Mildred Doran, a Michigan schoolteacher.

Next went the *Dallas Spirit,* with Captain William Erwin, a World War I ace, and navigator Alvin Eichwaldt. Then it was time for Schluter and me to go.

"All set?"I yelled back to Paul, crouched behind our fuselage fuel tanks.

"Honolulu here we come!" he yelled back, and I gave her the gun. Sand tugged at the disc wheels of our little monoplane. The propeller whirred furiously, but nothing happened. Half a dozen mechanics rushed out and threw their weight against the wing struts. Slowly the *Aloha* began to roll, coming to life as we bounced down the sandy runway and struggled into the air.

We flew along San Francisco's waterfront, gaining very little altitude as we passed through Golden Gate. Dead ahead I already could see trouble—a blinding white fog bank shrouded the Farallone Islands. I remembered what the government weather forecaster had told me—there would be a high-pressure ridge out over the mid-Pacific and the sky would be clear beyond that. The cloud deck, he assured me, wouldn't extend clear to the ocean.

Schluter had given me a compass heading to fly and instructions to stay within one hundred feet of the water so he could work his celestial navigation problems as we flew west. I remembered my promise to fly the way Paul wanted and stayed down low.

We didn't know until later what happened to the others, and it was just as well; I might have turned back then and there. Goebel, who left after us, of course made it to the islands and won first-prize money, but the others ran into serious trouble almost from the start. The *Oklahoma* had gone back to Oakland, her belly ripped open during her rough takeoff. The *Dallas Spirit* had given up and returned for similar reasons. *El Encanto* and the *Pabco Flier* had crashed on takeoff, and so only four planes were left.

The *Golden Eagle,* which made such a beautiful start, simply vanished at sea, and to this day her fate is unknown. *Miss Doran* returned to Oakland with engine trouble, but Pedlar decided to take a chance and start off again. The decision cost them their lives—the last ever seen of that ship was on its second takeoff, Mildred Doran blowing kisses from the window. So Goebel, the only instrument pilot, and I, though neither of us were aware of it, were the only ones left in the Dole Race, droning west into the approaching night of terror, one I will never forget.

I had faith in my ship, and even when the fog bank closed in around me, I believed we'd make it. I held close to my heading of 248 degrees, hour after hour. Schluter had suggested a great circle route, changing course one degree left every two hours, but he forgot to tell me to change course during the flight, and I didn't. I found out why later. Schluter had been unable to get a sun-sight through the broken clouds, and when night closed in, even Polaris, the North Star, was hidden from view.

There was plenty for me to do to keep busy, and so I had little time to worry. My ears were tuned to the beat of our Wright engine, and from time to time I had to switch over the crossfeed to use up fuel evenly. The rest of the time I sat there hypnotized, staring at the earth-inductor compass. The *Aloha,* a $15,000 plane I'd bought from Vance Breese, flew well, and there was nothing to do but fly west toward Honolulu, where Peg, my wife, was waiting for me.

I remembered back to our barnstorming trips across the country and in the Islands, and how bravely she'd helped out by clambering over the wings to draw customers. I thought of our first venture, in San Diego, when a good crowd showed up to watch, but nobody had courage enough to fly with me. So I gave a ballyhoo pitch about flying safety, then told them I was going to go up to 1,500 feet, stop the engine and glide back deadstick to land right in front of them. I wasn't too sure I could do it, but I did. It worked like a charm.

If my engine quit now, I knew, there would be no place to land down below; nothing but miles and miles of open sea. It was warm in the open cockpit of the *Aloha,* directly behind the engine, but the orange glow from the exhaust stack was blinding; even when the moon rose four hours after sunset, I could see nothing but a dim, almost imaginary horizon. The fog bank had pressed us lower and lower until we were completely enveloped in it. I know now what must have happened to the *Golden Eagle* and *Miss Doran,* and even then I wondered how they were making out.

Blind flying without instrument knowledge or a radio beam to follow was deadly and foolish; we were at the mercy of the elements. Vertigo and mysterious sensory illusions began to confuse me as I groped my way through the blackness. Even my old reliable seat-of-the-pants technique couldn't be trusted; we could be flying upside down and not know it until too late.

My hands were sweaty as I nervously gripped the stick, straining to see ahead. Suddenly there was a jolt; we had struck the ocean! Shocked, I struggled with the controls as I felt the *Aloha* slam hard against a big whitecap, a blow that threw spray over the ship and almost wrenched the controls from my hands. Fighting against panic, I hauled back on the stick and rammed the throttle full open. I caught a quick glimpse of angry waves reaching for us as I pulled up into the choking fog once more. My one hope lay in climbing on top of the cloud layer, which now stretched clear to the Pacific's surface, but any instrument pilot will tell you what it means to maintain control with only a needle, ball and airspeed to guide you, particularly if you aren't familiar with the trick of it.

We got past one thousand feet, then two thousand, and three thousand, and still no moon or stars. Then, passing four thousand feet, everything went crazy. I heard the scream of the wind, but couldn't figure out what the instruments were telling me. We were in a graveyard spiral dive! It was then that I took a chance and kicked the *Aloha* into a deliberate spin, simply because I knew how to get out of a spin.

We managed to recover, and three more times we went through the same agonizing experience, until somehow, by sheer desperation, I taught myself to fly by instruments. It was the world's quickest instrument flying course! I learned to disregard the nauseating vertigo that wanted to pull me off into those sickening spiral dives, and I discovered that I could keep the turn needle indicator centered by moving the stick gently to one side or the other.

Through the rest of the night I flew for my life—and Schluter's—with agonizing concentration on the needle, ball and airspeed, until at last a misty whiteness unfolded into a beautiful dawn. We were still alive and flying, but

Art Goebel, flying for news-reel cameras, buzzes under Pasadena's Colorado Street Bridge with two wing-riders, Gladys Ingle and Shirley Calishak, on top wing.

COURTESY ART GOEBEL

where were we? Schluter had not had a chance to check our position during the entire flight, but by dead reckoning I knew that we had flown long enough to be over the Islands by 8:30 A.M., Honolulu time. I looked at my watch. It was 9 o'clock!

I clipped a note to the clothesline and slid it back to Schluter. "Which way from here?" I pleaded. We were running low on gas, and I wondered if we'd picked up a tailwind and overshot our target.

Paul's note came back. "Circle until noon," it read.

I stared at it unbelieving. Fly in a circle for three more hours, when we were running out of gas? Then it dawned on me what he had in mind. It would be fatal to rush headlong in any direction, because we were completely lost; Hawaii could be on any of 360 different courses. We had to have a celestial fix, and the only way for Schluter, a ship's navigator with limited experience at that, was to get a sun shot at high noon.

It was deadly, flying in circles hour after hour, draining one auxiliary tank after another and perhaps throwing away any chance we had of making a landfall. But any plan was better than none, so we continued flying in circles until the main tank ran dry. The engine sputtered and almost quit before I could grab the wobble pump. A trickle of gas came from the outboard wing tanks.

Noon came. Paul, his face sweating, shot the sun as I circled once more, barely above the breaking whitecaps. Finally he clipped a note to the clothesline and sent it forward. I read it; Hawaii was two hundred miles away—to the south! That is, if Schluter was right!

I knew what had happened; by not changing course during the long night, and sticking to 248 degrees, we had gone too far north. Wondering if our fuel would hold out, I headed the *Aloha* south, leaned out as much as possible and waited. Every towering cumulus cloud became Diamond Head, and more than once I cried out, "Land ahead!"

At last I spotted a cloud that did not change shape—Oahu! Weaning every ounce of gas in a shallow letdown, I streaked in across Kahuku from the north, and over the Koolau range. I finally cut my power and glided noiselessly down the slope and, with a final roar, gunned across Wheeler Field, scattering the thousands of waiting people who were staring the other way, looking for ships that would never come in.

I pulled up and around in a fighter approach, then eased down and landed —28 hours 16 minutes after leaving Oakland. We had exactly four gallons of gas left—enough for one more hop around the field!

Schluter and I crawled out, exhausted but grinning. We embraced each other, intoxicated with the thrill of being alive. We felt certain we were the last in, having spent so much time circling north of Hawaii.

"The *Woolaroc* is over there," someone told me. "Goebel got in two hours ago. No one else is here."

Schluter and I looked at each other and understood. Goebel had flown on top of the weather the whole way and won first prize with a fine, professional job. For a barnstormer, it was something of a miracle that I had come in at all. The Dole Race was over, and Honolulu got its publicity—at a cost of ten lives.

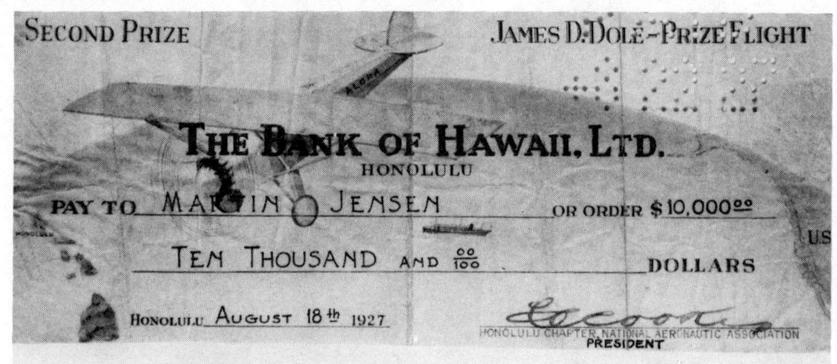

Jensen won this $10,000 check for coming in second in Dole Race.

First to die had been George Covell and Dick Waggener, two Navy officers who crashed to their deaths at Point Loma, near San Diego, while bringing their ship, a Tremoine Hummingbird up to Oakland for the Dole Race. Next was Arthur Rodgers, who failed in an attempt to bail out of his Bryant monoplane when it went out of control near Los Angeles, also en route to Oakland. The others were lost at sea—Augie Pedlar, Lieutenant Knope and Mildred Doran in *Miss Doran;* Jack Frost and Gordon Scott in the *Golden Eagle,* and Bil. Erwin and Alvin Eichwaldt in the *Dallas Spirit.*

Peg and I returned to the mainland and took the *Aloha* on a cross-country tour—this time with plenty of landing fields underneath! The $10,000 second-prize money helped to pay for the ship, and ahead of us lay another ten years of living the life of gypsy fliers.

Since then I've designed helicopters and twin-engine executive ships and helped engineer giant Douglas transports, but the call of barnstorming is still strong. I'm working on a new design now, a little two-place job with swept wings and four engines, and who knows? Maybe I'll go barnstorming again with that!

8/THE OLD AND THE BOLD

"THERE ARE OLD PILOTS and there are bold pilots, but there are no old, bold pilots," goes a barnstormer's adage. The same thing may be said of the "gypsy moths," roving bands of parachute jumpers who even more than men with wings surely knew that one misstep meant sudden death.

Now and then, in some remote airfield hangar, at some obscure county fair, you'll still find a few living old-bold ones, last of the breed of thrill-seekers, men who cheated death a dozen times, plus a few women daredevils also living on borrowed time.

Around rigging tables at scores of modern skydiving DZ's (drop zones), you'll still hear gossip about their amazing feats, and how they did the impossible and lived to tell about it. And about others who didn't. Much of the lore is legendary; most stories get better with retelling. But many were true.

There was dynamic Eddie Angel, brother of the famed barnstorming jungle pilot Jimmy Angel, a World War I Royal Flying Corps hero who joined the gypsy fliers in the 1920's and operated a cow-pasture squadron of death known as Jimmy Angel's Flying Circus. After a wild day's antics, the Flying Angels lured the suckers back to the field at night to witness Eddie's spectacular "Dive of Death"—a free fall plunge down the sky from five thousand feet while holding a pair of big flashlights.

"When I could see the ground," Eddie once explained to me, "it was time to pull the ripcord."

Then there was Bobby Rose, first of the great Hollywood stuntmen, who made women faint by falling backward off the top wing of a Jenny during a plane change. People shrieked in horror as Bobby tumbled down the sky head over heels toward certain death. Finally a 400-foot static line jerked open a hidden chute and Bobby floated earthward with cheers in his ears. "I used to plead with Sky-Hi Irvin to invent a ripcord and make my life easier," Bobby sighed recently. (Leslie Irvin did eventually become a successful parachute maker.)

Another old timer, Ed Unger, in 1967 celebrated his ninety-fifth birthday after a lifetime dedicated to barnstorming as a balloonist and parachutist. It was Ed who taught Sky-Hi Irvin to jump, and he still claims to have originated the idea for the manually controlled back pack parachute in use today.

Unger, Rose and many another old-time parachute jumper wintered in Venice, California, for two reasons—the weather was salubrious, and there it was that the early newsreel companies shot most of the early aviation thrillers. Few aerial stuntmen of the 1920's bothered with parachutes. They were cumbersome to lug around while clambering over the outside of Standards and Jennies while a camera ship flew in close, getting the action on film.

Duke Krantz, who originated the standing loop, scoffed at the idea of a "jump sack." After all, centrifugal force held him in place when he stood on the top wing all through a loop! Another barnstormer with disdain for the parachute was Walter Hunter of the Hunter Brothers Flying Circus. At a time when parachute jumps were standard fare at air meets, this team of four Oklahomans drew the biggest crowds by advertising their specialty—a "Thrilling Death Leap from an Airplane WITHOUT a Parachute!" The Hunters, a rugged band of grease-stained gypsy fliers, capped their show by dragging low across the field in front of the grandstand, Albert, Kenneth and John waving from the cockpits of their Standard biplane and Walter swinging by his knees from the axle. He'd let go and drop headfirst—right into a haystack.

Many a character used to drift into the old Venice Field hangar looking for stunt work during the Depression years. Of these oddballs, the strangest to Sam Greenwald, an International Newsreel cameraman, was a full-blooded Cherokee Indian, Chief Whitefeather, who showed up one day in the early 1920's, thick, black hair braided down his back, looking for wampum.

"Half a dozen pilots and cameramen were playing poker in a corner of

Barnstormer Dick Grace settled down in Hollywood where the pay was good. He specialized in crashing planes.

Walter Hunter (left) with brothers Albert, Kenneth and John, made up the Hunter Brothers Flying Circus. Walter dropped from their Standard biplane without a chute— into a haymow. Picture taken at Scott Field, Illinois, April 1924.

the hangar when Whitefeather drifted in," Greenwald remembers. "He told us he wanted to hang by his pigtail from an airplane for $50."

To make sure that he could do it, they interrupted their game long enough to string the Indian up by his hair, which was tied to a rope thrown over a rafter. "We forgot all about him until that night," says Greenwald. "When we finally went back to cut him down, he was pretty mad, but in good shape."

The following day Frank Clarke, one of the top Hollywood stunt fliers, took Chief Whitefeather up two thousand feet over town. He swung happily in the wind while Sam shot movies of him from a camera ship. Clarke had visions of making a fortune by touring the country with Chief Whitefeather, the stunt to be sponsored by a hair-restorer company.

Whitefeather had other ideas. He wanted to try an even riskier stunt—a cutaway parachute drop with *ten* chutes. His idea was to open them one at a time, cutting away each one as it opened, then land with the tenth. Things went well as cameraman Joe Johnson ground away furiously, following him down the sky in another ship. But Whitefeather's sixth chute fouled; he plunged into a barley patch and was killed.

There was something quite dashing about carnival jumpers, handsome young fellows with pencil-line moustaches and a quick eye for pretty girls, heroes who lounged around small airports in white coveralls, white cloth helmets, and white sneakers. From habitually packing a thirty-pound parachute on their backs, they walked with a forward list, as if encountering a stiff headwind.

Such a barnstormer was Jimmie Goodwin, a veteran jumper and Hollywood bit player who added to his costume a pair of canvas batman's wings and a cloth stabilizer between his outspread legs to enable him to descend in spirals. Hank Coffin, a flier who used to take Jimmie up to jump altitude, claimed with a straight face that he once glided all the way down from ten thousand feet, over Big Bear Lake, to Alhambra Airport, seventy-five miles away! There are probably less than half a dozen batmen left today, and the work isn't steady, but before World War II no flying circus was complete without one.

Cliff Rose, another barnstorming jumper who knows what it's like to fly like a bird, made his first canvas wings at seventeen to become the world's youngest professional batman. When the regular performer didn't show up at a Long Beach air show, Rose bought some sailcloth and rigged an outfit that looked fine, but he wasn't too sure it would work.

At ten thousand feet over the airport Rose stepped out of an Aeronca, spread his arms and legs and felt a freedom he had never experienced before. He related the story to me this way: "I knew from the sound of the wind that I was diving faster than I ever had before in thirty standard jumps. I gripped the hand holds on two broomstick arms and pulled up, drawing the canvas taut. I spread my legs apart and there I was—a batman!

"I decided to try a spiral dive. Cautiously, I raised my right arm and dropped my left, then lifted my head, so that I was a sort of human corkscrew. Below me, the earth began to spin like a top. I was flying!

"I tried a loop next. Both arms extended rigidly, I lifted my head and arched my back. I tumbled over backwards with an easy mushing feeling. From the ground, they said I looked sensational!

"I was so busy trying out my wings I almost forgot about my parachute. In alarm, I realized I was so close to the airport I could see the yokels' faces. I yanked hard on the D-ring, too close for comfort. I hit the ground seconds later."

To prove he was really flying and not just falling erratically, Rose added an extra canvas pocket and filled it with red flour, then painted the blue sky with streaks of slashing red.

There is a bad joke about the fellow who bought a parachute and was told by the salesman that he could bring it back and get a refund if it didn't work, but it really happened to Rose.

"I'd organized a jump team billed as the Cliff Rose Death Angels, and I was making a long free fall. I waited almost too long to pull, trying hard to zero in on a forty-foot ground target. As luck had it, the chute streamed and failed to blossom. I barely had sky left to hit the emergency chute, and while I missed the spot, I lived to take the big twenty-eight footer back to the complaint desk. They cheerfully refunded my money."

Among other famous batmen were Clem Sohn, the first man to use canvas sails for body gliding (and first to be killed doing it); Leo Valentin, France's famed "human bird," who also died a victim of his crazy contraption, and Roy "Red" Grant, who made headlines by sailing above Niagara Falls from the United States to the Canadian side and back on homemade wings.

Not all barnstorming jumpers were killed by falling. There were other danger zones around small-town airports that took their toll of lives. Gladys Roy, a pretty wing-walker and jumper with nerves of steel, thrilled many an

Cliff Rose, without a parachute, jumps from an airplane.

audience with her leaps into the wild blue yonder, which ended with her landing daintily and reaching for her compact. In Youngstown, Ohio, on August 15, 1927, Gladys lost her life—by accidentally walking into a spinning propeller.

Perhaps the world's greatest raconteur and authority on *Americana barnstormiana* is the irrepressible Louis "Speedy" Babbs, a still-active old-boldster who as much as anybody enjoys hangar flying tales "spun from whole cloth for the benefit of younger pilots, especially those who recently soloed."

Speedy will tell you about the twin brothers who in the '20's almost set a cross-country record of twenty hours in a Jenny, East coast to West. One brother, in a blaze of publicity took off from New York in a Jenny, after a $5,000 prize. On the side of the ship was emblazoned its name: *Soul of Africa*. He waved farewell to the newsreel camera ship and winged off into the night—landing in a Pennsylvania cow pasture after dark. At dawn, the other brother arrived in Seattle, having taken off in another Jenny from a small town fifty miles to the east. Things went well until the award banquet, when newsreels of the departure and arrival were shown the admiring audience. The last scene showed Jenny No. 2 setting down at Seattle's municipal airport, the pilot climbing out smiling broadly. Behind him, in large letters on the fuselage, he had painted, *Sole of Africa*.

Speedy remembers an Army pilot named Weaver, stationed in the Canal

84

Zone. who had several forced landings over the ocean. Each time he hit, his ship had nosed over. Finally he figured a way to lick the problem—the next time his engine quit he would glide down inverted. When his ship nosed over, there he would be, right side up!

Wild as some of the hangar flying tales were, Speedy heard few that could top his own personal experiences at cow-pasture dedications in the late 1920's and early 1930's. In his career as a stuntman Babbs broke only fifty-six bones, a few at a time, and so considers himself pretty indestructible. He started out to be a mere pilot, learning to fly in 1925 with Waldo Waterman, a pioneer birdman, but he couldn't raise enough money to buy an airplane. He invested in a $35 parachute instead.

"I helped dedicate more cow pastures than you can shake a control stick at," says Speedy, "not only doing parachute jumps and wing-walking, but some fancy rope ladder stunts, hanging from a Canuck by my feet and picking up a flag from the ground, and so forth."

His first "break" came at the dedication of San Bernardino's Tri-City Airport in Southern California on October 13, 1928. "I was supposed to ride the top wing of Bob Crooks' Curtiss Oriole, and while upside down in a loop, trickle off, fall free and open my chute as close to the ground as possible. It had never been done before."

Perched on the top wing, Speedy's skinny body nevertheless created such drag that Crooks barely got off the ground, flying *beneath* power lines you'll find at the edge of any small airport. "I had to duck to miss 'em!" he says.

Climbing finally to 1,700 feet, as high as the ship would go, Crooks dove for speed and mushed up through a loop. At the top Speedy slid off, and when he got low enough to hear the spectators screaming he reached for the rip cord. It wasn't there. It had fouled underneath his chest pack.

"With the same motion I jerked the rip line on my spare chute. It opened with a jerk, broke three shroud lines and split from the center to the lip, and then I hit the ground. I hit so hard I believe my tracks are still in the center of that airport!"

Speedy went to the hospital with two crushed vertebrae that left him paralyzed for weeks, but when he got out he went right back to the same airport—and opened a school for jumpers.

"I also did some more wing-walking, and once my dad was watching me. This lady next to him says, 'That man must be crazy up there! When he gets down I'd like to see what he looks like!' Dad says, 'Lady, stick right with me and I'll introduce you. He's my son!' "

With his back not fully healed, Speedy decided to forego parachuting for a while and take up something tamer. So he bought a motorcycle and had a big steel mesh ball erected at Ocean Park Amusement Pier, next to Venice, where he set a world record by looping the loop on the motorcycle inside the "Globe of Death" five hundred times in a row.

It was 1932 when Speedy made further headlines with a wild Fourth of

July pyrotechnic parachute stunt that literally backfired and darned near killed him. And grabbing headlines wasn't easy that week, for big things were happening everywhere: Franklin Delano Roosevelt had just won the Democratic Party's nomination for President in Chicago, Amelia Earhart had made a forced landing near Los Angeles, Jimmy Mattern and Bennie Griffin were off on their round-the-world flight, Jean Harlow and Paul Bern were honeymooning in Hollywood and Norma Talmadge was suing Joseph Schenck for divorce.

All this was brushed aside on July 5 when the Santa Monica *Outlook* bannered the story: HUMAN SKYROCKET SERIOUSLY BURNED!

Speedy had made a deal with the Venice Pier manager to jump from an airplane at night and set off a bunch of fireworks on his way down. He went from airport to airport, looking for a pilot to take him up, but was turned down cold.

"I found out later that the CAA had threatened to ground any pilot who took me up," he recalls. "Finally I found a guy who would do it—he needed the money to lift a mortgage on his plane."

Airport dedications drew barnstormers like flies. This is Union Air Terminal near Hollywood.

Speedy's angel of mercy was Chuck Sisto, a barnstormer who later made headlines himself as an airline pilot by accidentally, when his trim tabs stuck, doing half an outside loop over Texas with a DC-4 loaded with terrified passengers.

Sisto flew Speedy up to eight thousand feet over Venice Pier in the early evening and found the pier obscured from view by a three-thousand-foot cloud layer. Babbs decided to jump anyway, needing the money, so he crawled out to the wingtip, waved good-bye to Sisto and did a pull-off.

Hanging from his feet was a hundred-pound gunny sack full of star bombs and powerful red-white-and-blue flares, and a big inner tube for a life preserver. "I thought I'd light a couple of star bombs and see how the fuse was timed for distance," he relates. "The first bomb I lit dropped sparks inside the sack and all hell cut loose. The inner tube caught fire and blew out, and I set it swinging so that it would only burn me as it passed below my feet. The star bombs were blowing up inside the sack and my clothes caught fire. The only way to unload that sack was to reach inside and throw the stuff out—I'd left my jackknife at the airport.

"Many of the pieces exploded in my hands, and I could smell my own flesh burning. But I had to keep on going—in the bottom of the sack was a large detonating bomb that was to be the finale. I knew that if that bomb exploded while still in the bag it would also blow me into little bitty pieces. How I did it, I'll never know, but I finally got it overboard . . . and you know something? It never did go off!"

The irony of the occasion was that all this action took place high above the cloud layer. Says Speedy, "No one saw it but me. My only regret was that I was the only one who saw that magnificent display!"

Meanwhile, back on the ground, people got tired of waiting and started to go home, which prompted the worried pier manager to call the airport. When Sisto informed him that Speedy had jumped half an hour before, rescue boats hurried out and picked him up, more dead than alive, two miles offshore.

Speedy was amused to read his own obituary in the Los Angeles *Times*, whose report of his death was, as Mark Twain once put it in a similar circumstance, greatly exaggerated.

On December 17, 1933, on the thirtieth anniversary of the Wright Brothers' first flight at Kitty Hawk, Babbs dreamed up another spectacular that almost cost him his life—the "Sky-Hook-Pendulum-Cloud-Swing." Speedy threaded three parachutes onto eight hundred feet of rope, one atop the other, with the nebulous idea of doing a pull-off at eight thousand feet and swinging down the sky in giant half-mile arcs while burning smokepots, so the folks on the ground could follow him.

For added insurance, he fortunately carried an emergency pack. The thought of landing at the end of a half-mile downswing unnerved even this cool character.

Things started off well. He crawled out onto the wing and clipped the rope

Gladys Roy, wing-walker, was killed when she walked into spinning propeller.

SECURITY FIRST NATIONAL BANK COLLECTION

to his harness, then released the triple chute rig, which finally jerked him into space.

"After falling long enough for the slack to be taken up, nothing happened," he remembers. "I grabbed the rope to relieve the shock on my harness and pulled in a few feet of it. The rope felt strangely slack, so I made a few more overhand grabs and there I was, holding the frayed end! It had broken about twenty feet above me!"

With more than three thousand feet of the sky already used up, Speedy lived up to his name by grabbing for the D-ring. But a horrible thought flashed across his mind: his life now depended on an old canopy that had been repeatedly immersed in sea water and hadn't been repacked in months.

"It opened like a charm, though," he says. "To keep my act from being a total flop, I lit the smokepots anyway, then spun the chute all the way to the ground—right in front of the grandstand, naturally."

Speedy's other chutes meanwhile drifted to earth miles away, and before he could locate them, one had been stolen. The next day he tried the stunt again with the two remaining chutes and got involved in still another adventure. Due to a stiff ocean breeze, he had decided to jump out over the Pacific Ocean, hoping to be blown back to the airport by the time he reached terra firma, but a lower wind-shift tricked him.

"Through force of habit I put the soles of my shoes together and took a drift sight between them. I was drifting out over the bay, instead of in toward land! I whipped out my sheath knife, cut loose from the two chutes and dropped, keeping my eye on the mountain ridge inland. When I had dropped to a level I was sure was dead air I popped the emergency chute. I took another drift sight; I was slowly drifting shoreward toward the airport,

Wing-walker Gladys Ingle makes plane-to-plane change without a ladder. She climbs from top wing of Bon McDougall's Jenny to Art Goebel's lower wing.

COURTESY IVAN UNGER

though still too far out over the bay for comfort. But my luck held and I landed within twenty steps of the water's edge."

Eventually Speedy recovered all three chutes, but he never again tried the "Sky-Hook-Pendulum-Cloud-Swing."

Another of Speedy's great ideas that didn't work was something he dreamed up and sold to Paramount News. ("When I was broke I could always sell Paramount News a stunt," he says.)

"First I sold Spud Manning on the idea. We were to go up and jump out of a blimp—and swap parachutes on the way down."

Due to a misunderstanding, the camera ship pilot thought the blimp was only up for a rehearsal (*nobody* would attempt such a fool stunt without practice!), and so the boys waited up there at four thousand feet in the big sausage while the camera crew went to lunch.

"Our heavy chutes and harnesses got mighty uncomfortable, so we took them off and played a game to kill time . . . see who could lean out the door the farthest. That wasn't too much fun, so we tried hanging by our hands, dangling our bodies in space, until the pilot, turning green, pleaded with us to put our chutes back on."

When the photo ship finally got off the ground, Spud and Speedy jumped together but missed, by a wide margin, coming close enough to trade chutes.

Speedy: "The next day I tied a mile-long linen string to my harness and rolled it into a ball, which I threw to Spud as we floated down. He was supposed to reel me in to where we could unbuckle and swap harnesses. But the string broke."

On the third try Lady Luck was either with them or against them, depending on how you want to look at it, for errant winds kept them from risking their lives in a foolhardy stunt that could have been the death of both of them.

Spud Manning, in fact, was one of Speedy Babbs' four partners to die in action, drowning in Lake Michigan when his plane ran out of gas during the 1933 World's Fair. The first of them to go had been Jimmy Young, who lost a struggle against a windstorm that blew his parachute out to sea. Next was a jumper named Curly Wells, another barnstorming gypsy moth. His last partner, Jimmy Pate, waited too long to pull the rip cord on a delayed drop and left his mark in Natchez, Mississippi.

Incidentally, Jimmy Pate's jump pilot was J. O. "Doc" Dockery, an old-time barnstormer the author fondly remembers from World War II days when Doc ran a cow pasture operation at a small field near Stuttgart, Arkansas, an advanced Glider Corps training base. Due to the usual sort of wartime foulup, the AAF had provided 1,600 student glider pilot trainees with only *four* CG-4A boxcar gliders. To keep their hand in, the students used to rent Piper Cubs from Doc Dockery and buzz freight trains chugging along the Missouri & Arkansas Railroad tracks. A favorite game, to the discomfort of the engineer, was to try to land on a moving flat car on a straight stretch of track.

COURTESY SPEEDY BABBS

After fracturing his spine, Speedy Babbs temporarily gave up exhibition jumping for something safer—motorcycle riding.

One of Speedy's pals who worked for the Paramount News thrillers was a carnival escape artist named Joe Campi, who came up with the remarkable idea of having himself nailed inside a box slung under the belly of a Hisso Jenny. He was supposed to be dropped from eight thousand feet, then pull a release rope that would let the box come apart so he could parachute down.

"Well," says Speedy, "Joe dropped fine, but due to air resistance, the release jammed. Joe literally kicked that box to pieces and got out just in time."

Another of Speedy's crowd was Bob Coy, whom he calls "the best precision jumper I ever saw drop out of the sky."

One time down in Pensacola, Florida, Speedy recalls, "the local television studio asked Bob if he could land on the lawn at the studio and then walk right in for an interview. He said he could and we got him the necessary clearances, but the airport control tower neglected to tell the Navy what we were doing. It was at the time of the Bay of Pigs incident and Fidel Castro was threatening to bomb Florida cities. There were lots of clouds that day, but I found a hole and dropped Bob through it. He opened up and slipped his chute to land about fifty feet from the door of the TV studio, right on the sidewalk.

"As he'd planned, Coy rolled up his chute, tucked it under his arm and ran inside, right into the camera lens, for a video tape that was to be broadcast later that day. By coincidence, somebody with a vivid imagination driving past saw Coy drop out of the sky and run into the studio. He stopped and grabbed a phone and called the Navy base, sure it was a Castro invasion, reasoning that invaders would naturally take over communications first. Well, they scrambled every plane they could, but by that time I'd landed. We both caught hell."

The last time Babbs and Coy worked together, Coy jumped in a high wind at Key West, Florida, and broke his leg when he slammed into his own parked truck. The next day Speedy went to the hospital to cheer him up, but he couldn't get in; it was after visiting hours, a nurse told him. So he turned his collar around and tried again with another nurse.

"She almost broke her arm helping me when I told her I was the Reverend Babbs. Bob and I bowed our heads in silent prayer, and I slipped this fifth of one-hundred octane Southern Comfort between his sheets. That night Bob got high and fell out of bed chasing a nurse, his leg cast still tied to the ceiling. They found him that way the next morning, sleeping happily."

Some parachutists, it occurred to Speedy, are born to lose. Others adopt a philosophy of life and death that makes it possible to really enjoy things like the sport of skydiving, which, incidentally, Spud Manning invented when he was a barnstorming gypsy moth. Speedy worked out his own philosophy, after a near-fatal spill on his first motorbike prompted his father to give him a stern lecture about risking his neck foolishly.

"I told him we all have to die sometime, and anyone who is afraid to die, dies a thousand-and-one-deaths, while those who don't worry about it only die once, so we might as well have a bit of fun before the coming event!"

His friend, Bob Coy, it turned out, was not a born loser. In the summer of 1967 he was one of only two survivors rescued from the storm-lashed waters of Lake Erie following a tragic mass-parachute jump that took fourteen lives. An overcast sky had hidden their watery grave from the jumpers until it was too late in their free fall to alter their fate.

9/SOUTH OF THE BORDER

THE LEGEND HAS GROWN that barnstorming was strictly an American institution, as homespun as apple pie, maple syrup, hamburgers. This has led, in fact, to a derisive image of gypsy fliers as a wild bunch who turned handsprings for hamburgers and somersaults for sandwiches.

To go back to 1919 for a moment, listen to the pitch of a copywriter for the Aircraft Engineering Corporation who saw the skies full of American eaglets: "The most encouraging sign of the future of aviation is the large number of former army and navy fliers who are coming back into flying for sport, and as a profession.

"The war was responsible for the rapid training of many thousand men—the best blood of the nation!—who are generally recognized as the most skillful, resourceful and daring pilots in the world.

"When these men left the service, their interest in flying did not cease. On the contrary, they are more eager to fly than ever before. The call of the air is irresistible. They will fly, henceforward, just as in past years they played polo and owned speed boats and rode to the hounds. . . ."

You believe it?

The way Benny Howard remembers barnstorming was not exactly as a picture of members of the polo and yacht club set too bored with life to bother with anything less than a flying yacht, or following the hounds in a Christmas Bullet.

"We used to put mothballs in our gasoline, supposedly to make the engine run better," Benny recalls. And not the least of the dangers of flying during Prohibition was getting shot at by hillbillies who took you for a "revenooer."

The first flush of excitement over taking a ride in an airplane wore off

LOS ANGELES COUNTY MUSEUM

Frank Hawks barnstormed Mexico, flying payrolls over bandit country.

quickly, and after the first year of scouring the back country for rubes willing to take a chance on a hop over the old barn for a five or ten-dollar bill, those gypsy fliers who did not starve to death or kill themselves soon began eying the illicit liquor traffic as a more lucrative and exciting way of life.

One of this breed was a lean, hawkish Texas barnstormer named Slats Rodgers, who started life in Dark Hollow, Mississippi, and knew little or nothing of polo, yachting or other "blue-blooded" activities of the social set to whom advertising copywriters mistakenly tried to sell an aviation boom.

Slats produced a book once, with a Texas newsman who helped him spell four-letter (and bigger) words, a colorful writer by the name of Hart Stillwell. To Stillwell's credit, the story of Slats Rodgers, called *Old Soggy* after a home-built plane Slats flew as a boy, is one of the more memorable accounts of early barnstorming. It has become a part of American folklore. The way Slats told the story to Hart Stillwell, satisfying the customers who wanted to get their kicks in his Jenny became increasingly difficult:

"They got to wanting loops and other stunts, and finally lots of them wouldn't go up if you *didn't* stunt with them. It suited the hell out of me. I

94

might have stayed with flying, I mean hauling passengers, if it had stayed the way it was. I was making money fast. But here came the gypsy barn-stormers from other parts of the country, horning in. It was easy to see that kind of money wasn't going to be around for many years. People would pay to go up once, twice—then they wouldn't pay any more. Before long you ran out of customers. So I kept in mind that I was hauling passengers mainly to build up the right kind of reputation for my ship. I wanted her to be a lady, above suspicion."

Slats, of course, had his eye on the Mexican border; there was cheap booze on the south side and booming oil towns on the north. Slats found he could pick up a case of Scotch for $20 in Mexico and get $50 a bottle for it from thirsty oil-field roughnecks any Saturday night.

Like Benny Howard, Slats soon found his Jenny the target of sniper fire, but this time the anti-aircraft batteries weren't manned by moonshiners, but by Texas border patrol officers who were quick to realize that Slats no longer was just an innocent passenger-hopper. He switched to night-flying, landing on remote dry lake beds lighted by flares made from wine bottles filled with kerosene.

Coming back, Slats improvised a unique cargo-drop that enabled him to make round-trip flights out of Mexico without running the risk of apprehen-sion in Texas. What he did was roll the bottles (or sometimes watches—

"Mein erster Bruch!" (My first crack-up!) World War I German pilot Max Holtzem walked away from this crash landing.

reputedly Swiss) in a sheepskin bedroll tied to the lower wing. He simply buzzed the drop zone and cut his cargo loose, right into the arms of the waiting customers.

One time Slats got greedy and paid the price; there was a big Saturday night celebration at an East Texas boom town and he was commissioned to bring in fifty gallons, a consignment worth a small fortune. Slats took a chance and lost; the Federals were waiting for him when he landed. They quickly tossed him in jail, locking up the liquor in a storeroom for evidence when he came to trial for rum running by air.

Waiting for his trial in the Dallas pokey, Slats once more convinced officials that he was simply a misunderstood aviator who had become the victim of circumstances, mostly economic. The jailer felt sorry for Slats and made him a trusty. Then—if you can believe Rodgers—he swiped the key to the storage room and made off with the evidence against him, selling it through the window bars to passersby at a profit.

Slats went on to become one of a group of barnstormers who settled down at Love Field, near Dallas, for the cold winter months and toured north when the geese began honking overhead. For a while he stunted in a fancy Travelair, preferring the life of a gypsy flier to the perils of bootleg flying. A favorite stunt of Slats was to bounce his wheels on the runway in front of the grandstand and then pull up through a loop. Several times he misjudged his distance, but each time he cracked up a ship, his fame—and price—grew proportionately.

Looking back at those good years, Rodgers remembered: "I guess we were a strange lot, those of us who flew those old traps every Sunday at the field. Maybe we were sort of a mixture of the cowhand of the Old West, the hot rod driver of today, and the real gypsy. We thought we were as free as the birds when we got into the air, just as the old-time cowboy thought he was as free as the coyote. We deliberately missed death by inches, and we played the sucker wherever we could find him, which meant roaming the face of the earth like a gypsy."

If Slats Rodgers fancied himself some kind of a heroic figure, he wasn't alone. A good many of the barnstormers affected a studied individualism that represented a kind of mass protest against growing regimentation, if you care to look at them in the same way that the leather-jacketed motorcyclists or beaded hippies of a later generation are viewed. True, the world then was not nearly so complex, but there was sufficient reason for rebellion, and rebel they did, many of them heading south of the border into the life of romance and love and music and new adventure they believed awaited them in Mexico, in Central America and in South America.

Frank Hawks, who had been a wartime flight instructor at Love, Taliaferro and Brooks Fields, took his discharge after the armistice and cleared out of the country, flying oil-field payrolls over bandit country in a Hisso Standard

Princess Elly Jonescu and Hans Geberth ride wings of Max's Salm biplane in this thriller.

on the Mexico City-Tampico run, along with Jimmy Angel, who had flown with the Royal Flying Corps.

Even before World War I, barnstormers had sought their fortunes in Mexico, when flying jobs were as scarce as hens' teeth in the United States. Some, like Mickey McGuire, flew for Pancho Villa during the Mexican Revolution, and others, like Didier Masson, flew for the *Nationalistas*. Earning up to $100 a day, they sometimes allowed themselves to be "shot down" by the other side so their buddies would have a ship to fly on retaliatory raids. It didn't make much difference which side they were on, for their homemade bombs, carried in their laps in flimsy wood-and-wire pushers, seldom did any damage.

Unlike the other barnstorming mercenary pilots, Hawks decided against joining in the prolonged Revolution, as he had many friends on both sides and didn't want any of them to get hurt on his account. So, until things quieted down, he temporarily gave up flying and worked as an oil field driller.

In 1921, under contract to the Mexican Government, Hawks put together a flying circus to help celebrate the national centennial. He hired Augie Pedlar, an expert mechanic and wing-walker, and played a leading role in making Mexico air-minded in its early post-revolutionary days.

Some years later another troupe of barnstormers, broke and hungry, found themselves stranded in Tampico. They didn't have Hawks' understanding of the Latin temperament and so decided to sell their four Lincoln Standard biplanes to a Tampico banker who saw an opportunity of using them for shuttle flights into the booming oil fields at Tuxpan, ninety-three miles across the green jungles. Thus was born on August 20, 1942, the oldest Latin-American airline, Compañía Mexicana de Aviación.

Hawks finally took a job with Mexicana and flew down to Tampico from Mexico City so many times he got to know every cactus on the route, but the day came when he covered the entire two hundred miles without one glimpse of the ground, on top of a solid overcast. On a routine trip, Hawks knew, there were rugged mountain peaks up to nineteen thousand feet high to climb over before starting to let down over the rain jungle of the coastal plain, but on this trip he had no way of knowing when it was safe to come down.

"I've faced death many times during the twenty years I've been flying," he recalled afterward, "but in thinking over all of those close calls, none of them could hold a candle to this one."

What haunted Hawks was that he felt sure he had picked up a tail wind, but he didn't dare drop down into the clouds until he was certain he was over the worst part of the mountains at Pachuca, normally a one-hour flight from Mexico City in his 75-mph Standard. For good luck he flew another fifteen minutes, then another five for insurance. He sucked in his breath, throttled back, and settled down into the blinding whiteness of the cloud layer, half-expecting a mountain peak to loom ahead of him at any moment.

Minutes passed and still Hawks was in the soup, getting lower and lower. Finally he saw something beneath the lower wing—water! He groaned. There was no telling how strong the tail wind had been. He could be as much as a hundred miles out over the Gulf of Mexico! Hawks quickly circled around and flew back over the course he'd followed, hugging the waves.

For what seemed like hours he nursed the Hisso along, saving fuel, until his engine began to sputter and threatened to quit. Straining his eyes to see ahead, he finally made out the distant shoreline. He landed at his home base with less than a gallon of gas left.

Jimmy Angel led an even more adventurous life during his Mexico barnstorming days. Like Hawks, he took a job flying payrolls over bandit country. He once told the author of this book the story of how a passenger in the front seat of his Jenny suddenly turned around and pulled a gun on him, five thousand feet over the Barranca de Cobre country.

"I'd brought him along as a guard, but I guess he got some ideas of his own," Jimmy recalled. "I was supposed to land somewhere down there, and I knew that the minute I did I was dead. He obviously meant to shoot me once we were on the ground, and take off with the gold. Well, it was his life or mine, so I rolled inverted and stuck my foot against the stick and shoved. The nose swung up hard. He popped out of the cockpit, bounced off the wing, and the last I saw he was grabbing air on his way to eternity."

One time Jimmy faced a ticklish situation that called for some fast thinking: a revolutionary leader tried to hire him to bomb the palace in San José, Costa Rica. Jimmy's pretty young bride, Marie, was staying in the hotel next to the palace, waiting to catch a plane back to the United States the next day. If he turned down the job, some other pilot would get it and could easily blow up the hotel by mistake, so Jimmy agreed to fly the mission.

Stalling as long as he could, he tinkered with the engine, cleaning the plugs and fooling with the carburetor, until the revolutionary began getting suspicious. Finally Jimmy told him he'd have to wait until dawn, which would be better because everybody would be asleep and that way they could more easily wipe out the opposition.

"*Bueno!*" snapped the leader, stalking off to attend to other matters. The next morning Jimmy was off the ground at first light, smiling to himself; during the night he'd got word to Marie to clear out in a friend's airplane. He buzzed over the palace, where President Teodoro Picado was sleeping, unaware how close he'd come to death. Jimmy zoomed off into the wild blue and headed south to continue his adventurous barnstorming career in Panama and Venezuela, where he operated the Jimmy Angel Interamerican Aviation Service.

His subsequent search for a lost gold mine atop Auyán-tepuí, an awesome mesa in Venezuela's Gran Sabana country, is the tragic story of a man who followed a dream unto death, the elusive prize always just beyond reach. Briefly, it is the story of an old prospector named Bob Williamson who showed Jimmy a hatful of huge gold nuggets and offered to share his secret if Jimmy would join him.

Jimmy landed his plane atop the distant mountain at Williamson's direction, and sure enough, there it was—a mother lode!

Before Jimmy could go back and set up a claim, the old prospector was killed in a Panama waterfront fight, and never again could he locate it, although he spent more than two decades trying. He did find a magnificent waterfall, however, the highest in the world (3,280 ft.), which today bears his name—Angel Falls. In 1956, he was off again with new backing, heading back to Venezuela with fresh hope of finding his lost mine. Instead Jimmy found death, in a simple accident in Panama, which occurred when he tried to land in a stiff crosswind.

In 1968, a renewed effort to discover the secret of Auyán-tepuí Mesa was made with the organization in Los Angeles of the Lost World Scientific Expedition, headed by two veteran parachuting barnstormers, Dave Burt and James C. Hall, and an associate, George Waltz. This time the search was conducted by helicopters and with the more scientific goal of discovering new flora and fauna in this region made famous by Sir Arthur Conan Doyle's book, *The Lost World.*

His middle name was Orville and the Arapahoes called him Tall Feather, so it was natural that Edward O. DeLarm, a lad from the Big Wind River Reservation in Wyoming, should become the first full-blooded Indian birdman and end up in a real wild west adventure, as the first barnstormer to invade remote Patagonia in South America.

In Redwood City, California, DeLarm qualified for his pilot's license by flying with Sy Christofferson, a wartime builder of Jenny trainers. He then worked his way up as a respected test pilot for several aircraft manufacturers. After the Wall Street crash in 1929, DeLarm found opportunity wasn't knocking loud enough to do a drifting pilot much good, so he set out on the day before Christmas of that year on one of the longest overwater flights in history to seek his fortune below the equator.

Piloting a Sikorsky amphibian on a delivery flight from Roosevelt Field, New York, DeLarm winged southward to Cuba. There he gassed up and continued on to the Greater Antilles via Haiti, Santo Domingo, Puerto Rico, the Virgin Islands, the British West Indies, Trinidad, and from there up the Amazon to Para, Brazil, and across Uruguay to Buenos Aires. The flight blazed a trail that would open regular passenger and mail service, but DeLarm's Indian blood was restless, and on September 20, 1930, he moved on to greater adventures, taking off from Buenos Aires with six passengers on a 1,100-mile charter flight across the continent to Concepción, Chile.

DeLarm was amazed to read in the logbook that his craft was the same Fokker F-7 in which Amelia Earhart had winged across the Atlantic in 1928 as a passenger with Wilmer Stultz and Louis Gordon—the *Friendship*. It was then owned by one of DeLarm's passengers, all of whom represented themselves as just a bunch of homesick Chileans. What the pilot did not know was that they were armed revolutionaries plotting a takeover of the Chilean government, and that their baggage, 1,100 pounds of it, was mostly firearms and ammunition.

No sooner had they landed at Concepción than a small army surrounded the amphibian. This, it seemed to DeLarm, was not in the best tradition of welcoming committees, particularly for a flying machine named the *Friendship*. He was even more upset when they marched him off as a prisoner to await the outcome of a trial of his passengers for attempting to oust forcibly the incumbent government the hard way. They were found guilty, of course, and sentenced to fifteen years imprisonment in a leper colony on a Pacific island, there being no death penalty. Later they escaped to France.

DeLarm, considered an accomplice to the others in the dastardly plot, was held on a battleship during the trial and was later transferred to a military prison. He remained there for thirty-seven days, until one night when he noticed that the guards were celebrating heavily. When one fell into a drunken slumber, DeLarm lifted his keys, opened the doors and walked out to freedom. For the next thirteen days he made his way across the Andes, following uncharted trails and living off the land. He finally reached Zapala, Argentina, and hopped a train back to Buenos Aires, reminding himself to check the passenger manifest more carefully on his next charter flight.

One of the most remarkable barnstormers in the entire Western Hemisphere was a self-exiled German national named Max Holtzem, who came to South America after the armistice for a simple but compelling reason—he loved flying, and under the terms of the Treaty of Versailles, flying was *verboten* in Germany.

He had done pretty well, as combat heroes go, as a pilot with the Bavarian Jagdstaffel 16, but even before that Max had been making aviation history as one of the real pioneers of flight in the Kaiser's conservative empire. A youth of twenty in 1912, Max built his own airplane, which resembled a Blériot monoplane but was underpowered, having a Delfosse rotary that could barely lift it off the grassy fields near his home in Cologne.

Against the wishes of his parents, Max enlisted with the Imperial Army Air Service in 1913, and with the outbreak of war the following year was assigned to a Tauben squadron, flying over the front lines in Luxemburg and France. "Dueling in the air in those days was unknown," Max related to the author, "and the Huns and poilus saluted each other whenever they passed each other in the air."

In 1916, Holtzem wrecked so many ships that Max's fellow pilots were wondering which side he was on. Here's what happened when a loop in a Pfalz didn't work out.

COURTESY MAX HOLTZEM

To liven things up, Max decided to play a trick on his friendly enemies. "I attached an extension pipe to the end of my Very signal pistol, and as I passed the next French airman I showed him the profile of my 'gun' and fired it. The smoke and fire scared him immensely, to my great amusement." He was less amused the next day when the French pilot took a pot shot at him—with a real gun.

Because he showed great promise as a pilot, Max was commissioned as a test pilot at the Pfalz Pursuit Aircraft Factory at Speyer, where with Gnome rotary engines, the Germans were turning out carbon copies of French Morane-Saulnier monoplanes.

By 1916, with the factories humming along, the German Air Force needed pilots more than it did planes, so Max was transferred again, this time to the sole Bavarian air base, at Schleisheim, near Munich, as chief (and only) instructor.

"I alone was authorized to take up a brand-new model, the 14-cylinder, double-row Gnome Pfalz single-seater," says Max. "This aircraft, with its castor-oil burning rotary 'Schnurpser' engine, was a very special ship, to be used only for the defense of Munich."

When the Grand-Inspector of the Bavarian Air Corps arrived in his flashy car one day to inspect the hot little Pfalz fighter plane, Max spotted in the car a beautiful dashboard clock that he decided would look just great in the Pfalz panel, so while the visitors looked over the Pfalz, Max removed the clock. The Grand-Inspector raised hell when he discovered the theft, but after things cooled down, Max installed his prize in the cockpit, quite proud of himself.

He wasn't so proud the day he told his student pilots to watch from the ground while he demonstrated a low-level wingover in the single-seater Pfalz. He took off, climbed high over the snowy Bavarian countryside, then came buzzing along a road toward the flying field and pulled up sharply in a climbing turn. The engine misfired at the critical moment, and instead of diving back the way he had come, under control, the Pfalz shuddered, shook and spun in.

"I *walked* to the ambulance," Max recalls proudly.

As an acrobatic instructor, Max first had to teach himself how to loop and roll and do falling leafs before demonstrating the stunts to his students, many of them mustachioed Turks who flew airplanes the way they rode horses—full speed ahead. "I got cold feet the first time I tried a loop," he remembers. "I went into a bad tailslide—the Pfalz had no stabilizer or fin—and then I tried it again, successfully. I did twenty in a row."

Switching to the Albatros, Max adopted a personalized insignia on the side of his ship calculated to scare the enemy to death—a blazing comet with the name HOLTZEM slashed across the top. Back in combat for the long-awaited German offensive in March of 1918, Max was very discouraged to see his commanding officer, Lieutenant Heinrich Geigl, a hero with twenty-one kills, ram his Albatros head-on into a Sopwith Camel. "The British plane disintegrated, while the Albatros seemed unharmed. Then it began a steep turn, and one wing fluttered off before it plunged to earth at terrific speed."

Max didn't kill anybody in the shooting war, not even himself, but he did frighten the wits out of the good burghers of Cologne two days after the armistice by leading a flight of three Jasta 16 Fokker D-7's back from the front to his old home town.

"Down there I saw the beautiful spire of the Cathedral, the bridges, all the landmarks I knew and loved. They had come through the war unscathed, but now everybody thought Cologne was under attack!"

Max led the formation in trail over the rooftops and beneath the bridges that spanned the Rhine, then high-tailed out to the airport and landed, returning by taxi. It was a glorious flight, and the last he ever made in Germany. "The end of the war brought an abrupt end to all flying for German aviators," Max recalls. "But I had different plans. Because I loved flying and struggled so hard to become an aviator, I could not bear to give it up. My decision was to leave Germany if it meant that I could still fly. I went immediately to Argentina, in South America, and there flew at the military airdrome, El Palomar, near Buenos Aires."

Max was once again his old self, though he admits, "I was detested by some members of the French Mission who had come to the Argentine to sell their old war planes." There also he met Lawrence Leon, an American representing the Curtiss factory, "who was showing the Argentineans the beauty of his high-priced, but low-powered Jennies."

Max went to work building up a name for himself as a crack acrobatic pilot and formed South America's first flying circus, the Circo del Aire. It was the start of a whole new career for him, very much in the same way that returning Allied fliers had gone into barnstorming in the United States and Europe. And for a while he had the continent virtually to himself, or at least the sky above it.

Max did not get the idea for the Circo del Aire from the notorious Flying Circus of Baron Manfred von Richthofen. "The Baron prohibited aerobatics," he said. "He was a straight killer. He never once made a looping." Max did, though, often with a pair of fearless wing riders sitting on top of his Salm biplane, which was a copy of a German trainer built in Argentina and powered with a 100-horsepower Rhone rotary. One of his stunters was Eugen Geberth, a destitute wartime buddy who followed Max to South America and joined the show. The other was Princess Elly Jonescu, a Romanian beauty and multilinguist who had found life boring in postwar Europe.

From 1921 until death parted them, Holtzem, Geberth and Princess Elly roamed from *ciudad* to *ciudad* and were billed by the Latin press as *Las Suicidas*. "People came from miles around to see us killed," says Max. Until Geberth finally obliged them by slipping and falling from a monoplane near

Max (right) didn't mind getting his hands dirty. Hans Geberth (left) and Princess Elly Jonescu, were wing-walkers. Henri Roger (next to Holtzem) was his "impresario."

Princess Elly Jonescu was Max's girl wing-walker.

Santiago, Chile, in November, 1924 (Max wasn't flying him that day), the Circo del Aire held a perfect record for aerial lunacy.

Most of their performances were staged at enclosed *sportivas* and race tracks, and one of Max's big crowd-pleasers was the old car-to-plane change that originated with the Fox-Movietone Newsreel stunt team, the Thirteen Black Cats, up in Hollywood. Because the cars had difficulty getting up speed to match Max's flying machine, he tied sandbags to the bottom of a ladder and swept low overhead, trusting to the stuntman to leap and catch it on the fly. Now and then one walloped the car driver on the head.

Max much preferred doing simple loops with Princess Elly and Geberth on the wingtips, because the ship was better balanced that way. These maneuvers were not done high in the sky, but virtually at ground level so that the spectators had to pay to get inside and watch Elly swooping past the grandstand below the roof tops.

Max did have trouble with Geberth, who insisted on using a long ladder on car-to-plane changes because the Gnome covered him with oil on a short one. On skidding turns, Geberth swung far out from beneath the landing gear until he could almost grab the wing skid. Once Geberth came close to having the ladder sawed in two by the propeller.

After Geberth's death, Princess Elly drifted off to seek new adventures, and in 1928 Max headed north to accept a job as a test pilot for his old friend Tony Fokker at Teterboro, New Jersey. Within two years the Depression had made it impossible for barnstormers to make a good living, and after a final show in Newport, Rhode Island, Max retired from active flying and moved to California to settle down. Now in his sunset years, he enjoys thumbing through several scrapbooks or threading a movie projector with a flickering old newsreel sequence of the Circo del Aire swooping low over an Argentine race track, back in the days when flying was so different. . . .

10/HERE COMES THE CIRCUS!

THE NEXT TIME you board an airliner, sit back in your comfortable, cushioned seat and settle down to watch a first-run movie or listen to music via the stereo sound system, pause a moment to reflect how it all began, back in the days of the barnstormers.

"Ladies and gentlemen," your airline captain will say over the closed-circuit speaker, "we are about to depart for a brief flight at an altitude of forty thousand feet. The weather is clear all the way, and we hope you enjoy your flight."

On May 25, 1924, hundreds of people traveled out to the airport in Lincoln, Nebraska, where Daredevil Lindbergh had learned to fly two years before. Gaudy posters had fluttered over town announcing a stupendous air show featuring races, stunt flying and parachute jumping. Besides the Lincoln Standard biplanes, five Army De Havillands from Fort Riley, Kansas, were there to introduce folks to the sky.

In that single afternoon, to help launch air travel, three Standards took 408 people into the air for the first time at three dollars a head. One pilot, Bob Cochrane of Casper, Wyoming, hopped 184 people. "At no time during the afternoon," reported *Aviation* magazine, "was there a lull in obtaining passengers. The new price of three dollars for five minutes seemed very popular, because in a number of instances the same passenger remained in the ship on landing and rode as often as four times in succession."

Much of this nationwide air-mindedness was traceable not to ex-military gypsy fliers who roamed the countryside scaring the wits out of passengers, but to astute promoters of flying circuses who realized that the stuntmen were only up there to attract attention. Once the yokels got to the airport, the job was to entice them into the sky for money.

Iron Hat Johnston was featured performer at Air Pageant with Colonel Roscoe Turner, Gladys O'Donnell and Hollywood Trio.

Slats Rodgers used to be highly amused at the way his passengers worked up enough nerve to get off the ground for the first time: "Here would come some middle-aged wife traipsing behind her middle-aged husband, and they would fight anyone who wanted to horn in ahead of them, but they were scared to death all the time.

"About the time it was their turn, the man would want to back out. But the woman would argue him into it. Almost every time it was that way—the man was the one that had to be talked into going up. The woman did the talking."

Perhaps the one person who talked more people off the ground than anyone else was Ivan R. Gates, an old-time promoter of exhibition fliers before World War I, who originated the great Gates Flying Circus that air-lifted more than one million people between 1922 and 1928. Among Gates' early star performers had been Didier Masson, who later flew in the Mexican Revolution and with the Lafayette Escadrille in France, and Art Smith. The latter was a spectacular stunt flier who thrilled early air show audiences with pyrotechnic displays on night flights, during one of which he almost blew the tail off his pusher when his mechanic accidently tied giant firecrackers instead of flares to a strut.

Recognizing the great potential of aviation during the early postwar years, Gates teamed up with two West Coast barnstormers, Clyde Pangborn and Lowell Yerrex, and toured virtually every state in the nation. Lieutenant Pangborn, who had been a flight instructor at Ellington Field near Houston, Texas, was a careful pilot, and during his association with Gates hopped more than 125,000 passengers without a serious mishap.

Originators of the dollar ride, the Gates Flying Circus flew five-place Standards with Hisso engines and steel ladders bolted to their sides so passengers could be gotten in and out faster. One Gates pilot, Bill Brooks, broke all records by flying 980 passengers in a single day at Steubenville, Ohio! To do this, Brooks maintained a remarkable marathon schedule, taking off, circling the field and landing to empty his ship and take on a fresh load of passengers being herded through like cattle by the cigar-smoking Ivan Gates.

Passenger flying was big business to Gates, who paid his pilots up to $1,500 a month in order to get the best qualified men. It has been said that the Gates Flying Circus turned out more famed pilots than the Army and Navy put together.

There was excitement too, along with the big push to get passengers airborne: there was the time a movie actress named Rosalie Gordon agreed to make a jump from Pangborn's ship with an old pull-out chute, but it got hung up in its container lashed between the wheels. Milton Girten, who was riding with them, clambered down to pull Rosalie back up, but the force of the wind was too much. Freddie Lund, also a wing-walker, swung over from another ship onto Pangborn's wing and crawled down to give Girten a hand. No luck. Both Girten and Lund were small men, agile but not muscular. So Lund, who was also a pilot, climbed back up and motioned Pangborn to go help Girten while he flew the ship. That way they did get the poor girl back into the cockpit, and all four landed safely, making more headlines for the Gates Flying Circus, of course.

The first man to rig his Jenny for inverted flight, Pangborn once piloted it for a distance of two miles upside down, and after that he painted his name on the top wing. After the Gates Flying Circus broke up in 1928, Pangborn formed his own circus, the Flying Fleet, which came before he launched a new career as a world flier.

Almost as well-remembered as the Gates team of barnstormers was the Doug Davis Flying Circus, a group of deep-South gypsy pilots who banded together in 1924 when they realized their profits could soar if they put on

Airline pilot Harold Johnson flew with National Air Show on his days off. Here he loops a Ford Trimotor at ground level.

Eleven Thousand Pounds Starts a Loop

a full-size air show instead of each doing a solo act. On their very first performance as a team, they took in $1,400 from the well-satisfied citizens of Opalaca, Alabama.

Davis, a wartime pilot trainee who came close to washing out a number of times at Kelly and Brooks Fields in Texas, started barnstorming in Atlanta, flying his clipped-wing Jenny from Chandler Race Track, today the site of one of the nation's busiest airports. When a competitor, Beeler Blevins, moved in, Davis moved out and began working the back country, where he introduced aviation to untold hundreds of country folk.

The Doug Davis Flying Circus stormed the countryside for a full year before running into fresh competition, this time in the form of a pretty Birmingham, Alabama, girl, Mabel Cody. Mabel was the first lass to risk her neck in a speedboat-to-plane change and was so well liked that she decided to organize her own show. This became the Mabel Cody Flying Circus, naturally, and it gave Doug Davis a headache. Whenever he led his squadron of war-surplus ships into a Southern town, it seemed that Mabel's troupe had already been there, skimming off the cream.

Mabel's pilots included a former Navy airman named Slim Culpepper and a wing-walker named Bonnie Rowe, who frequently had to put his parachute in hock. When it came time to put on his jump act, Bonnie would often have to sit up all night in a hotel room sewing bed sheets together for a makeshift canopy.

Curly Burns, Mabel Cody's promoter, was the first to see virtue (and profit) in billing the troupe's pilots as war heroes. Passengers paid more money to ride with this hero-star—whoever he happened to be on a particular day. So well did Mabel Cody's outfit do financially that Doug Davis flew over one day and tossed her a note in a bottle, suggesting that they bring the two shows together into one gigantic enterprise. Mabel turned him down cold. In addition, she began flying into towns Davis had already circularized, playing one-day stands to catch the bulk of his business.

There is the story of how the two flying circuses met one day over a small town and for a full hour fiercely battled each other all over the sky. True or not, there was intense rivalry until the Doug Davis show, adopting Mabel Cody's own tricks, began working itineraries she had already pitched with full page newspaper ads. Finally the two did join forces, and with a new sponsor, the Curtis Candy Company, were known henceforth as the Doug Davis Baby Ruth Flying Circus.

With the coming of the Depression years following the 1929 stock market collapse, the era of the gypsy flier drew to a close. Already, the Air Commerce Act of 1926 had shot down many of the baling-wire brigade of barnstormers who had not yet killed themselves in machines that couldn't come near passing an airworthiness inspection. It was a time for overhaul, for a new approach to country aviation, with some semblance of sanity in flying and

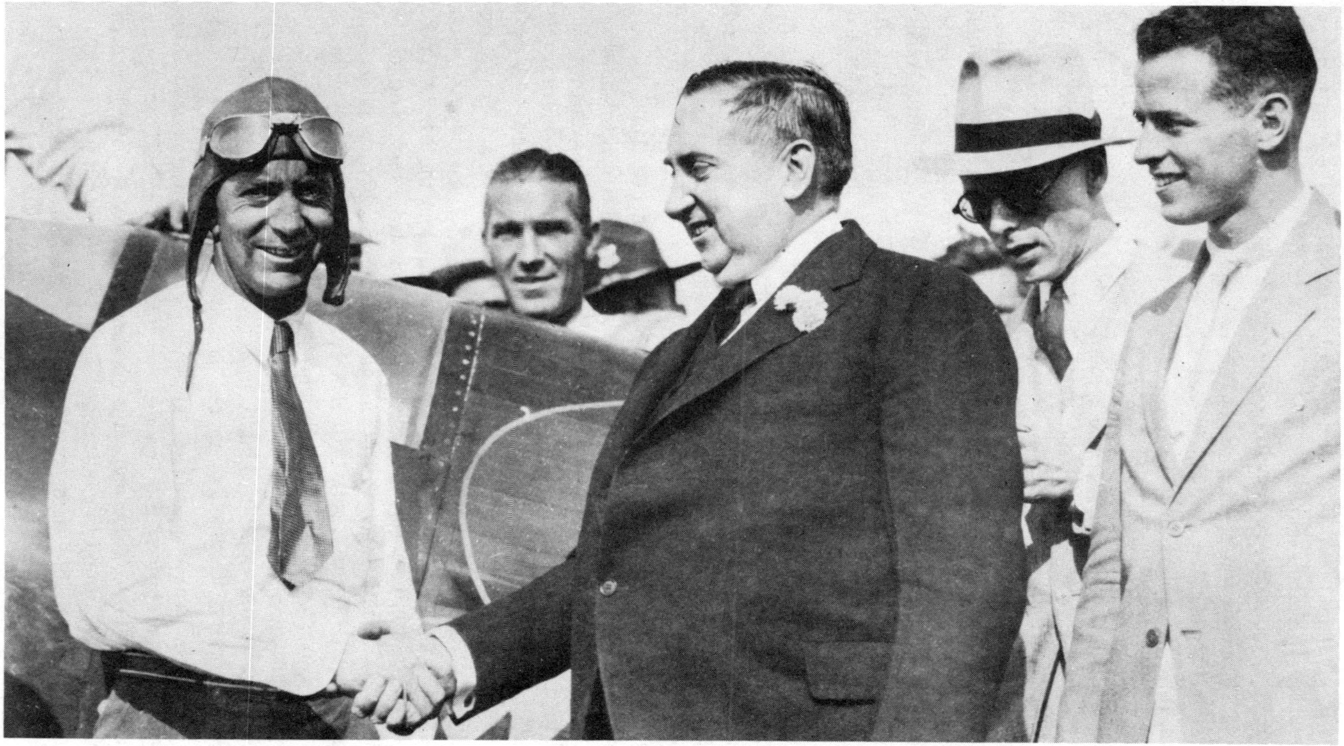

BENDIX CORP. PHOTO

Doug Davis, head of Doug Davis Flying Circus, gets handshake from Vincent Bendix after winning 1934 Bendix Race.

licensing of plane and pilot to protect the public as well as the barnstormers themselves from needless tragedy.

Before we look at the new breed of barnstormer, the pilot in trim uniform who would give aviation a much-needed air of respectability, let's acknowledge one last time those old-bold heroes who soon would be flying west to vanish from the American scene.

There was Loxla Thornton, a heroic Alabamian who lost both arms in a railroad accident and yet barnstormed a Jenny with the aid of a steel forearm and hand and a versatile right shoulder. Strapped into a cockpit by his mechanic, he carried hundreds of passengers without accident until Department of Commerce inspectors grounded him. And there was Charles "Smiles" O'Timmons, a one-legged, one-armed parachutist remembered by aviation historian Russ Brinkley, who once got Smiles a job wing-walking in a show when the regular stunt man failed to show up. All went well until Smiles' artificial leg went through the Jenny's wing, fouling the control cables. In desperation, he pulled off his pants, unbuckled the leg and dove for the cockpit, to the great hilarity of the audience.

There was Major R. W. "Shorty" Schroeder, noted ex-Army pilot, who barnstormed a while before becoming chief pilot for Henry Ford and later a United Air Lines executive. And there was Lawrence Fritz, another World War I veteran who went broke barnstorming and became chief pilot for Bill Stout, designer of the Ford Trimotor "Tin Goose."

Colin "Boots" LeBoutillier, who never used his given first name, Oliver, went into barnstorming as a Royal Air Force hero who had helped to shoot down the famed Red Baron, Manfred von Richthofen. After the war he then drifted west to Hollywood to become an early member of the wild bunch known as the Motion Picture Pilots Association. Boots flew with Paul Mantz and Frank Clarke as wingman with the famed Hollywood Trio stunt team, which etched graceful patterns across the sky with special red, white and blue smoke LeBoutillier had invented. Others to join the MPPA ranks from the barnstorming circuit, besides Clarke, included Frank Tomick, a wartime March Field flight instructor, Ira Reed, an ex-Navy pilot, and "Colonel" Roscoe Turner, who won his commission as personal pilot of California's Governor "Sunny Jim" Rolph.

Colonel Turner, one of aviation's most colorful figures, was among the first to recognize the value of a natty uniform to impress customers, particularly the ladies. Resplendent in red jacket, green whipcord breeches, riding boots and waxed mustache, Turner looked exactly the way folks thought a pilot should look. He flew the way they liked, too, well enough to win the Bendix Trophy twice and set a number of city-to-city speed records nobody has gotten around to challenging.

Turner first went into barnstorming on borrowed money and with the rank of Lieutenant in the Air Service, and from the start knew how to please the crowds well enough to earn as much as $1,000 a day performing such spectacular stunts as "Falling a Mile in Flames"—a vertical dive performed with a smokepot. Teaming up with another barnstormer named Arthur H. Starnes, Turner formed the Roscoe Turner Flying Circus, which guaranteed to perform:

ONE WING-WALKING SHOW, INCLUDING THE
"SWING OF DEATH"
A PARACHUTE JUMP
AN AEROPLANE ACROBATIC FLIGHT, LOOPS, SPINS,
WINGOVERS, WHIP STALLS, ROLLS

Later on, Roscoe bought an eighteen-passenger Sikorsky, the first built in America, and for several years made broadcasts from the sky, staged aerial pink teas for society women and carried charter parties wherever they wanted to go. In Hollywood, Turner leased his Sikorsky to moviemaker Howard

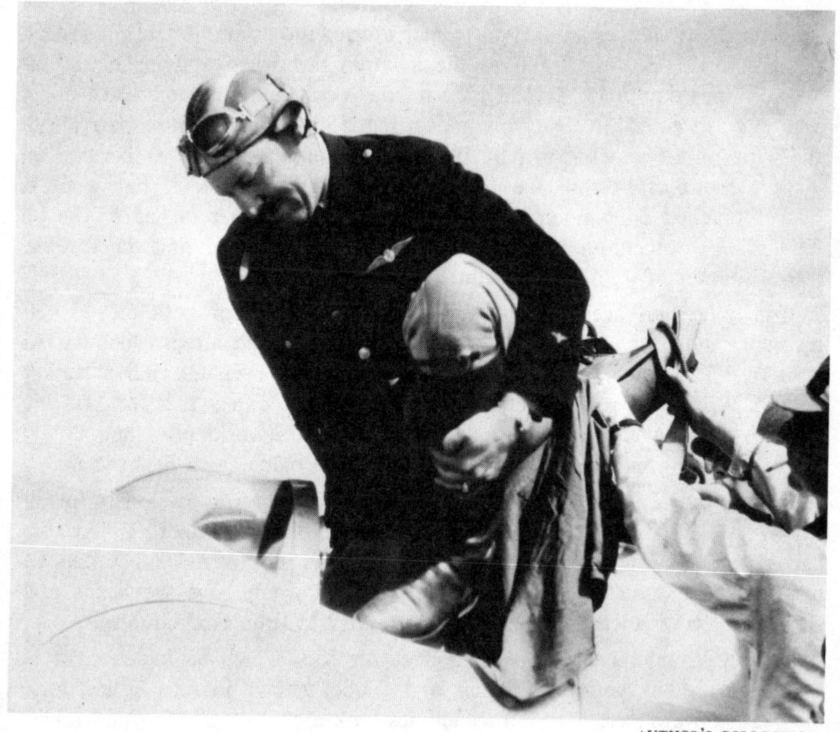

AUTHOR'S COLLECTION

Colonel Roscoe Turner wore fancy uniforms with wings on chest, sleeves, collar, belt. He was a top racing pilot.

Hughes, who had it painted to resemble a German Gotha bomber for the great war film, *Hell's Angels.* If you saw the picture, maybe you remember the thrilling scene in which the Gotha, trailing smoke, spins to earth and crashes. Al Wilson, the pilot, found himself in a real, not a make-believe crisis when he heard a wing spar snap. He bailed out and lived, but a prop man named Phil Jones, operating the smokepot in the rear of the craft, failed to jump and was killed.

Other barnstormers went on to become aircraft company executives: Carl Squier was the first president of the Lockheed Aircraft Company; Alan Loughead, the founder of Lockheed when it was a small Santa Barbara, California, firm; and Tubal Claude Ryan, founder of the Ryan Aeronautical Company, who used one of Donald Douglas' Cloudsters to fly beer across the border from Mexicali to San Diego.

Enterprising Cliff Henderson, who staged the famous Cleveland National Air Races in the 1930's, bought three brand-new JN-4Ds in their crates, traded one to barnstormer Fred Kelly in return for a short course in flying and used the others to hop passengers who had been guaranteed a free ride with every car they bought from Henderson's Nash agency, a sideline business.

Wiley Post, another great barnstormer, started out as an oil-field roughneck in Oklahoma. He was bitten by the aviation bug when a flying circus hit Wewoka one day in 1919. When the regular parachute jumper broke his leg, Wiley volunteered to take his place for $50, leaping into space from the wing of a Canuck flown by Berl Tibbs. For a while Wiley became an airport tramp, following the barnstormers on their circuits, but when a gusher touched off a boom at Seminole, Oklahoma, he returned to the oil field. There something happened to change his life—the first day on the job, a splinter of steel blinded him in one eye.

Ironically, that accident got him back into flying, for with the $1,800 insurance money he bought a wheezing old Canuck and hired a flier named Sam Bartel to teach him to fly it. Wearing a black eyepatch and a leather helmet, Wiley Post was launched on a barnstorming career that would bring him world fame as a high-altitude flier, around-the-world pilot and finally as the man who met death with Will Rogers in a plane crash in Alaska.

One of the last, but not the least colorful, of this army of loners in the sky was Lieutenant Ralph Vaughn, who was one of Arizona's first two barnstormers. Vaughn and Charlie Mayse had discovered an untapped resource—the Apache Indians *loved* flying and would give the beads and silver trinkets off their backs to go up for a visit to Rain God country.

At first Vaughn operated with a Kreutzer K-3, which he had bought to set up an air link between Phoenix and Globe, a town in Apache territory. To drum up business, Vaughn sometimes took along a pair of wing-walkers and parachute jumpers, Suicide Slim and Reckless Rosie. In the 1930's, Vaughn bought a highwing Stinson R monoplane and until his retirement

Will Rogers (left), the great humorist, was favorite of barnstormers like Tex Rankin.

worked the Indian country throughout Arizona, Colorado, New Mexico and Utah with great success.

One by one, death thinned the ranks of the remaining independent barnstormers, who toward the end of the Roaring Twenties were finding the skies too crowded with competition or, in order to keep eating, tempted fate just once too often. There was Second Lieutenant Ormer Lester Locklear, a World War I flight instructor at Barron Field, Texas, who one day in luckier times was flying along with a student when his radiator cap blew off. Casually, Locklear clambered up on top of the Jenny's center section and stuck a rag in the hole to keep the scalding water from blowing back into the cockpits.

Other Army instructors at Barron heard about this and soon were trying to outdo each other with wing-walking stunts that petrified their students. Finally Locklear and two other mentors, Lieutenant Milton "Skeets" Elliot and Lieutenant Shirley Short, formed an exhibition team and went barnstorming, following Locklear's honorable discharge on May 5, 1919.

In January of that year, Locklear had made the first public plane change

117

These nine shots show the great Ormer Locklear at work. A former Air Corps instructor, Lieutenant Locklear was a nerveless wing-walker. He finally lost his life in film stunt plane crash.

from one moving craft to another, over Love Field, Dallas. He then drifted
west to become one of the madcap motion-picture pilots and stuntmen,
until his death on August 2, 1920. Locklear was flying, with Skeets as a
passenger, for a big scene for the movie *The Skywayman* that called for a
night tailspin over Cecil B. DeMille Field, which was surrounded by blazing
searchlights. Frank Clarke, who was there, remembered: "The cameras
caught the silver wings as they twisted groundward. It was a beautiful scene,
a dramatic shot. It seemed as if Locklear were outdoing himself in daring.
And then in one minute the make-believe of pictures changed to tragedy in
reality. . . . The plane hit the ground with a roar and shattered to bits. The
mangled bodies of Locklear and Elliot were taken from the twisted mass.
The battery of lights had been too strong and had blinded Locklear."

The wreckage, of course, was snapped up by Arrigo Balboni, the flying
junkman, who added it to his collection of plane carcasses on Riverside
Drive in Los Angeles. Balboni, who was born in Renazzo, Italy, and who
learned to fly with the United States Army Air Service, once barnstormed a
Jenny all the way around South America, up the East Coast of the United
States, and finally into Alaska, before settling down. By the 1940's, with a
new crop of Air Force pilots cracking up ships, Arrigo opened a bigger
airplane junkyard closer to March Field, at a desert location he named
Balboni, California (population 1).

Curiously, although perhaps one-tenth of America's population of a hundred million in the 1920's was introduced to flying by barnstormers (some esti-

120

mates run to twenty per cent)—a far better record than the jet airlines can claim today, as a matter of fact—people were slow to accept scheduled flying, in the latter part of that hell-for-leather decade, as anything but a blue-sky dream. In 1926, a reporter for the Pekin, Illinois, *Daily Times* interviewed Dale Seitz, a barnstormer whose Safe and Sane Flyers troupe was operating in a field at the edge of town, near the Quaker Oats plant, and was told: "The time is rapidly approaching when any city that wishes to hold its own must have a municipal airport." Flying, Seitz assured the readers, "is a great experience, unlike anything else in the world. It does not make one dizzy, like standing on a tall building; rather, it affords a view of the city, rivers, lakes and fields that far surpasses the best maps ever made."

The next year, Seitz's Safe and Sane Flyers were cashing in on the Lindbergh boom, advertising, "Lindy crossed the ocean! You can see South Bend and your home from the air—enjoy a safe AIRPLANE RIDE!"

The Safe and Sane Flyers, who flew Wacos, Standards and other early OX-powered biplanes, naturally had their own stunt people, like "Daredevil Joe" LeBoeuf of Kankakee, Illinois, billed as "the world's oldest parachute jumper" who had been "diving out of airplanes and balloons for thirty-three years."

Dale Seitz' emphasis on flying safety was an indication of changing times. The image of the heroic gypsy flier was passé, and in his place was the crisp, efficient airman, a man you could trust. One of the first flying circuses to adopt

Dale Seitz' Safe and Sane Flyers barnstorming troupe brought aviation to grassroots communities in Midwest.

With handbills and newspaper advertisements, the Safe and Sane Flyers circularized towns, luring crowds to the performing fields.

Pilots Bert Brown and Ted Fordon organized successful National Air Show, adopted fancy military uniforms, hired best-known pilots on barnstorming circuit.

this new image was the Fordon-Brown National Air Show, which boasted it was the only troupe of fliers to hold a Class One Bureau of Air Commerce waiver and $100,000 liability insurance with a property damage clause, in the event one of their pilots flew into the grandstand or hit a barn.

Created by two fliers named Bert Brown and T. N. Fordon, the National Air Show scouted the countryside for the most popular stunt fliers, hiring men like Milo Burcham, Dick Granere, Harold Johnson, Joe Jacobson, C. W. "Flash" Whittenbeck and Buddy Batzell, a parachute jumper. "The management," they announced, "has its fliers measured for special uniforms. The appearance of these ace fliers is neat. British-type uniforms set off with Sam Brown belts sound the death-knell of the oil-spattered, overalled fly-by-night aerial exhibitionists of the postwar barnstorming days. This is a new era in aeronautical entertainment."

"At the sound of an exploding bomb," they advertised, "a band strikes up a military air, leading the parade past the reviewing stand. Mounted state troopers add color and dignity to the occasion. Directly behind the band marches the troupe of world-famous fliers which is the Fordon-Brown National Air Show. The band swings into the National Anthem, a bomb bursts overhead, a flag floats slowly to earth. And as the music fades away, another bomb explodes, an airplane roars down the field and surges aloft, and the show is on!"

And what a show! Instead of OX-Jennies, there was Flash Whittenbeck plunging through outside loops in a Great Lakes biplane, Milo Burcham snapping inverted in his Boeing P-12 fighter (how Burcham got it for civilian use he won't tell) and Joe Jacobson acrobating in his little white Howard racer (he takes it up to seven thousand feet and dives at the earth, engine wide open; down he comes at close to four hundred miles an hour, and then a few feet above the grass Joe pulls out in a screaming zoom!).

The finale was Buddy Batzell's grand parachute drop from the dizzying height of fourteen thousand feet: "After the first excited cry of recognition the crowd is deadly quiet. His body picks up speed, to 140 mph. Twelve thousand, ten, eight, six, four, two—will he *never* open his chute? Still he falls. Then there's a sharp crack! At that low altitude you can hear his twenty-eight-foot silken canopy whip open. He has fallen two miles in practically nothing flat. The show is over."

Not all barnstormers of the 1930's were out after speed records and passenger dollars. On September 14, 1933, a West Coast pilot, F. Myrten (Iron Hat) Johnston, won the dubious distinction of official recognition by the National Aeronautic Association for the nation's slowest plane flight: he averaged 37.008 mph over a 98-kilometer course in an Aeronca "flying bathtub" fitted with floats.

Johnston was an air show favorite with his skilled handkerchief pickup with a wingtip, takeoffs from the top of a touring car and endurance flights in which he grabbed five-gallon cans of fuel from an aide in a speedboat or automobile.

And then, finally, came the barnstorming transports—big, three-engine Ford "Tin Gooses," Fokkers, and Boeing Trimotors. People by the thousands began taking to the air, not to go somewhere but to get off the ground, up into the sky, just to ride around and look at the beauty of the country spread out below and realize they were in on the start of something marvelous, a revolution that some day would make air travel as safe and convenient as riding a train—and faster!

Brigadier General Leslie G. Mulzer, an airman who spent three years with the Strategic Air Command and served as 15th Air Force Commander at Colorado Springs, home of the Air Force Academy, was among those who came up from Jennies to Wacos and finally to Ford and Stinson Trimotors, as part of a glorious career "flying anything that would fly."

Howard Fisher Maish, holder of Transport Pilot License No. 68, was another ex-World War I barnstormer who graduated to Ford 4-ATs in 1930, barnstorming 175,000 miles across the United States, Canada, and Mexico for two years, making people airliner conscious.

One of the last to add glamour to barnstorming in the Trimotor era was Ben Gregory of Kansas City, who rigged his three-engine Ford transport with giant searchlights and with neon tubes outlining its great tin wings. This beast he called the *Ship from Mars,* and as he looped and rolled it through the sky, he pulled a lever that squirted kerosene into the exhaust stacks. Despite the shock of seeing huge flames and smoke plumes trailing from the engines, an estimated six hundred thousand people took a chance and went up to see what airline flying was like.

The greatest of the multi-engine barnstormers by far were the brothers Rolly, Don and Art Inman, whose Inman Brothers Flying Circus introduced a whole generation of Midwesterners to the glory of whistling across the sky in "America's largest Trimotor, the 80-A Boeing Clipper." It was Art Inman, in fact, who first came up with the name "Boeing Clipper" and who was obliged to sue the plane company for the right to retain it after they adopted it for their big flying boats that pioneered the Pacific Ocean runs.

More than one innocent would-be airline passenger made the mistake of climbing aboard the Inman Clipper, belting himself into one of the wicker seats and asking, "What time do we get to Chicago?" Art, resplendent in military uniform and silver wings, smiled and explained that it was only a local hop.

A confirmed realist, Art Inman more recently remembered with a laugh that the whole philosophy of barnstorming at the end of the 1920's and in the 1930's was aimed at building public confidence and used the tactic of making a farce of danger. "All rides are guaranteed to get you back in one piece!" Inman's barkers told the crowds. "Your money back if you get killed!"

Being a pilot was only a small part of what it took to be a barnstormer in those days, according to Art. "He had to be a politician, ballyhoo artist, con man, roustabout mechanic, showman, fence-mender and, of course be able to get a ship off the ground and back in one piece."

The Inman Brothers came along at a time when, strangely enough, local airport operators who should have welcomed the big business they drew either ordered them off the field or made them sign excessive thirty-day leases they couldn't afford. When the powers-that-be at Des Moines Municipal Airport pulled that stunt on Art, they chose the wrong man. One night Art flew over the field with his 80-A Boeing Trimotor idling, then aimed the monster straight down, engines screaming and landing lights stabbing the sky. It was right after the Orson Welles "Martian Invasion" broadcast scare, and the Iowans were certain an L.G.M. (Little Green Man) was at the controls.

Inman had already arranged with city officials to provide a motorcycle escort for a giant portable searchlight he and his brother Rolley had rigged

Don Inman died in this 1935 plane crash in Florida.

up. The airport officials were outraged, but the Inman Brothers Flying Circus did a land-office business that wouldn't stop.

Art, Rolley and Don Inman were not the last of the first barnstormers, but the first of the last. Rolley learned to fly under the tutelage of the great Speed Holman and was something of a master at the controls of their first barnstorming plane, a Lincoln Standard biplane. Luckily for his passengers, Rolley's skill once paid off when Art, landing with a good payload at Newton, Iowa, noted with a shock that the right wheel of Rolley's LS-5 had spun off just as he left the ground. Thinking fast, Art grabbed a loudspeaker and ordered everyone out onto the field. At the end of his hop, Rolley went furiously beet-red when he saw the rubes jamming his landing place, then paled, understanding the reason, when a mechanic held up the wheel. Rolley flew around until the crowd cleared off and then made a perfect one-wheel landing.

Another time early in the Inmans' career on a night flight from Waterloo, Iowa, a town well-named, he considered later, it was Art who faced disaster. His Stinton Trimotor, loaded with paying passengers, pierced a sudden line squall fraught with turbulence that came close to ripping the wings off. When he finally found the field, only the baleful glare of auto headlights marked the runway instead of the powerful searchlight that was invariably lugged from town to town. Angrily confronting their assistant, Clifford Pitts, to demand an explanation of why he hadn't turned on the big beam, Clifford

126

shrugged and spat. "You told me, Art, always cover it with the tarp if it rains, no matter what."

Well, there were good days and there were bad days in barnstorming, Art recalls. There were gay times they had with Kitty, a toothless old lion who weighed nine hundred pounds and who wrestled impressively with Art's wife as part of their circus routine, and there was the sheer joy of flying, taking up passengers and showing them what it was all about.

Rolley Inman, brother of Art, barnstormed until World War II. He was killed in 1944 plane crash.

Kitty, pet lion, poses with Inman Brothers Flying Circus troupe.

Sometimes Art was hard put to outwit the freeloaders who parked their Model T Fords outside the field and watched the air show for nothing. At such times he had a helper start a fire out on the field and set off a dynamite charge. A few moments later one of his brothers arrived on the scene, driving an ambulance, its siren screaming. The rubes then flocked through the gates like lemmings, to find out what had happened.

Rolley came close to death once when he flew the Ford Trimotor inverted, causing battery acid to spatter all over him. His bushy eyebrows kept him from being blinded, but he "put on the greatest acrobatic show of his life, jumping around the cockpit," Art remembers. That happened at Wichita, Kansas, a town hostile to outside barnstormers and one where Inman found another way to break the monopoly on the local airport. He ran full-page ads in the newspapers, creating such a controversy that everybody in town came out to see what kind of a nut he was.

For all their frolicking, the Inman Brothers are remembered today for the expert and efficient way they went about their business. They dressed in neat, well-pressed uniforms and flew with such skill that some six hundred thousand Midwesterners were lured into the sky.

Death finally caught up with Don Inman, ironically as a passenger in another barnstormer's ship, on February 10, 1935, just four days after he had been grounded by the Department of Commerce (for not wearing a chute when he took up Carl Hall, the circus parachutist, for a jump). Rolley was killed nine years later, on June 19, 1944, in Maine, when a military C-54

Inman Brothers Flying Circus had its own portable searchlight for night shows. Once, in 1938, they used it to help firemen put out fire in Centerville, Iowa. Inmans did such things to help break down opposition to itinerant flyers.

Ray Dugan, Inman Brothers Flying Circus mechanic, to help bring out the crowds, doubled as wing-walker.
COURTESY ART INMAN

transport he was ferrying on an Atlantic crossing hit a mountain peak during a rainstorm. Art, who was flying another ship ahead of him, had been shocked to hear Rolley get an incorrect weather briefing by radio. "He never had a chance," Art says sadly.

Art Inman, the sole survivor of the Inman Brothers Flying Circus, remembers that only World War II ended barnstorming, although today you'll find a few stray pilots out beating the bushes for passengers on weekends. The author of this book remembers helping out a friend, Alvin Algee, who flew weekend passengers from a small airport in Compton, California, right after the war. One of the passengers was a little old lady who settled down in the back seat of Algee's Aeronca.

"Don't be scared when we lift off the ground," I reassured her. "Just relax and enjoy it."

At five hundred feet altitude, climbing over a highway, I glanced down and saw a convoy of military trucks inching along, bumper to bumper. There was a sudden, startling noise behind me, not unlike the chatter of a machine gun. I glanced around in alarm. The little old lady, her finger cocked, yelled, "Got every damn one of 'em!"

130

Inman Brothers Flying Circus' Ford Trimotor over Wichita.

Interior of Ford Trimotor was the ultimate in flying luxury: it had seats.

Art Inman, incidentally, became something of a World War II hero when, on July 1, 1941, he ferried the first Lockheed Lodestar across the Atlantic Ocean from the United States to British West Africa to spearhead the gigantic flow of lend-lease planes that helped divert the Nazi thrust into Africa. Today he lives not far from bustling Los Angeles International Airport, the

In the late 1930's the Inman Brothers airlifted more than half a million people in big Boeing Clipper and Ford Trimotor transports, introducing them to airline-type travel.

thunder of jets a constant reminder of the barnstorming days when he helped America to grow wings.

Stop by at any airport today and you'll likely find some old-timers around, anxious to swap tales about the good old days of barnstorming, whether in the United States, Mexico or Canada. Bob Arabsky, an up-and-coming young bush pilot living in Winnipeg, Manitoba, reports the story of a charter pilot who was hopping passengers back into moose country, along with a case of Scotch. While waiting for a storm to pass, they opened a bottle. By the time the weather cleared they all fell into the aircraft, the pilot fumbling for the starter. One of his passengers got out to pull the blade through by hand just as the pilot hit the switch. But someone had forgotten to remove the canvas engine cover, and suddenly the guy outside found himself being whipped by lashing ropes. It sobered them all up pretty quickly.

Speaking of Canada, it was back in 1932 that a former Royal Air Force pilot named Wilfred Reid "Wop" May helped out the red-coated Royal Canadian Mounted Police in tracking down a fugitive known as the "mad trapper of Rat River."

The mad trapper, a man named Albert Johnson, led the Mounties on a twenty-nine day chase over the frozen Arctic wasteland. He was wanted for robbing Indian fur traps along the Rat, and, in an arrest attempt, shooting a constable in the chest. During the month-long manhunt, Johnson shot and killed a Mountie, then took off his snowshoes and disguised his tracks through the snow by combining them with those of a migrating moose herd.

The Redcoats finally appealed to "Wop" May, who went out looking for him in a ski-equipped Bellanca, which was armed with tear gas bombs, dynamite and 30-30 ammunition. Flying more than a thousand miles over the frozen tundra, the Canadian barnstormer finally picked up Johnson's trail near the Eskimo village of Aklavik, where a posse flushed him out and shot him. The dead trapper was found to have a grisly bag of gold teeth in his possession, plus $2,000 in cash.

The girls, lest we forget them, played their part in the barnstorming days when flying was a glorious adventure, and one should not overlook the cool, steady birdwomen—chicks, if you prefer—who added glamour to the skyways. Today's Powder Puff Derby had its origins in the barnstorming era, when the nation's top ladybirds competed in the First National Women's Air Derby, Santa Monica to Cleveland, in 1929. There was Louise Thaden, Bobbie Trout, Patty Willis, Marvel Crosson, Blanche Noyes, Vera Walker, Amelia Earhart, Marjorie Crawford, Ruth Elder and Florence "Pancho" Barnes. Marvel Crosson was killed, but the others found their way to Cleveland, with Louise Thaden the winner.

Aviatrix Helen Richey in 1934 became the first female airline pilot in America when she flew the right seat of a Central Airlines Ford Trimotor from Washington to Detroit.

Ethel Dare, the Flying Witch (Margie Hobbs), thrilled crowds in 1920's with ninety- five plane changes.

COURTESY ETHEL DARE

Gladys Ingle of Hollywood made hair-raising plane changes, flying with the famed Thirteen Black Cats for the newsreels, but perhaps the most daring of the ladies was a lovely teenager, Margie Hobbs, who today lives in Miami, Florida, and from that sunny city can look back on her stunt career as a barnstormer billed as "Ethel Dare, the Flying Witch."

Margie was a familiar figure around the old Ashburn Field near Chicago, where she first got the yen to scamper out on the wing of a Jenny and swing herself up onto another ship flying overhead, the first woman to make a plane-to-plane change in mid-air. In all, she performed the dangerous feat ninety-five times, hoping to make it an even hundred, before the pilots, a superstitious lot, refused to take her up any more.

One of her pilots, George W. Parmley, today remembers the time their circus was playing the Illinois State Fair in Detroit in 1920: "There was another plane-changer on the bill who was to perform first, before Ethel Dare. His two pilots made several unsuccessful approaches and on the third try he managed to grasp the ladder and leave the lower plane. But his hold was insecure and he lost his grip and came hurtling to the ground, falling about six hundred feet, end over end.

"We all stood watching the tragedy, and decided that Margie should be excused from performing that day, but she would have none of it. So we took off and she put on one of the best performances of her career."

Parmley's flying partner in the act was a man named Elmer Partridge, whose career dated back to 1910. Says Parmley, "After we finished with show business he decided to build an airplane for himself and incorporate his own ideas. He built this plane in an old and very small barn in Homewood, Illinois, and to everyone's amazement it not only flew but flew very impressively."

With this unusual plane, which featured an enclosed cockpit, Partridge became one of the first barnstormers to try to make the final transition to airline flying. This was on June 7, 1926, when he attempted, in the midst of a raging thunderstorm, to pioneer the dangerous Chicago-Twin Cities Air Mail Route, CAM-9.

Four pilots had already crashed to their deaths, and eight planes had been demolished in 1920, the year the Post Office Department started mail service over the route as a feeder line to the newly established transcontinental airway system. The run was shut down until Congressional passage of the Kelly Bill, an act designed to stimulate the airmail service. Then it was awarded to a bearded earlybird named Charlie Dickinson, a man who made his fortune in the seed business and spent it on planes. It was Dickinson who helped the noted designer Matty Laird build his great racers, such as Jimmy Doolittle's *Super Solution,* in Wichita.

An inveterate flier, Dickinson looked not unlike Santa Claus whipping across the sky, his whiskers flying in the wind, and as historian Henry Ladd Smith once observed, "The resemblance did not stop there."

On that June day in 1926 when Dickinson began operations on CAM-9, his mail fleet consisted of Partridge's three homebuilt Laird biplanes with OX-5 motors, and a Laird Swallow. Partridge climbed into his enclosed cockpit, signaled to the mechanic to turn the prop, then taxied out onto the runway at the Minneapolis field. He gunned the engine, lifted the tail and chugged off into the blinding storm, disappearing from sight. Moments later his ship was seen spinning down from the clouds. In the crash Partridge was killed, his lovely plane destroyed.

Within three months, all of Dickinson's pilots quit and all but one of his ships were wrecked. In August, 1926, he gave notice he was through and turned over his operation to Colonel L. H. Brittin, an officer of the St. Paul Chamber of Commerce, who, when he could find no other takers, decided to run the hard-luck airline himself. The line was called Northwest Airways, and it goes down in the history of aviation as the first barnstorming airline, one that lives today as Northwest Airlines, one of the world's major air carrier systems.

Brittin, an ex-Army colonel who knew absolutely nothing about flying, decided at the outset to hire only ex-barnstormers, wing-walkers and parachute jumpers. "We started with the idea that men who had barnstormed their own planes for years, through winter and summer, in the northwest territory we planned to serve, were the best men to fly our ships in that same district," he explained. "We had nothing against Army-trained pilots—they are fine fellows and fine fliers—but we believed that the men who had learned to nurse Wright pushers through the air, patch up wartime Jennies with haywire and keep them going and care for their own property to keep it flying, had peculiar gifts that we might utilize."

For his first pilot, Colonel Brittin picked Speed Holman, who, like Lindbergh, had started out as a parachute jumper, barnstorming the farming country of Minnesota, Montana and the Dakotas. Others who helped get Northwest going were Homer Cole, Chad Smith and Walter Bullock. Bullock logged more than thirty thousand hours in the air from the time he learned to fly back in 1916 at the old Curtiss Flying School at Newport News, Virginia.

They're still around, many of those great old barnstormers, so don't underrate those fatherly airline captains with graying temples who sit in air-conditioned cockpits and flash you across the sky at better than 600 mph.

There's Walter Hunter of American Airlines, the boy who hung by his knees from a Standard biplane and dove into haystacks. There are Dick Rossi and Bob Prescott, of the Flying Tiger Line, who barnstormed the Orient and flew with the Chinese in Burma, before America entered World War II. And there are Clay Lacy of United and Mira Slovak of Continental, dedicated transport drivers who spend their weekends out in the boondocks, flying open-cockpit racing planes close to the good sweet earth where all fliers come from, and to which they must someday return.

FOR FURTHER READING

THE AIR DEVILS
Don Dwiggins
J. B. Lippincott Co., 1966

THE AMERICAN HERITAGE HISTORY OF FLIGHT
Alvin M. Josephy, Jr., Editor
American Heritage Publishing Co., 1962

BARNSTORMING
Martin Caidin
Duell, Sloan & Pearce, 1965

THE CHALLENGING SKIES
C. R. Roseberry
Doubleday & Co., Inc., 1966

HEROINES OF THE SKY
Jean Adams & Margaret Kimball
Doubleday, Doran & Co., 1942

ONCE TO EVERY PILOT
Captain Frank Hawks
Stackpole Sons, 1936

SKY STORMING YANKEE
Clara Studer
Stackpole Sons, 1937

THE WORLD IN THE AIR
Francis Trevelyan Miller
G. P. Putnam's Sons, 1930

A NOTE ABOUT THE AUTHOR

DON DWIGGINS' interest in aviation began when a barnstormer with "a waxed mustache and beautiful boots" took ten-year-old Don up in a Jenny. He has been a flying enthusiast ever since.

At college, Mr. Dwiggins studied journalism and in his spare time researched and wrote a history of American ballooning in the 1800's. In World War II, he enlisted in the Glider Corps and later switched to the RAF as a flight instructor for three years.

When peace came, he bought seventeen surplus planes to start a non-scheduled airline that carried movie crews and equipment to filming locations. But this venture never became profitable and Mr. Dwiggins went back to journalism, first as aviation editor of the *Los Angeles Daily News,* then of the *Los Angeles Mirror.* In recent years, he has been associated with the Walt Disney Studios and Los Angeles television studio KKTV. He is also the author of several notable books about various aspects of aviation, among them a biography of Paul Mantz, the famous Hollywood stunt flier.

INDEX

PILOT'S DIGEST OF FAA REGULATIONS
New Revised 2nd Edition

PILOT'S DIGEST OF FAA REGULATIONS

2nd Edition

BY JOHN L. NELSON

Printed in the United States of America

Reproduction or publication of the content in any manner, without express permission of the publisher, is prohibited. No liability is assumed with respect to the use of the information herein.

Library of Congress Cataloging in Publication Data

Nelson, John Lewis. 1926-
    Pilot's digest of FAA regulations.

    Includes index.
    1. Aeronautics—Law and legislation—United States.
I. Title.
KF2400.Z9N4   1981   343.73'097'02636   80-28682
ISBN 0-8306-9621-0
ISBN 0-8306-2295-0

Foreword

In today's world Federal Aviation Regulations are a basic part of the business of flight. They exist for reasons of safety in air transportation—safety to the pilot, the passenger, and the person on the ground over whose land we travel. The FAR's exist as a means of achieving equity between people in the use of a national resource—our airspace. They are the aviation "law of the land."

Written in the language of the lawyer, Federal Aviation Regulations often require study and interpretation. A purpose of this text is to briefly restate in simple narrative form those regulations commonly employed by the general aviation pilot. However, a more important objective is to outline the manner in which our regulatory system operates. A system does exist for the general aviation pilot to make his voice heard. By understanding the regulatory process we may participate directly in the formation of future Federal Aviation Regulations.

The author wishes to thank the many people who made this book possible, especially the Phoenix FAA General Aviation District Office and TRACON, Barbara Nelson for all of the artwork, Dick Roberts for many helpful suggestions, and daughter Sandy for the final typing. Sincere appreciation is extended to Cessna Aircraft Company, Piper Aircraft Corp., Beech Aircraft Corp., Bellanca Aircraft Corp., Mooney Aircraft, Evans Aircraft, King Radio Corp., and EAA for photographs and reference material. The author is indebted to the Electrical and Electronic

Engineers, Inc., Washington, D.C. for permission to quote from the text, *The Federal Airways System* by William Jackson and likewise to The Michie Co., Charlottesville, Virginia for legal reference from *The Law of Aviation* by Rowland Fixel. Last, but far from least, sincere thanks are directed to editor Joe Christy for supplying guidance in writing as well as editing of the final product.

John L. Nelson

Contents

Chapter 1
Our Regulatory System

Federal Aviation Regulations are minimum standards, broad guidelines, desirable procedures, and definitions pertaining to the administration and preservation of our National Airspace. Webster has defined regulation as *"the art of reducing to order."* And so it is with Federal Aviation Regulations; safe, efficient, and orderly use of the National Airspace directed to fairness and service to the community is the primary objective of our aviation regulatory process. This objective is founded upon democratic principles which recognize public needs as being dominant, and fairness as a stronger requirement than efficiency.

Currently there exist more than 180,000 aircraft operating throughout a network of 12,500 civil airports in the continental United States. Predictions indicate that our air transportation system will expand to 206,000 civil aircraft within the next few years and double in services to the user. A fair and democratic regulatory process is essential to the continued growth of travel by air.

A "PREFLIGHT" BRIEFING

This text is directed to a brief common language explanation of and rationale for the various Federal Aviation Regulations utilized by the student, private, commercial, and airline transport pilot. Material has been selected from many parts of the Federal Aviation Regulations and condensed to a digest form of the subject;

regulatory intricacies and special situations have necessarily been deleted. Should the reader require exact legal definitions, descriptions, or precise detail, it will be necessary to refer directly to the Federal Aviation Regulations (FAR's) for such.

A compromise considered necessary in preparing this document is the limitation of material to Visual Flight Rules (VFR). The subject of Instrument Flight Rules (IFR) is a specialty which is beyond the scope of a discourse on the general subject of FAR's.

A particular objective of this text is to outline the manner in which Federal Aviation Regulations evolve and the large measure of public participation incorporated in the process. In the year 1903, the whole of the National Airspace was shared by two persons, brothers Wilbur and Orville Wright. Today, that same airspace is put to use by more than 800,000 pilots. As a companion in the enjoyment and utilization of our National Airspace we can now ask only our fair share. The challenge at hand is *direct participation* by the airspace *user* in rule making procedures to both *define* and *protect* that "fair share."

HISTORICAL BACKGROUND

Perhaps the first recorded example of an aviation regulatory process-in-action was provided by brothers Jacques and Joseph Montgolfier (acting as air traffic controllers) in providing a takeoff clearance for three stout-hearted aeronauts (a duck, a rooster, and a sheep) for a "once around the patch" balloon flight of eight minutes in the year 1783. Possibly a need for regulation stemmed from early balloonists utilizing rocks for ballast, thereby making flight hazardous to persons on the ground. Perhaps the 1861 actions of Lowe in sending the first aerial communication from his balloon signalled the need for regulating the activities of aerial vehicles as observation platforms. Whatever the case may have been, the regulatory process *preceded* the invention of the heavier-than-air craft by some four years for in 1899 representatives attending the first Hague International Peace Conference acted to prohibit the launching of projectiles and explosives from balloons.

Following the invention of the airplane and during the period 1905 to 1911, serious consideration was given to registration of aircraft, certification of pilots, and also whether or not property rights and claims of sovereignty of airspace permitted free circulation of aircraft. The first recorded instance of government contol of pilot certification and regulation of operations in the United States is found in the Connecticut Act of June 8, 1911.

With the growth of aviation, particularly after World War I, a public need was sensed to control traffic in the vicinity of airports. Initially, the control of aircraft was by men waving arms or flags to indicate to pilots circling the field that a landing was safe—or a takeoff was permissible. Subsequently, colored directional lights were developed so that they might be focused on a specific aircraft. Pilots learned to look for these lights to give them landing or take-off information.

In 1930, at Cleveland, Ohio, an airport traffic control tower was put into operation. This installation proved so successful that several larger municipalities erected towers and staffed them with personnel using light guns and low powered radio equipment. By 1936, approximately 20 cities had followed Cleveland's lead and established similar radio-equipped control towers.

The continuing growth of aviation brought about the need for uniformity in operating procedures. In response, the Civil Aeronautics Act of 1938 was passed which provided for the development of safety provisions related to civil aeronautics. Initially, airport control towers continued to be owned and operated by municipalities. However, personnel employed at these airport traffic control towers were certified by the Civil Aeronautics Authority (CAA) as to theoretical knowledge, physical qualifications, and experience requirements.

Early in 1941, due to the increasing aviation activity brought on by World War II, Congress assigned to the CAA the responsibility for operating certain control towers. Most of the municipally operated traffic control towers were taken over by the CAA on January 1, 1942, with the intention that these towers would be operated by the CAA only for the duration of the war. With the CAA's assumption of these activities, standards were prepared for equipment and operations; tower operators became employees of the Federal government. At that time there were no regulations which required a pilot to comply with instructions issued by a tower controller. Consequently, enforcement of an airport control procedure by the Federal Government was not possible. This condition led to the development of air regulations establishing control zones for governing operation of aircraft operating therein. For the next 20 years, the CAA remained the primary government regulatory agency for U.S. aviation.*

The tremendous growth of aviation during World War II, the introduction of jet aircraft, and the need for a modern electronic

---

*History of Control Tower Operation and CAA; Jackson, *The Federal Airways Systems*.

system of navigation to accommodate high speed aircraft resulted in a corresponding need for modernization of the aviation regulatory process. A tragic collision of a TWA Super Constellation and a United Airlines DC-7 over the Grand Canyon on June 30, 1956, catalyzed the situation and prompted action. The Federal Aviation Act of 1958 resulted, which established the Federal Aviation Agency as the primary regulatory body for both *military and civil aviation* (Figs. 1-1 through 1-8) in the United States.

In 1966 the Department of Transportation (DOT) was created for the purpose of developing national transportation policies and programs conducive of a fast, safe, and efficient national transportation system. The Federal Aviation Administration was placed under the guidance of the Department of Transportation along with all other federal transportation agencies.

THE SIGNIFICANCE OF FEDERAL LAW

While aviation became a matter for state regulation as early as 1911, the enactment of the Air Commerce Act of 1926, the Civil Aeronautics Act of 1938, and the Federal Aviation Act of 1958 have made aviation law in the United States, its territories and possessions, the law of the land. Exclusive control of aviation by federal law has become increasingly necessary by reason of the increased use of navigable airspace by federally certificated commercial air carriers flying scheduled operations. Furthermore, ground based radio navigation equipment, weather service, air-to-ground communications, instrument landing system, and the like are provided for by the federal government. Other compelling factors are the extensive use of aircraft by the Army, Navy, and Air Force; the use of commercial aircraft by the U.S. Postal Service in the performance of a national function; the use by foreign aircraft of airspace in the United States; and the protection of airspace over military, Naval, and Air Force installations.

It is a well-established principle of constitutional law that where the exercise of control by the federal government is necessary and imperative, and the subject is national in character (such as the regulation of interstate commerce), and furthermore requires uniformity of regulation, the federal government becomes paramount and exclusive. Federal control applies to *intra*state as well as *inter*state traffic. In such cases where there is a state law and also a federal act on the same subject, the latter will control. The United States Supreme Court stated the following with respect to federal control of air commerce:

12

Fig. 1-1. Where altitude and speed are needed the Mooney Turbo 231 gives both. A 210 hp Continental 6 provides a maximum speed of 200 knots.

Fig. 1-2. Latest of the Piper line, the Turbo Arrow IV employs a T-tail for additional speed.

Fig. 1-3.   A classic light twin, the Cessna 310 and its turbocharged counterpart feature seating for six adults and cross-country speeds up to 207 and 237 knots respectively.

Fig. 1-4. Top of the Cessna singles line, the pressurized Centurion cruises at over 200 knots at 20,000 feet.

Congress has recognized the national responsibility for regulating air commerce. Federal control is intensive and exclusive. Planes do not wander about the sky like vagrant clouds. They move only by federal permission, subject to federal inspection, in the hands of federally certificated personnel and under an intricate system of federal commands . . . . Its privileges, rights and protection, so far as transit is concerned, it owes to the federal government alone, and not to any state government.

Regulation of air commerce and air transit in the airspace above the United States is conceded to be a national responsibility, and the control and regulation of air commerce and air traffic of all kinds in the United States is recognized to be a federal function.* In general, state aviation law is limited to matters of taxation, regulation of air carriage wholly within a state, law concerning liability for damage to persons and property on the ground by aircraft, and the power of eminent domain over airports.

THE AVIATION GOVERNMENTAL COMMUNITY

Democratic freedom is a matter of making choices, which includes the freedom to share with others in setting up possible choices. Democratic freedom also implies restrictions in that no society will allow unlimited choice. Laws, as well as unwritten codes, may restrict the freedom of the individual in favor of a

---

*The Significance of Federal Law; Fixel, *The Law of Aviation*.

Fig. 1-5. After three decades, the Beech Bonanza is still the classic single-engine general aviation aircraft. This model V35B cruises 172 knots at 6000 feet, carries four, has a range of 716 miles, and is certificated in the utility category at full gross weight.

community standard. The balanced liberty of the United States, sometimes referred to as "ordered freedom" contains the elements of justice, order, and restraint. Justice is the principle by which each man is assured the things that belong to him; order is the principle by which peace is maintained; and restraint is the exchange of unlimited personal freedom for freedom of society as a whole. Our flying freedom embodies all three basic principles of political freedom:

Justice: The National Airspace is shared equally between general aviation, military aviation, and commercial air car-

Fig. 1-6. Typical of the corporate aviation fleet, the Beechcraft Hawker 600 offers performance comparable to large transport jet aircraft.

Fig. 1-7. The Piper Turbo Aztec F features two Lycoming engines of 250 hp each and cruises at 215 knots at 22,000 feet.

riers. (Reflect for a moment as to how many other nations of the world grant as much airspace freedom to general aviation as that enjoyed within the United States.)

Order: A superb system of airways exist for use by general aviation, military aviation, and the air carrier.

Restraint: Aviation rules and regulations are the "price paid" by the individual to obtain his fair share of our National Airspace.

Understanding our aviation governmental community is the first step to understanding Federal Aviation Regulations.

Department of Transportation

The Department of Transportation (DOT) is the parent governmental organization whose responsibilities extend throughout all forms of transportation in the United States. DOT directs the efforts of the United States Coast Guard, Federal Aviation Administration, Federal Highway Administration, Federal Railroad Administration, Urban Mass Transport Administration, and other like transportation agencies. In addition to administrative functions, DOT is responsible for scientific and technological research to advance our national transportation capability in safety, effectiveness, economy, noise abatement, telecommunications, and transportation of hazardous materials (Fig. 1-9).

Fig. 1-8. High performance is the mark of the Super Viking. The Viking cruises at 165 knots (204 knots at 20,000 feet for the turbo version). Takeoff distance is an amazingly short 460 ft. and landing distance only 575 ft.

Federal Aviation Administration

The Federal Aviation Administration (FAA) is the primary governmental authority responsible for the conduct of aviation in the United States. The Federal Aviation Act of 1958 charges the FAA with: regulating air commerce to promote its safety and

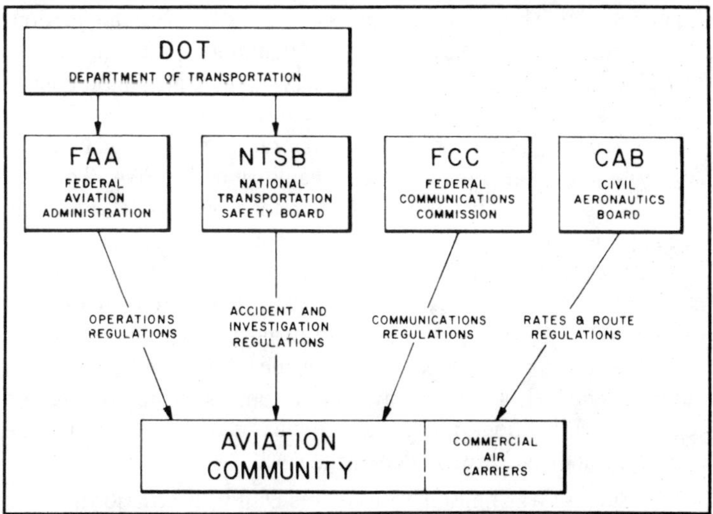

Fig. 1-9. Under the Executive branch of the U.S. Government, the FAA and NTSB are the principal elements of general aviation government.

development; achieving the efficient use of the navigable airspace of the United States; promoting, encouraging, and developing civil aviation; developing and operating a common system of air traffic control and air navigation for both civilian and military aircraft; and promoting the development of a national system of airports. The FAA thus issues and enforces rules, regulations, and minimum standards for the operation, maintenance and manufacture of aircraft, as well as rating certification of airmen. The agency provides a system for the registration of aircraft, engines, propellers and appliances as well as a system for recording aircraft ownership. The FAA may, from time to time, reinspect any civil aircraft, aircraft engine, propeller, and the like as well as re-examine any civil airman. If, as a result of a reinspection, it is determined that safety or public interest so requires, the FAA may amend, modify, suspend, or revoke aircraft type certificates, production certificates, airworthiness certificates, airmen certificates, and other such authorizations as granted by the agency.

Another function of the FAA is to conduct research and development programs directed to systems, procedures, facilities, and devices needed for air navigation and air traffic control. Other research involves the development and testing of improved aircraft, engines, propellers, and avionics appliances.

The FAA is responsible for the location, construction, installation, maintenance, and operation of our federal system of air navigation. The agency operates and maintains communications equipment, radio teletype circuits and equipment, radio navigation equipment, radar equipment, and equipment at air traffic control towers and Air Traffic Control (ATC) centers.

The management of air traffic operating within our National Airspace system is a primary responsibility of the FAA. To carry out this responsibility the FAA develops air traffic rules and regulations and allocates the use of airspace. It provides for the security control of air traffic to meet national defense requirements. The FAA coordinates with foreign governments in matters of aviation, administers federal aid airport programs, develops specifications for the preparation of aeronautical charts, publishes current information on airways and airport service, and issues technical publications for the improvement of flight safety. The following facilites are provided by the FAA:

Airports (Washington National and Dulles International)
Towers and Combined Station Towers
Flight Service Stations

Very High Frequency Omniranges (Radio Navigation)
Airport Surveillance Radars
Air Route Surveillance Radars
Air Route Traffic Control Centers
Instrument Landing Systems

The FAA is divided into several regional offices (including Alaska and Hawaii), an aeronautical center at Oklahoma City, and an experimental center at Atlantic City.

National Transportation and Safety Board

The National Transportation and Safety Board (NTSB) was created as a separate branch of the government by the Department of Transportation Act of 1966. The Safety Board has the authority to investigate, determine the probable cause, and issue reports on all civil aviation accidents; make final cause determination, and report the facts and circumstances related thereto. Authority to investigate accidents of a routine nature is delegated by the NTSB to the FAA. When acting in cooperation with or on behalf of the NTSB, the FAA determines whether or not the aircraft and flight crewmen were properly certificated, the extent to which FAA air traffic control may be involved, whether or not government operated navigation and communication equipment was performing to specified standards, etc. By so doing, FAA lends technical expertise to the various phases of accident investigation and analysis. The determination of probable cause, however, *remains the responsibility* of the NTSB. This ensures impartial judgement of responsibility. Being a separate government agency (*not* a part of the FAA), NTSB is obligated to examine all aspects of an accident as an independent observer. NTSB is thus able to assign fault to the FAA as well as to an individual or an organization.

In addition to accident investigations, the NTSB conducts special studies and makes recommendations on matters of aviation safety and accident prevention. Related to the airman, the board reviews (upon request) the suspension, amendment, modification, revocation, or denial of a pilot's certificate. Thus, the board acts in the manner of a "higher court," providing the airman a second opportunity to appeal FAA decisions relating to his certificate.

Federal Communications Commission

The Federal Communications Commission (FCC) is responsible for the licensing and regulation of radio broadcasting stations and, in addition, licensing of radio telephone operators. FCC regulations require an airman to have at least a restricted radio telephone operator permit to operate a licensed radio station. Furthermore, FCC regulations require any aircraft which contains a transmitter have an appropriate station license. The aircraft station license must provide for transponder and distance measuring equipment as well as communications transmitters and radar equipment. The FCC, through its Field Engineering Bureau, performs monitoring, inspection, and investigative activities. Periodically the FCC monitors aircraft transmissions to assure that transmitter frequencies are within tolerances and that stations (aircraft) have a valid and current station license.

Civil Aeronautics Board

The Civil Aeronautics Board (CAB) grants authorizations for commercial air carriers to engage in interstate and foreign air transportation over assigned routes. Similarly, it issues permits to foreign air carriers authorizing them to engage in air transportation between the United States and foreign countries. The Board has jurisdiction over tariffs, rates, and fares charged to the public for air transportation. Regulatory action of the CAB is directed to the air carrier and has limited impact on the general aviation pilot.

TYPES OF REGULATIONS

Aviation regulations may be categorized in general as *safety based* and *precedural* regulations. Of these, regulations devoted to public safety constitute by far the majority of the FAR's. Safety based regulations provide minimum standards for the certification of student pilots, private, commercial, and airline transport pilots. Air traffic control operators, aviation mechanics, parachute riggers, flight instructors, and other specialists related to the aviation industry are governed by safety based certification regulations. The certification program is a major regulatory action to promote professionalism throughout the aviation community.

Another class of safety based FAR's is associated with the definition of and operating procedures for our more than 280,000 miles of federal airways. The fundamental purpose of these FAR's is to prevent collisions between aircraft by segregating air traffic in lanes and altitudes.

Legal procedures constitute the second broad category of regulations. FAR's of this type describe means for accomplishing a function such as rule making or enforcement. Examples are aircraft registration and title recording, federal aid to airports, and procedures for the conduct of hearings. Regulations of the NTSB dealing with accident investigations and hearings are largely procedural in nature.

HOW REGULATIONS COME INTO BEING

Fundamentally, regulations come into being because a *public need exists*. Without a public need there would be no regulations and no regulatory system. Public needs are numerous in quantity and diverse in scope; many result simply from population growth. Consider the following typical examples:

Population Increase: The rapid increase in travel by air has resulted in congestion at major terminals (Washington, New York, Chicago, Los Angeles, Atlanta, and San Francisco). Air traffic studies performed by the FAA and other agencies conclude that the probability of mid-air collisions will become intolerable if an appropriate action is not taken. The result: regulatory action by the FAA to control *all* aircraft in the vicinity of major air terminals; in this case the establishment of Terminal Control Area (TCAs).

Public Safety: Studies of aviation accidents often disclose cause factors which prompt regulation as a means of reducing accidents. Recent studies performed by the NTSB have revealed that weather is a major cause of general aviation accidents. The result: regulatory procedures are modified to increase the emphasis on the study of weather as a part of pilot training.

National Defense: Pilot training in modern high speed jet aircraft constitutes a hazard for general aviation and air carriers alike if conducted in common airspace. The result: regulations are enacted to segregate military training operations from general aviation and air carrier traffic. Military Operation Areas (MOA's) on aviation maps exemplify this type of regulation.

Modernization of Standards: Increased use of the airways and technical advancements in general aviation aircraft necessitate upgrading minimum standards for pilots and air crewmen. A modern single engine aircraft such as the V35B Bonanza is in many ways more sophisticated than many combat aircraft of World War II. Similarly, general aviation jet aircraft such as the Learjet or Falcon Fan Jet approach the commercial air carrier in equipment sophistication and performance. The result: FAR-61, Certification of Pilots and Flight Instructors, is modernized. Proficiency requirements are increased for all levels of the aviation pilot community.

The Basic Rule-Making Process: Initiated by a need, a regulation begins life as a petition. An interested person may petition the FAA to issue, amend, or repeal a rule or request an exemption. Petitions must be submitted in duplicate to the FAA, Washington, D.C., 20590; and set forth the text or substance of the rule or amendment proposed, or the rule from which an exemption is sought, or specify the rule that the petitioner seeks to have repealed (Fig. 1-10).

The petitioner must include a complete description of the action being sought and, in the case of an exemption, reasons why safety would not be adversely affected. The FAA may also initiate a rule-making action. In doing so the FAA considers the recommendations of other agencies of the United States and the petitions of interested persons. Petitions for rule-making and related responses are available to the public from the office of the General Counsel of the FAA.

Action on petitions for rule-making may take a number of courses. If the FAA determines that the petition discloses adequate reasons, a Notice of Proposed Rule-Making (NPRM) is issued, or a final rule is adopted, or, if in the public interest, an exemption is granted. If the FAA determines that the petition does not justify instituting rule-making procedures or granting the requested exemption, the petitioner is notified accordingly. For example, consider a significant rule-making procedure such as the modernization of FAR-61. In this case, the FAA, acting on behalf of many segments of the aviation community, formulates an initial version of the rule. The proposed rule is advertised to the aviation community by the Federal Register, the aviation news media, and personal notification (where a party may be directly affected). The formal means for advertising NPRM's is via the Federal Register. A Notice of Proposed Rule-Making includes:

1. A statement of the time, place, and nature of the proposed rule-making proceeding;
2. A reference to the authority under which it is issued;
3. A description of the subjects and issues involved or the substance and terms of the proposed rule;
4. A statement of the time within which written comments must be submitted and the required number of copies; and
5. A statement of how and to what extent interested persons may participate in the proceedings.

As an example of this process, the Notice of Proposed Rule-Making for FAR-61 reads in part:

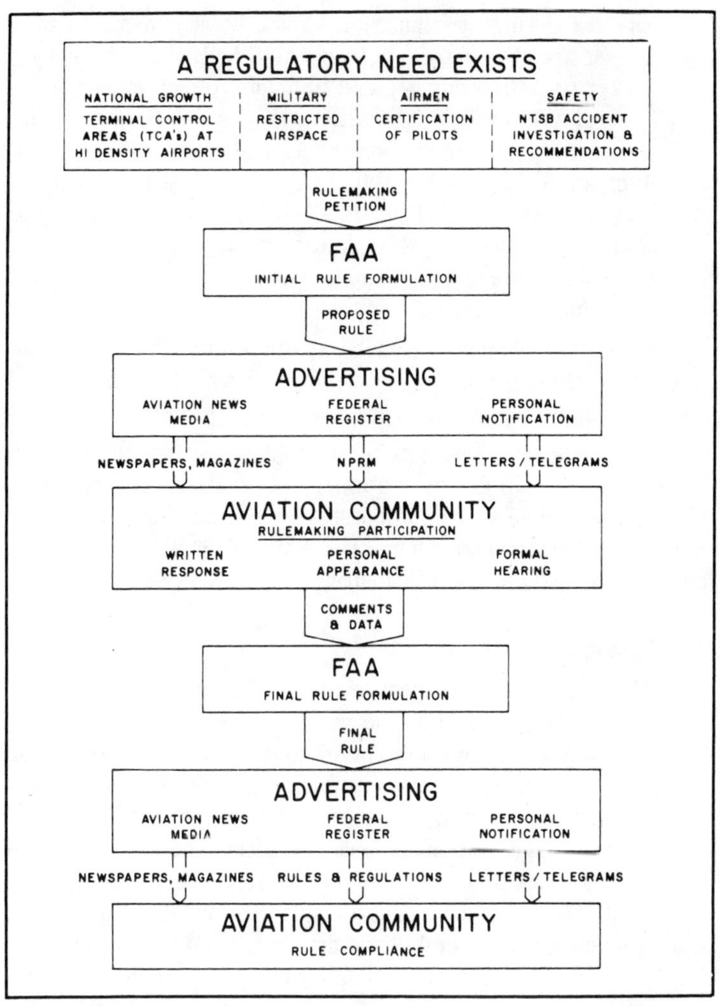

Fig. 1-10. Rule making begins with a public need; proposed regulations are advertised in a formal manner and public opinion solicited. Regulations result after interested parties have forwarded opinions (by letter or in person) and the FAA "weighs" the returns. The objective of the process is to formulate laws that benefit the community in general. In certain special instances aviation laws result from direct congressional action. Laws governing Emergency Locator Transmitters and noise abatement are examples.

The Federal Aviation Administration is considering amending Part 61 to revise the standards for issuing pilot and flight instructor certificates and ratings. Interested persons are invited to participate in the making of the proposed rule by submitting such written data, views, or arguments as they may desire. Communications should identify the regulatory docket or notice number and be submitted in duplicate to: Federal Aviation Administration, Office of the General Counsel, Attention: Rules Docket, GC-24, 800 Independence Avenue, S.W., Washington, D.C., 20591. All communications received on or before . . . will be considered by the administrator before taking action on the proposed rule. The proposals contained in this notice may be changed in the light of comments received.

Interested parties do have an opportunity to participate directly in the rule-making procedures. Part 11 of the federal Aviation Regulations states:

Each interested person is entitled to participate in rule-making proceedings by submitting written information, views, or arguments. In addition, he may comment on the original information, views, and arguments submitted by other persons, if, after receiving them, the administrator considers it desirable.

Part 11 goes on to state:

The rule-making procedure also includes any further procedural steps that best serve the purposes of a particular proceeding. For example, interested persons may be allowed to make oral arguments, participate in conferences between the administrator or his representative and interested persons and organizations, appear at informal meetings presided over by a designated FAA official at which a stenographic transcript is made, or participate in any other procedure whenever it is desirable and appropriate to assure informed administrative action and adequate protection of private interests.

Is your voice as an individual heard? The answer is indeed yes! The Federal Aviation Administration *is* sensitive to the opinions of the aviation community. Typically, the results to NPRM's as published in the Federal Register read:

1. Numerous comments have been received in response to the notice of proposed rule-making and changes have been made in the regulation in the light of such comments . . . (FAR 21-2, P5, FR 8464).
2. Finally, in response to some comments, the language of the proposal was changed to specifically include within the term

"surface of the airport . . . —etc." . . . (FAR 91-43, P65, FR 9640).

3. Upon considering the comments submitted by the air travel clubs and other interested persons, the FAA has determined that the form of the proposed rules should be changed by . . . (Far 91-60, P85, FR 12887).

After public comment and supporting data has been received, the FAA acts to formulate and issue the final rule. Advertising is by the Federal Register and the aviation news media.

VARIATIONS ON A THEME

The diverse nature of Federal Aviation Regulations leads to variations in the process of rule-making. Rule-making authority is often directed to regional FAA directors, particularly related to airspace assignments and Airworthiness Directives. Rule changes of a routine nature do not require advance public notification. For example, action to modify a transition area at Twenty-Nine Palms, California, reads in part:

Since this change is minor in nature and imposes no additional burden on any person, notice and public procedure hereon is unnecessary.

Similarly, the alteration of a control zone at El Toro, California, as published in the March 18, 1972 issue of the Federal Register reads:

Since this action is less restrictive in nature than currently designated airspace and imposes no additional burden on any person, notice and public procedure hereon is unnecessary.

Upon occasion, regulations such as Airworthiness Directives require immediate attention to a particular problem. As a typical example an Airworthiness Directive related to a rotor blade system reads:

Since it was found that immediate corrective action was required, notice and public procedure thereon was impractical and contrary to the public interest and good cause existed for making the airworthiness directive effective immediately—.

Still another variation on the theme is the issuance of an Advance Notice of Proposed Rule-Making (ANPRM). The purpose of such notification is to propose several courses of rule-making actions and request public reaction to the proposals. This technique is advantageous in shortening the rule-making procedure by inviting public reaction at a very early stage in the rule formulation process.

HOW TO KEEP CURRENT WITH REGULATIONS AND PROCEDURES

Aviation is rich in literature. The history of aviation has been thoroughly documented. News media widely distribute current events. Even our children speak knowledgeably on future explorations in the fields of aviation and space. What then, about the subject of rules and regulations? Why is the question so often asked, "Where can I obtain a set of regulations—how can I keep current on changes?" The question is indeed valid for in the abundance of aviation literature, Federal Aviation Regulations comprise but one small part. The next few paragraphs are devoted to a step-by-step procedure which answers that important question, "Where can I obtain information on regulations—how can I keep current on changes?"

Step 1—Publications Catalogs

The initial step to be taken in searching for information is to obtain catalogs which list current aviation publications. The Government Printing Office (GPO) is an excellent source of such material. The following four catalogs are recommended:

The catalog *FAA Publications* contains selected material which is of interest to the public, to pilots, and the aviation industry. Revised editions are issued periodically. The catalog may be obtained by writing:

> Department of Transportation
> Distribution Unit, TAD-484.3
> Washington, D.C. 20590

In your letter, request the text *FAA Publications*. Be sure to include your return name and address, written legibly. A self-addressed mailing label is recommended.

Publications of the National Transportation Safety Board is a listing of printed material of general interest to the public and the transportation industry. It contains a record of all publications issued by the Safety Board. The pamphlet is available upon request by writing:

> National Transportation Safety Board
> Publications Section
> Washington, D.C. 20591

The FAA issues *Advisory Circulars* to inform the aviation public in a systematic way of non-regulatory material of interest.

These excellent documents supplement regulations by providing related material on pertinent subjects such as wake turbulence, aircraft fuel management, radar capabilities and limitations, altitude-temperature effects on aircraft performance, etc. To obtain information on Advisory Circulars, write:

Department of Transportation
Distribution Unit, TAD-482.3
Washington, D.C. 20590

Request Advisory Circular 00-1, the *Advisory Circular System* and Advisory Circular 00-2, *Advisory Circular Checklist*. These documents are available free of charge.

The FAA makes available to the public on a no-charge loan basis films, filmstrips, and audio slide packets on subjects ranging from aviation careers to weather. To obtain information, write to the following address for the publication *FAA Film Catalog*:

Film Library, AC-921
Federal Aviation Administration
P.O. Box 25082
Oklahoma City, Oklahoma 73125

Step 2—Federal Aviation Regulations

FAR's are published in a series of volumes. To obtain your pilot certificate it is necessary that you study material contained in FAR's 1, 61, 91 and NTSB regulation 830, as an absolute minimum. For persons entering aviation on a commercial basis, the information contained in FAR's 11, 13, 23, 43, 71, 95, 97, and 135, is recommended. Volumes may be purchased from:

Superintendent of Documents
U.S. Government Printing Office
Washington, D.C. 20402

To expedite getting your publications promptly, send a check, not cash, in the exact amount of the purchase. List the exact name of the publication you desire. Enclose a self-addressed mailing label if you have no order blank.

NTSB regulations pertaining to the notification and reporting of aircraft accidents are also required as a part of your general knowledge when applying for your pilot certificate. This information is distributed free of charge by writing:

National Transportation Safety Board
Publications Section
Washington, D.C. 20591

Fig.1-11. Some of the many government publications for keeping current; available at no cost are publication listings of the FAA and NTSB, Exam-O-Grams, and most Advisory Circulars.

Request Part 830, *Rules Pertaining to the Notification and Reporting of Aircraft Accidents, Incidents, and Overdue Aircraft, and Preservation of Aircraft Wreckage, Mail, Cargo and Records*.

Step 3—Keeping Current

Keeping current is to a great extent a matter of individual need (Fig. 1-11). Let us assume that you are preparing for your FAA written and wish to be current in subjects related to the examination. The FAA publishes *Exam-O-Grams* which are nondirective in nature and are issued solely as an information service to individuals interested in airman written examinations. *Exam-O-Grams* are distributed free of charge by writing:

Department of Transportation
Federal Aviation Administration
FSTD, Operations Branch
P. O. Box 25082
Oklahoma City, Oklahoma 73125

If you are studying for your private or commercial examination, request the *VFR Examo-O-Grams*. In the event you intend to obtain your instrument rating request the IFR *Exam-O-Grams*.

For the average general aviation airman, the FAA offers two publications available on a subscription basis as a means of keeping current. The first of these is the *FAA Aviation News*. This monthly magazine contains information on rule changes, piloting techniques, special environment flying, aircraft maintenance, etc. A second publication which provides current information in greater depth is the *Airman's Information Manual*. Part I, *Basic Flight Manual and ATC Procedures*, is issued quarterly. This manual contains information for both the VFR and IFR pilot. Either publication can be obtained by writing:

> Superintendent of Documents
> Government Printing Office
> Washington, D.C. 20402

Let us assume you are in a business that demands absolute currency of Notices of Proposed Rule-Making, Rules and Regulations, etc. Then the answer is a subscription to the *Federal Register* (Fig. 1-12). The *Register* contains regulatory information from the Office of the President, as well as the many executive agencies of the government. The *Federal Register* is distributed only by the Superintendent of Documents, U.S. Government Printing Office, Washington, D.C. 20402.

Need an immediate updating on regulations or procedures? Try the telephone book. Under United States Government, the listing: Department of Transportation, may contain the telephone number for either the Federal Aviation Administration Flight Standards District Office (FSDO), General Aviation District Office

Fig. 1-12. The Federal Register is the formal government publication for advertising regulatory material. Notices of Proposed Rule Making, Rules and Regulations, Airworthiness Directives, and listings of Advisory Circulars are published.

31

(GADO), or Flight Service Station (FSS). These agencies will be able to provide current information on regulations and procedures.

Other ways of keeping current? The aviation news media, magazines, and newspapers perform an excellent service by publicizing current news about regulations, interpretations of precedures, and examples of legal actions related to flying. State aviation agencies often issue newsletters which contain current event information on FAR's. Flight instructors, ground school instructors, and fixed base operators are still another way of keeping current on regulations. In reality regulatory information is abundant and, for the most part, either free or obtainable at a minimal cost.

If you elect to obtain information from the government expect to practice patience. The Superintendent of Documents receives anywhere from 10,000 to 60,000 requests for publications *each day*. A wait of six to eight weeks may take place before you receive your documents. In some instances the process can be speeded up by obtaining documents from your local GPP bookstore.

FUTURE TRENDS

Looking to the future of aviation one can only say that its "rate-of-climb" will be phenomenal. During the next few years combined general and air carrier aviation activity will result in vast increases in traffic to be handled by FAA facilities. Terminal area facilities logged 56.2 million aircraft operations in the year 1970. This total is expected to grow to approximately 130 million operations in 1982! Air carrier and general aviation flying will account for all of the growth in operations as military activities are expected to decline at airports with FAA traffic control towers. As for the general aviation fleet, the total number of aircraft in the year 1971 was 131,000. This number is expected to increase to 206,000 by 1982! Instrument operations are expected to increase from 17.4 million to 32.3 million in 1982, with general aviation as the major contributor to the increase. Pilot briefings, which constitute the largest volume of services and which are primarily for general aviation pilots, are expected to increase over three times their current level. No matter which statistic is examined, the answer is the same—dramatic growth in air transportation.

To accommodate the needs of the future a major air traffic control modernization plan has been enacted. New terminal and enroute radar equipment is being designed to provide more accurate longer range air traffic control coverage. New controller computer and display equipment is being installed. A microwave instrument landing system has recently been implemented. This system will eventually permit the landing of air carrier aircraft in conditions of zero visibility and zero ceiling! Air carrier data links will become operational in the not too distant future. Technical answers and equipment are being evolved to handle the predicted 1990 air traffic load. What will be the likely effect on Rules and Regulations?

Predicting the future is difficult at best. However, some events appear certain. The floor of Positive Control will be lowered. The number of Terminal Control Areas will increase and so the number of aircraft under ATC direction. The Microwave Instrument Landing System will require development of new procedures for employing the facility. Increased emphasis on the use of transponders is certain. Radar services for the VFR pilot will be increased and improved. Rules, regulations, their number and content will change and increase proportionately. Such is the price of air safety and an equitable distribution of our National Air Space. With this preamble let us begin study of Federal Aviation Regulations.

Chapter 2
The Pilot

Flight is an event that renders a sense of satisfaction unique in life's quota of experiences. The world of three dimensional motion offers new vantage points for viewing both the earth and the sky. As with anything worthwhile a significant effort is required to master the theory, practice, and discipline of flight. The sky is fickle in nature, often placid, sometimes hostile. Mastering this medium requires understanding, a bit of skill, and a large measure of determination.

To be certificated as a pilot an applicant must meet the minimum physical standards and aeronautical skill requirements listed in the federal aviation regulations. Three steps are involved in the process. First an applicant must take a physical examination given by an FAA aviation medical examiner. The purpose of this examination is to assure that no defects exist which may be detrimental to flight. Second, an applicant must take a written examination as a measure of his theoretical aeronautical knowledge. Third, an applicant must demonstrate his flight proficiency by a flight test. As we shall see in this chapter the regulations do not require a person to be a physical Superman or a mental Einstein. Average abilities are quite adequate to qualify for a pilot certificate.

MEDICAL STANDARDS

Medical standards are divided into three classes. To qualify for a student or private pilot certificate an applicant must pass the requirements for a third-class medical certificate. A third-class medical certificate is valid for a period of 24 calendar months and represents the minimum physical standards set forth by FAR. The certificate expires on the last day of the calendar month in which it was issued. For example, let us assume that a private pilot obtained a third-class medical certificate on Nov. 15, 1981. The certificate would then expire at midnight on Nov. 30, 1983.

A second-class medical certificate is required for commercial pilots and flight instructors. This certificate denotes slightly higher minimum physical requirements as logically should be expected of pilots involved in flight for hire. A second-class medical certificate is valid for a period of 12 calendar months after the date of issue whereupon it serves as a third-class medical certificate for an additional 12 calendar months. Therefore an applicant who qualified for a second-class medical certificate on Nov. 15, 1981 may render commercial aviation services until Nov. 30, 1982. Should he wish to continue commercial operations it will be necessary that he renew his second medical prior to this date. However, should he wish to fly only as a private pilot he may do so until Nov. 30, 1983, at which time he must renew his medical if he is to continue flying.

The first-class certificate is the most rigorous of the FAA physical examinations. This certificate is intended primarily for airline transport pilots. A first-class certificate is valid for a period of six months as a first class certificate, six more months as a second-class certificate, and twelve additional months as a third-class certificate (24 months in total). The medical certificate is a required companion to the pilot certificate. Both *must* be carried at all times when performing the duties of a pilot.

Third-Class Medical Certificate

To be eligible for a third-class medical certificate, an applicant must have distant visual acuity of 20/50 or better in each eye separately, without correction; or if the vision in either or both eyes is poorer than 20/50 and is corrected to 20/30 or better in each eye with corrective glasses, the applicant may be qualified on condition he wears glasses while acting as an airman. He must be able to hear the whispered voice at three feet, have no acute or chronic diseases of the internal ear or disturbances in equilibrium. In addition there must be no established medical history or diagnosis of personality disorders, psychosis, alcoholism, drug dependency, epilepsy, unexplained loss of consciousness, convulsive disorders, or other such items. The applicant must have no established medical history of myocardial infarction: or angina pectoris or other evidence of coronary heart disease that may reasonably be expected to lead to serious heart problems. A history of diabetes that requires insulin or similar agents for control may be disqualifying.

Fundamentally the third-class medical is a conventional physical examination whose purpose is to determine possible organic, functional, or other such defects which may make an applicant unable to safely perform the duties of an airman. A further purpose of the certificate is to give reasonable assurance that such will continue to be the case for a period of two years.

Second-Class Medical Certificate

The second-class medical certificate is similar to a third-class certificate except for tolerances being somewhat tighter. To be eligible for a second-class medical certificate an applicant must have distant visual acuity of 20/20 or better in each eye separately without correction; or at least 20/100 in each eye separately corrected to 20/20 or better with corrective glasses, in which case the applicant may be qualified provided he wears glasses while exercising the privilege of his airman certificate. In addition to normal fields of vision and the ability to distinguish aviation red, green, and white, the applicant must have certain other eye requirements related to bifoveal fixation and vergencephoria. An applicant for a second-class medical certificate must be able to hear the whispered voice at 8 feet with each ear separately. Other requirements are similar to those of the third-class certificate.

First-Class Medical Certificate

The first-class medical certificate embodies all of the re-quirements of the second-class certificate with additional emphasis on sight, hearing, and heart condition. An applicant for a first-class medical must have distant visual acuity of 20/20 or better in each eye separately, without correction; or at least 20/100 in each eye separately corrected to 20/20 or better with corrective glasses, near vision of at least $V=1.00$ at 18 inches with each eye separately, normal color vision, normal fields vision, and no acute or chronic conditions of either eye that might interfere with the applicant's ability to perform his function as a pilot. The wearing of glasses is permitted. The applicant must have the ability to hear the whispered voice at a distance of at least 20 feet with each ear separately; or demonstrate a hearing acuity of at least 50% of normal in each ear throughout the effective speech and radio range as shown by a standard audiometer. After age 35 an applicant is subject to an electrocardiographic examination to determine his heart condition. Blood pressure limitations are as shown in Table 2-1. Blood pressure limits are specified only for the first-class medical certificate. However, as a general guide for the second and third-class medical certificates a maximum reading of 170/100 is typical. The person's age, weight, and total physical condition is considered in cases of elevated blood pressure readings.

Medical Limitations

A medical certificate may be issued to an applicant who does not meet the medical standards required by FAR in certain instances. For example, an applicant who is moderately color blind may be able to read the standard aviation red, green, and white signals issued by a tower even though the medical examiners' cards were indistinct as to their color. In a situation of this nature the applicant may apply to the FAA GADO for a practical test of color blindness. A typical test is simply that of the applicant and a local FAA representative "reading" a group of light signals from a tower. If the applicant can successfully determine the color of the signals his operating limitations may be removed; however, should he be unable to do so it is likely that he will be prohibited from flight at night.

Denial of Medical Certificate

Any person who is denied a medical certificate may, within 30 days after the date of denial, apply in writing (in duplicate) to the Federal Air Surgeon, Attention: Chief Aeromedical Certification Branch, Civil Aeromedical Institute, Federal Aviation Adminis-

Table 2-1. First Class Medical Blood Pressure Limits.

Age Group	Maximum readings (reclining blood pressure in mm)		Adjusted maximum readings (reclin ing blood pressure in mm)	
	Systolic	Diastolic	Systolic	Diastolic
20-29	140	88	---	---
30-39	145	92	155	98
40-49	155	96	165	100
50 and over	160	98	170	100

tration, P.O. Box 25082, Oklahoma City, Oklahoma 73125, for reconsideration of that denial. If such action is contemplated by an applicant it is important that it be accomplished within 30 days after the date of the applicant's medical certificate denial. Following review by the FAA medical authorities the denial may be upheld, the applicant may be asked to supply additional medical evidence, or simply to demonstrate competence of airmanship (as in the case of a physical handicap). Assuming the latter the FAA may give a combined medical-private or medical-commercial flight test. If the pilot demonstrates competence his certificate may be renewed without waiver; that is, additional "medical" flight tests will not be required.

Medical Examinations: Who May Give

An FAA physical may be given only by an aviation medical examiner who is specifically designated for the purpose. Your family doctor may or may not be so designated. To obtain a list of qualified medical examiners, in any area, contact the FAA regional director or the FAA GADO. Most fixed base operators, flight schools, flight service stations, or state aviation authorities can provide a list of designated aviation medical examiners.

THE STUDENT PILOT

To obtain a student pilot certificate, a person must be at least 16 years of age, able to read, speak, and understand the English language; and qualify for at least a third-class medical certificate. A combination medical certificate and student pilot certificate will be issued, at your request, by the medical examiner upon the satisfactory completion of your physical examination. Student pilot certificates may be issued by FAA inspectors or designated pilot examiners if the applicant already possesses a valid medical certificate. Since a medical is required prior to solo flight it is really simplest to obtain a student pilot certificate by visiting your aviation medical examiner as the initial step in learning to fly. A combination medical and student pilot certificate is valid for a period of 24 calendar months. The certificate expires at the end of the calendar month in which it was issued two years hence. (For example, a certificate issued on January 2, 1981, will expire at midnight on January 31, 1983).

To solo is to taste adventure. A high point in any pilot's career, the initial solo requires adequate preparation. FAR's specify that, as a minimum, a student pilot be given training in aircraft pre-flight inspection, operating the aircraft engine, taxiing, takeoff, landing, traffic pattern procedures, level flight, turns, climbs, glides, stalls, and emergency landings prior to solo. Upon completion of these requirements and demonstration of competence in the control of an aircraft a flight instructor may endorse the student pilot's certificate that the holder is competent to solo. This endorsement must be made prior to the first solo flight and is an endorsement to solo *only in the make and model of aircraft so designated* by the flight instructor. A solo endorsement is also required in the student's log book. Except for special cases this solo log book endorsement is valid for a period of 90 days. After this time a student must fly with and have his log book endorsed again by a flight instructor in order to continue solo flight. These endorsements permit student solo only in a local area designated by the flight instructor. A student who is endorsed to solo in one type of aircraft may *not* solo in another unless his flight instructor has endorsed his student pilot certificate and log book to permit soloing in both aircraft types.

Solo cross country flight is perhaps the high point in student training. Regulations require that, prior to solo cross country flight, a student be given instruction (and demonstrate competence) in basic flight planning elements such as plotting courses, evaluation of weather reports, estimating time on route, and fuel required. In addition, competence in crosswind and simulated soft field takeoffs and landings, climbing and gliding turns at minimum safe air speeds, and cross country navigation by reference to aeronautical charts is required. Knowledge of safe operating procedures in simulated emergencies such as engine failure, loss of flying speed, marginal visibility, deteriorating weather, becoming lost, and other critical situations is necessary. Skill in conforming with air traffic control instructions by radio and lights, the proper use of two-way radio communications, VFR navigational procedures, and simple maneuvers by reference only to instruments is required prior to solo cross country flight. When a student demonstrates competence in the required skills, a flight instructor may endorse the student's certificate for solo cross country work. A log book cross country endorsement must be made by a certificated flight instructor prior to *each* solo cross country flight. A student must carry his log book on each solo cross country flight. The student's pilot certificate must also be carried on all solo flights. In the event operation of an aircraft radio transmitter is anticipated, a student must obtain an FCC radio telephone operator's permit. This permit can be obtained by sending the Federal Communications Commission an FCC Form 753-A. These forms are usually available at flying schools. Once obtained a radio telephone operator's permit is currently valid for life.

Student Pilot Limitations

Federal aviation regulations prohibit a student pilot from carrying passengers, or operating an aircraft for compensation or hire, or in the furtherance of a business. In addition, a student pilot is not permitted to make international flights. The purpose of the student pilot certificate is purely for training; extended flight privileges are reserved for those who pass the private pilot examination (or commercial, etc.).

The Written Examination

As a part of the effort in preparing for a private pilot's examination a student must pass a written test on the theory of flight, meterology, the Airman's Information Manual, FAR's, use of the flight computer, navigation, pre-flight planning, radio procedures, emergency procedures, and basic attitude flying using only aircraft instruments for reference. The minimum passing grade for the written test is 70. In general the written is given at the local FAA GADO and requires typically 3 to 3½ hours to complete. The student need supply only a navigation computer and plotter; text books or notes of any kind are forbidden. A portable calculator may be used provided it is battery-operated and contains no formulas. The written test is valid for a period of 24 months. In the event an applicant fails a written test he may apply for re-testing after 30 days after the date he has failed the test or upon presenting a statement from an instructor stating he has given additional instruction to the applicant and considers him competent for re-testing. Although ground school is not currently required by FAR many students find this an efficient means of preparing for the written examination. As a minimum the following study materials are recommended in preparation for the private pilot written examination:

1. *Student Pilot Guide* AC 61-12J
2. *Pilot's Handbook of Aeronautical Knowledge,* AC 61-23A.
3. *Private Pilot Written Test Guide,* AC 61-32C.
4. *Private Pilot Airplane Answer Book* 61-32CP
5. Federal Aviation Regulations Parts 1, 61, and 91.

6. Federal Aviation Regulations Written Test Guide AC 61-34B.

7. National Transportation Safety Board Investigation Regulations, Part 830.

8. *Medical Handbook for Pilots* AC 67-2.

9. *Private Pilot (Airplane) Flight Training Guide*, AC 61-2A.

All items are available from the Superintendent of Documents, Government Printing Office, Washington, D.C. 20402. Item 9 is specially recommended as a means of keeping track of flight training progress.

PRIVATE PILOT CERTIFICATE

To be eligible for a private pilot certificate an applicant must be at least 17 years of age, able to read, speak and understand the English language, hold at least a third-class medical certificate and within 24 months have passed a written examination for the rating of private pilot. In addition the applicant must have at least 40 hours of flight instruction of which 20 hours are solo time. Of the solo time at least 10 hours must be devoted to cross country flight. One cross country flight must include three landings, each of which is more than 100 nautical miles from each of the other two points. Three hours of night flight including 10 take offs and landings are also required.

Following the student's first solo cross country, FARs require the flight instructor provide at least three hours of training to include maneuvers previously learned, as well as additional maneuvers required for the private pilot test. While the regulations specify a minimum of 40 hours total flight time, a typical value for the average student is 53 hours. With the increasing complexity of flight we can expect this to approach 60 hours in the near future. The reason is simple. FAR's specify minimum hours only from the standpoint of aeronautical experience. The real requirement is proficiency, not hours! Proficiency limits are described in the FAA publication, *Flight Test Guide, Private Pilot, Airplane, Single-Engine AC 61-54A*. This publication is generally available at aviation flight schools.

When a person has acquired the necessary qualifications and has obtained a written recommendation for a flight test from an appropriately rated flight instructor he may apply for the private pilot flight test. The flight test can be taken at any FAA facility or from a designated pilot examiner. There is no charge for flight tests when conducted by an FAA inspector; however, designated pilot examiners are entitled to charge a reasonable fee. A private pilot test consists of three phases, an oral examination, a basic piloting technique, and a cross country test. FAR's require that an applicant perform the following procedures and maneuvers:

1. Phase I—Oral Operational Test. An applicant will be required to present and explain aircraft registration, airworthiness and equipment documentation, as well as airplane log books and airworthiness inspection reports. To determine that the applicant knows what performance and operating information is important he is required to demonstrate a practical knowledge of aircraft performance parameters, range, operation, weight and balance, etc. In addition, the oral test covers aircraft pre-flight procedures and the use of radio for voice communications.

2. Phase II—Basic Piloting Technique Test. The objective of Phase II is to evaluate the student's flying skill. A demonstration of pre-flight procedures, taxiing, normal and crosswind takeoffs and landings, climbs, level flight, descents, and flight at minimum controllable speeds is required. Stalls and stall recovery, 720 degree steep turns about a point, full stall landings, short and soft field takeoffs and landings, and engine out emergency landings are required.

3. Phase III—Cross Country Flight Test. The objective of this phase of testing is to determine whether the applicant can effectively prepare for a cross country flight in a reasonable period of time and furthermore conduct such in a safe expeditious manner using normally available aids and facilities. Before take-off for the flight test, the applicant will be requested to plan a cross country flight to a point at least two hours cruising range distance in the airplane to be used for the test. At least one intermediate stop will normally be included. The student is expected to procure pertinent available weather information, plot the assigned course, establish check points, estimate flying time, and fuel requirements. The use of the Airman's Information Manual for reference information and a flight computer for dead reckoning computations is expected. As a part of the flight test the examiner will request the student demonstrate cross country flying ability by following a designated course, the use of radio aids in VFR navigation, and instrument flight. During simulated instrument flight the student must be able to recover from the start of a power-on spiral, recover from the approach to a climbing stall, and execute normal turns of at least 180 degrees to within plus and minus 20 degrees of a pre-selected heading. Shallow climbing turns to a pre-determined altitude, shallow descending turns at reduced power to a pre-determined altitude, and straight and level instrument flight is required.

During a flight test the examiner acts in the capacity of an observer; the student is pilot in command. The examiner simply poses the problem, the student is responsible to demonstrate the correct response. A successful flight test is denoted by obvious mastery of the aircraft, the outcome of a maneuver must never be in doubt. A student who fails a flight test may apply for a re-test upon presenting a statement from his flight instructor that he has been given additional instruction and the flight instructor now considers the applicant ready for re-testing.

COMMERCIAL PILOT CERTIFICATE

To qualify for a commercial pilot certificate a person must be at least 18 years of age, be able to read, speak, and understand the English language, hold a valid first or second-class medical certificate, and have the aeronautical experience required (including an instrument rating). In addition, to take the commercial flight test, an applicant must have passed within 24 months the commercial pilot written examination and have a written recommendation from a flight instructor stating that he has qualified for a commercial flight test. An instructor's recommendation for a flight test is valid for 60 days.

Aeronautical experience required for the commercial certificate consists of a minimum of 250 hours of total flight time including at least 100 hours of flight time in powered aircraft. In addition, 100 hours of flight time must be as pilot in command including a minimum of 50 hours of cross country time. Cross country experience must include takeoffs and landings from two different airports under two-way radio instruction from an airport tower as well as one cross country flight which includes three landings, each of which is at least 200 nautical miles from each of the other two points. In preparation for the commercial flight test a student must have a minimum of 10 hours of instruction in a complex aircraft, 10 hours of instrument flight instruction, and 10 hours of flight instruction devoted to maneuvers required for the commercial pilot flight test. To qualify for night flight, an applicant must have at least 5 hours of flight time at night, including at least 10 takeoffs and 10 landings as the sole manipulator of the controls.

The commercial pilot flight test is conducted in four phases: an oral examination; basic flying techniques; precision flight maneuvers; and a cross country flight. The test is similar in principle to the private pilot examination except that a greater depth of knowledge, accuracy, and flying skill is expected. In addition, the applicant must demonstrate the required precision flight maneuvers. These maneuvers consist of gliding spirals, lazy eights, steep turns, chandelles, maneuvering at minimum controllable air speed, and stalls from all normally anticipated flight attitudes with and without power. Accuracy landings within 200 feet beyond a designated mark are also required. Proficiency requirements are listed in the FAA publication *Flight Test Guide, Commercial-Pilot-Airplane*, AC 61-55A.

A limited commercial certificate is available to those who have no need for an instrument rating. The limited commercial prohibits the carriage of passengers for hire in airplanes on cross country flights of more than 50 nautical miles or at night.

FLIGHT INSTRUCTOR CERTIFICATE

To be eligible for a flight instructor certificate an applicant must hold a commercial pilot certificate (including an instrument rating) without night flight limitations. In addition, the applicant must pass a written test on the fundamentals of flight instruction and the performance and analysis of flight training maneuvers. A first or second class medical certificate is required as well as a flight instructor written recommendation, made not more than 60 days before applying for the flight test.

The flight instructor examination consists of two phases; an oral examination and a flight test. Phase I requires that the applicant exhibit knowledge of flight instructor procedures and responsibilities, factors conditions and principles which control the learning process, and objectives and limitations of a lesson plan. As a part of this examination the applicant must prepare a lesson plan for flight instruction. Phase II, the flight test explores the full capabilities of the aircraft and the applicant including spins (or a log book entry thereof) and other precision maneuvers. Should a flight instructor wish to qualify to give instrument instruction, he must take the instrument instructor's written examination as well as a flight test devoted to instrument flight.

A certificated flight instructor is required to sign a student's log book for each period of flight instruction given and, in addition, record the name of the person to whom flight instruction has been given, the date, and the type of flight instruction involved. The flight instructor must keep a record of each person for whom he has signed a recommendation for a written test. Records are to be kept for a period of at least three years.

PILOT CERTIFICATES

Upon the successful completion of a flight test the pilot will be issued a temporary certificate which is valid for a period of 120 days. During this time the FAA Airman Certification Branch at Oklahoma City will issue a permanent pilot certificate. Pilot certificates for the private and commercial ratings are permanent in nature and are kept valid by the renewal of the associated medical certificate. At any time a person acts as pilot in command he must have in his possession a current pilot certificate and medical certificate. If requested a pilot shall present either or both to the FAA, the NTSB, or any federal, state, or local law enforcement officer for inspection. The holder of any certificate that is suspended or revoked shall, upon request, return it to the FAA. Flight instructor certificates and student pilot certificates automatically expire at the end of the 24th calendar month after the month in which they were issued.

To replace a lost or destroyed pilot certificate write to the Department of Transportation, Federal Aviation Administration, Airman Certification Branch, P.O. Box 25082, Oklahoma City, Oklahoma 73125. Include your name, permanent mailing address, social security number, date and place of birth, and any available information regarding the grade, number, and date of issue of the certificate. A check or money order for $2.00 payable to the FAA is required. In a similar fashion an application for a replacement of a lost or destroyed medical certificate is made by letter to the Department of Transportation, Federal Aviation Administration,

Table 2-2. Category, Class, and Type Nomenclature.

ITEM	CATEGORY	CLASS	TYPE
Aircraft	Transport Normal Utility Acrobatic Limited Restricted Experimental Provisional	Airplane Rotocraft Glider Balloon Landplane Seaplane	Cessna 150 Beech B19 Etc.
Pilot	Airplane Rotocraft Glider Lighter-than-Air	Single-Eng. Land Single-Eng. Sea Multi-Eng. Land Multi-Eng Sea Gyroplane Helicopter Airship Free Balloon	Cessna 150 Piper 140 Etc.

Civil Aeromedical Institute, Aeromedical Certification Branch, P.O. Box 25082, Oklahoma City, Oklahoma 73125. A check or money order for $2.00 is likewise required.

AIRCRAFT AND PILOT RATINGS

As a means of subdividing the various kinds of airplanes, FAR's use the words category, class, and type. Unfortunately, the same three words are also used for subdividing airmen ratings. Applied to an aircraft the word category describes the intended use of the vehicle with class being a descriptive modifier. Applied to a pilot the word category describes the fundamental airborne device involved with class as a descriptive modifier. To clarify the matter, Table 2-2 lists category, class, and type nomenclature applicable to the aircraft and to the pilot.

Type Ratings

A pilot's certificate describes the category, class, and type of aircraft the holder is rated to fly. For example a commercial pilot's certificate may read "Airplane, multi engine-land, Boeing 707." Thus the certificate clearly describes the category, class, and type of aircraft that the pilot is rated to fly. A private pilot may have a certificate which reads "Airplane, single engine-land". In this instance only the category and class of aircraft are specified. Type ratings are required for large aircraft (over 12,500 pounds), turbojet powered aircraft, and certain helicopters. The absence of a type rating on a certificate indicates the pilot is authorized to operate only small aircraft for which no type rating is required (Figs. 2-1 through 2-6). In order to obtain a type rating an applicant must hold or concurrently obtain an instrument rating, possess the required aeronautical experience, and demonstrate proficiency by a flight test. Since large aircraft often differ in their operating limits and techniques, type ratings are employed to require and test for pilot proficiency in the specific aircraft of concern.

Class Ratings

The class rating is an item of greater interest to the general aviation pilot than the type rating. For example, a pilot whose certificate reads "Airplane, single engine-land," wishes to transition to a multiengine-land aircraft. In this instance a class rating is involved. In order to obtain a class rating an applicant must show that he has received flight instruction in the class of aircraft for

Fig. 2-1. Another fast mover in the Cessna line, the centerline thrust Pressurized Skymaster cruises at 205 knots at 20,000 ft.

Fig. 2-2. The world's most popular trainer, the Cessna is available in Standard, 152 II, and Aerobat models.

which a rating is sought and has been found to be competent. In addition, he must pass a flight test appropriate to his certificate and the aircraft class rating sought. The flight test may be given by an FAA examiner or a designated pilot examiner and, for all practical purposes, is similar to the private or commercial flight test (depending upon the level the applicant seeks).

SOLO FLIGHT

A pilot may fly a small aircraft solo provided he meets one of the following conditions:

1. He holds a category and class rating for that aircraft.

2. He has received flight instruction in that category & class of aircraft and has been found competent to solo by a flight instructor.

3. He has soloed and logged pilot-in-command time in that category and class of aircraft prior to November 1, 1973.

To act as pilot in command of an aircraft that has more than 200 horsepower, or that has a retractable landing gear, flaps, and a controllable propeller, the pilot must receive flight instruction from a flight instructor to the extent necessary to prove competency. The flight instructor will then "sign off" the candidates log book accordingly.

CARRYING PASSENGERS

To carry passengers a pilot must have a category and class rating for the aircraft in which he intends to carry passengers as well as meet FAR requirements relating to recency of flight experience. FAR 61.57 states that, within the preceding 90 days, a

Fig. 2-3. One of the newest entries into the trainer field, the Beechcraft Skipper 77.

49

Fig. 2-4. With six aboard, the Cessna 421 Golden Eagle has a typical cruising range of 1118 miles.

pilot must have made at least three take-offs and three landings in an aircraft of the same category & class (and type if required) to act as pilot in command of an aircraft carrying passengers. If the aircraft is a tailwheel type the landings must be full stop. If he wishes to carry passengers at night he must have made the three landings and takeoffs within the past 90 days during the nighttime. Insofar as recency of experience is concerned, night is defined as beginning one hour after sunset and ending one hour before sunrise. The pilot is not required to perform takeoffs and landings both day and night in all types of aircraft in which he may carry passengers; night experience is transferrable. As an example let us take the case of a flight instructor that regularly gives instruction in a Cessna 150, a Cessna 172, a Cessna 182, and a Cessna 310. So long as he makes three takeoffs and landings in one of the four aircraft at night he is considered current to carry passengers in the other three types of aircraft during the day *or* night.

In addition to the 90 day requirement, a biennial flight review is required of all pilots not engaged in airline or commercial operations where the FAA already requires periodic flight checks. The review includes an examination of a pilot's knowledge of FAR's and flying skills appropriate to the pilot's certificate. The purpose of the review is to assure that at least once every two years each pilot rides with a competent instructor who can comment on his ability. The flight review is *not* a flight check! The person giving the review need only certify (in the pilot's logbook) that the pilot has successfully accomplished the review—an endorsement of the pilot's competency is not required.

Fig. 2-5. The four-place Beechcraft Sundowner 180.

51

Fig. 2-6. Piper Archer II carries four with only 180 hp.

PILOT LOG BOOKS

Flight time used to meet experience requirements for any pilot certificate or rating, or to meet recent flight experience requirements must be shown by a reliable record. The logging of other flight time is not required. Needless to say fraudulent or intentionally false entries to show compliance with certification requirements are prohibited.

CHANGE OF ADDRESS

In the event a pilot changes his permanent mailing address he is required to notify the FAA within 30 days. New address information should be forwarded to the Department of Transportation, Federal Aviation Administration, Airmen Certification Branch, P.O. Box 25082, Oklahoma City, Oklahoma 73125.

PILOT PROFICIENCY AWARDS

All pilots holding a private pilot certificate or higher and a current medical certificate are eligible to receive the Phase I Pilot Proficiency Award upon successful completion of the following recurrent training:

1. One hour of flight training to include basic aircraft control, stalls, turns, and other maneuvers directed to mastery of the aircraft.
2. One hour of flight training to include precision approaches takeoffs, and landings including crosswind, soft field, and short field techniques.
3. One hour of instrument training in an airplane or instrument simulator or training device.

4. Attendance at one accident prevention meeting or clinic conducted after July 15, 1979, held in cooperation with the FAA, and attendance attested to by an accident specialist or counselor.

Note: Requirements above must be completed after original pilot certification and must be completed within any 120 day period after July 15, 1979.

A pilot becomes eligible for the Phase II award by repeating the recurrent training program at least one year after completion of Phase I. The pilot may wait as long as two years (when his or her next Biennial Flight Review is required by regulation) to complete the training and become eligible for the Phase II award. A third repetition of the program earns the pilot the Phase III wings. Again the pilot must wait another year after receiving the previous award.

Chapter 3
The Aircraft

If ever variety was abundant, aviation was deeply blessed in its dawning moments. Early experimenters designed, constructed and flew (sometimes) an almost endless mixture of airplane configurations. Fuel sources included steam, electricity, diesel fuel, and gasoline. Building materials ranged from paper, wicker, and bamboo to duraluminum. Slowly, as the limits of technology increased, the evolutionary process of selecting the "fittest of the breed" laid the ground work for standardization of airplanes and motive power. The magnificent helium and hydrogen filled dinosaurs of the lighter-than-air age were relegated to extinction. The biplane gave way to the monoplane; the pusher to the tractor; and the radial engine bowed to the turboprop and turbojet. The flying wing designs of John Northrop (Fig. 3-1), the lifting body fuselage of Burnelli (Fig. 3-2), and the canard aircraft no longer traverse our airways. The airplane has become standardized in shape, motive power, quality, operation, and safety.

Title VI of the Federal Aviation Act of 1958 states that the FAA shall be empowered to regulate civil aircraft in air commerce by prescribing minimum standards governing the design, materials, workmanship, construction, and performance of aircraft, aircraft engines, and propellers as may be required in the interest of safety. Furthermore, the FAA is empowered to prescribe reasonable rules and regulations and minimum standards governing: (a) the inspection, servicing, and overhaul of aircraft, aircraft

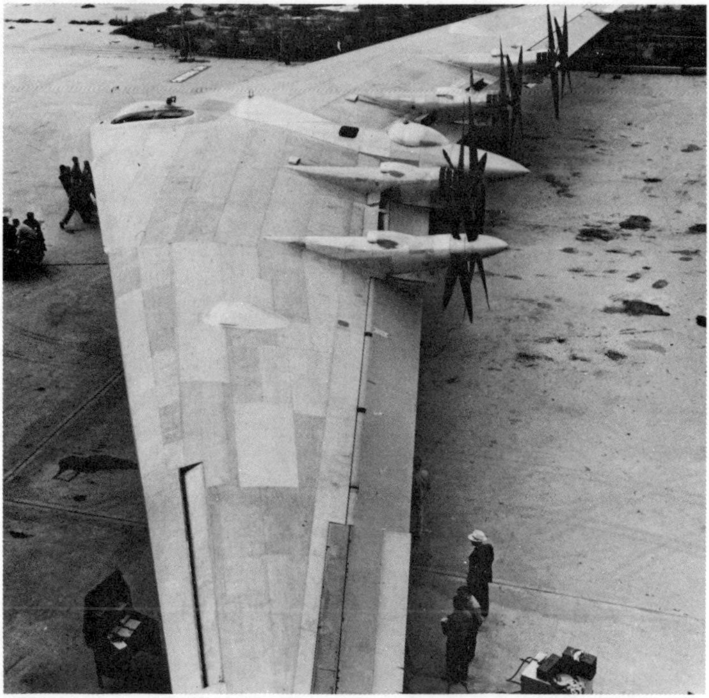

Fig. 3-1. The Northrop XB-35 flying wing. Powered by four Pratt & Whitney R-4360 engines, this unique aircraft employed tandem contrarotating propellers. The aircraft was designed as a military vehicle and first flew on 25 June 1946. A jet version, the YB-49, made its maiden flight 16 months later.

Fig. 3-2. A most unusual aircraft, the Burnelli CBY was built to a joint Canadian/British/U.S. specification. Powered by two R-1830 engines, the aircraft first flew in 1945; cruise speed 264 mph at 7000 ft. The 17,500 lb. aircraft once carried a payload of 27,500 lbs.

Fig. 3-3. The Beechcraft Baron 58TC is an excellent aircraft for high altitude fields and fast transportation over long ranges.

engines, propellers, and appliances; (b) the equipment and facilities for inspection, servicing, and overhaul; (c) periods for, and the manner in which inspection, servicing, and overhaul shall be made, including provision for examination and reports by properly qualified persons. Fundamentally, the FAA is granted "standardization" authority; an authority that extends from design concept throughout production, testing, and operation of an aircraft during its total life span.

GENERAL PROGRAM OF AIRCRAFT CERTIFICATION

The program of aircraft certification provides for the issuance of a *type certificate* which signifies a design as satisfactory, a *production certificate* which maintains a design through the process of manufacture, and an *airworthiness certificate* which denotes to the owner that his aircraft meets the requirements specified by the original type certificate. Since the airworthiness certificate is renewed by periodic maintenance inspections, the standards set forth by the original type certificate are theoretically retained throughout the life of the aircraft.

The purpose of the certification program is to govern standards required in the interest of public safety. This is accomplished by requiring *demonstrated proof of performance* to qualify for a certificate. Design techniques, material choices, and production processes are left to the manufacturer or individual so long as basic safety standards are met. The program provides for the individual in the process of constructing an experimental

56

aircraft of unique design as well as the manufacturer who may produce hundreds of look-alike aircraft (Figs. 3-3 through 3-8); either may obtain an airworthiness certificate for his final product.

TYPE CERTIFICATES

Type certificates are granted for aircraft, engines, propellers, and appliances (radios, instruments, etc). Aircraft type certificates denote weight and balance limits, propeller speed and pitch limits, stall speed limits, takeoff, climb, landing and spin characteristics. Controllability and maneuverability limits are specified as well as ground handling characteristics. Structural safety factors are specified for flight loads that act upon the various surfaces of the aircraft. Aircraft fuel, oil, cooling, induction, exhaust, power plant, instrumentation, electrical, hydraulic, lighting, oxygen, and other like subsystems are defined. In essence a type certificate is a *formal specification* which describes *all vital aircraft characteristics* including the properties of its subsystems.

Type Certificate: Normal, Utility, Aerobatic, and Transport Category Aircraft

Type certificates are granted for normal, utility, aerobatic, and transport categories of aircraft provided design data and test reports demonstrate that the product meets applicable airworthiness and aircraft noise requirements. The normal category type certification is limited to airplanes intended for non-aerobatic operation. Non-aerobatic operation includes maneuvers incident to normal flight, stalls (except whip-stalls), lazy eights, chandelles,

Fig. 3-4. The Cessna Conquest is typical of the corporate fleet.

Fig. 3-5. When it comes to large cargo doors and weight hauling the Cessna Stationair 7 is hard to beat.

Fig. 3-6. Cessna Skymaster features centerline thrust, an aircraft concept so successful that it was adopted by the Air Force as a military observation ship (0-2). With no minimum single-engine control speed, the Skymaster has a special FAA Centerline Thrust pilot Certificate rating.

and steep turns in which the angle of bank is not more than 60 degrees. The utility category provides for airplanes intended for *limited* aerobatic operation. Utility category aircraft may be certified for spins (if approved for the particular type of airplane), lazy eights, chandelles, and steep turns in which the angle of bank is more than 60 degrees. Some utility aircraft (including popular models) are extremely limited in spin recovery characteristics! Check with your local GADO to determine the exact limitations of your aircraft *before* entering into any maneuver that may result in an accidental spin! The aerobatic category is the only category provided for aircraft intended for use without restriction(other than those resulting from certification flight testing).

Another distinction between normal, utility, and aerobatic categories is the limit maneuvering load factor. For aircraft in the normal category that do not exceed a gross weight of 4,117 pounds, the positive limit maneuvering load factor is 3.8 G's. Aircraft in excess of this weight are permitted a lower positive G limit depending upon weight but not less than 2.1 G's. Utility category airplanes must accommodate a positive limit load factor of 4.4 G's. The positive limit load factor for aerobatic category aircraft is 6.0 G's. Negative load limit factors are required to be at least 40% of the positive load factor for the normal and utility categories and 50% of the positive load limit factor for aerobatic category aircraft. The limit load factors represent the minimum strength of the weakest part of the aircraft when subjected to G loadings from turbulence, steep turns, dive recovery, and the like. These loadings apply with flaps up only. With flaps down positive G loadings are reduced by a factor of 50% and negative G loadings are reduced to zero!

Type Certificate, Restricted Category Aircraft

Aircraft intended for special purpose operations may be granted a restricted category certificate if it can be shown that the aircraft is safe when operated under the limitations prescribed for its intended use. Such aircraft are not permitted to operate for reasons other than the special purpose for which the aircraft was certificated. Crew members are limited to persons essential to the special purpose function. Restricted category civil aircraft may not operate over densely populated areas, in congested airways, or near a busy airport where passenger transport operations are conducted. Typical of special purpose operations for which a restricted type certificate may be granted are:

Fig. 3-7. One of the most economical aircraft ever devised, the Mooney 201 delivers "mileage" of 19.3 mpg at 185 mph and 12,000 ft.

Fig. 3-8. The sleek E55 Beech Baron can carry a useful load of over 2,000 lbs.

1. Agricultural spraying, dusting, and seeding, and predatory animal control.

2. Forest and wild life conservation.

3. Aerial surveying (photography, mapping, oil, and mineral exploration).

4. Patrolling (pipeline, powerlines, and canals).

5. Weather control (cloud seeding).

6. Aerial advertising (sky writing, banner towing, airborne signs and public address systems).

Type Certificate, Provisional Category Aircraft

Provisional type certificates are *temporary* certifications issued for a limited time special purpose operation; 24 months for a class I certificate and 12 months for a class II certificate. The provisional type certificate is typically granted for:

1. Demonstration flights by a manufacturer for prospective purchasers; as in the case of a foreign built aircraft being demonstrated in the United States.

2. Market surveys by a manufacturer.

3. Flight checking of instruments, accessories, and equipment that do not affect the basic airworthiness of the aircraft.

4. Service testing of the aircraft.

Only persons who have a proper interest in the special purpose operation being conducted or who are specifically authorized by the manufacturer and/or the FAA may be carried in a provisionally certificated aircraft.

Other Type Certificates

The FAA grants type certificates for gliders, surplus military aircraft, and various imported products. Supplemental type certificates are issued to update an original type certificate or extend its use (as in the case of a new model of an existing aircraft). The holder of a type certificate is required to report any defect in any product or part manufactured that could result in a structural failure, flight control system malfunction, engine failure, propeller failure, fire, brake failure, or other hazardous conditions.

AIRWORTHINESS CERTIFICATES

Standard airworthiness certificates (Fig. 3-9) are granted aircraft in the normal, utility, aerobatic, and transport categories at the time of manufacture and remain current provided maintenance and alterations to an aircraft are performed in accordance with FAR 43 (Maintenance, Preventative Maintenance, Rebuilding and Alterations) and FAR 91 (General Operating Flight Rules).

Special airworthiness certificates are granted to restricted, limited, provisional, and experimentally qualified aircraft. These certificates remain in effect as long as prescribed maintenance operations are conducted with the exception of experimental certificates which are valid for one year after the date of issue or renewal (unless a shorter period is prescribed by the FAA).

The experimental airworthiness certificate deserves special consideration for under this certificate amateur built aircraft are certified airworthy. The growth of the amateur-built aircraft movement under the direction of the Experimental Aircraft Association is one of the truly significant events occurring in

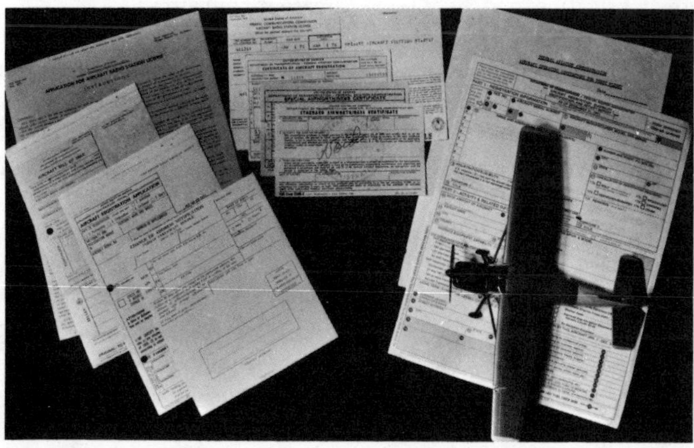

Fig. 3-9. Every aircraft must carry an Airworthiness Certificate, and a Certificate of Registration. An internal revenue tax (IRS Form 4638) is an annual obligation.

Fig. 3-10. Symbol of the Experimental Aviation Association, this fully aerobatic aircraft, the Acro Sport, is typical of amateur-built aircraft.

aviation today (Fig. 3-10). This revitalization of flying at the grass roots level was made possible by changes in federal law which extended airworthiness certification to amateur built aircraft (circa 1950). As a result, amateur built aircraft are competing with—and often exceeding—performance capabilities of production aircraft! At present, amateur built aircraft hold the world championship for aerobatics and numerous racing titles. It is interesting to note that NASAD (National Association of Sport Aircraft Designers) has adopted a system of *voluntary minimum standards* for the specification, fabrication, and flight test of amateur built aircraft.

Experimental airworthiness certificates are awarded amateur built aircraft on the basis that the major portion of the fabrication and assembly is conducted by persons for their own education or recreation. FAA inspections are required prior to enclosing the aircraft framework and following final assembly of the craft. Typically, 50 hours of flight test is initially required, to be conducted in a specified test area. The purpose of the flight test program is to demonstrate that the aircraft is controllable throughout its normal range of speed and maneuvers; furthermore, that the aircraft has no hazardous operating characteristics or design features. Assuming the aircraft meets minimum safety criteria from the standpoint of design and workmanship and successfully passes the assigned flight test program, an experimental airworthiness certificate will be issued. Experimental aircraft are not permitted to carry persons or property for compensation or hire. Furthermore, such aircraft may operate under VFR day-only conditions unless otherwise specified by FAA. Except for takeoffs and landings, operation of an aircraft with an experimental certificate is not permitted over a densely populated area or in a congested airway. Passengers must be advised of the experimental nature of the aircraft.

In addition to amateur built aircraft, experimental airworthiness certificates are granted for research and development purposes, demonstrating compliance with regulations, crew training, exhibitions, motion pictures, television, air racing, and market surveys. Experimental aircraft may employ non-conventional materials, unique fabrication techniques, and engines

Fig. 3-11. Typical of general aviation radio equipment which meets FAA Technical Standard Orders (TSO's) are the King KT 76 hi-altitude transponder, KT 78 low altitude transponder, and KX 175 Nav/Comm.

that are not type certificated. Amateur built aircraft have employed automotive engines, ground power unit engines, and, at least in one case, multiple go-cart engines for power.

Limited Airworthiness Certificates

Limited airworthiness certificates are specialized airworthiness authorizations typically granted for unique situations such as the conversion of an ex-military aircraft to civil use.

INSTRUMENTATION AND AVIONICS EQUIPMENT STANDARDS

Just as the type certificate is a standard for an airframe, an engine, or propeller, the Technical Standard Order (TSO) is the standard for aircraft instrumentation and avionics equipment (Fig. 3-11). The purpose of the TSO is to insure that an article will operate satisfactorily and accomplish its intended purpose under specified use conditions. In order to obtain a TSO authorization a manufacturer must demonstrate that his device, (a transponder for example) meets minimum performance requirements under the environmental conditions specified. A manufacturer of an article for which a TSO authorization has been issued may market the device providing that he maintains an adequate quality control system, a current file of complete technical data and records, permits the FAA to inspect the article as well as the manufacturer's facilities, and reports any failures, malfunctions, or defects that may result in a hazardous situation to the user. The TSO is *not* an ironclad guarantee of equipment quality and reliability. However, it goes a long way toward providing the dependable service all airmen desire of their avionics equipment.

At the present time only certified air carriers are required to equip their aircraft with TSO approved equipment. An exception is the transponder. Since improperly operating transponders raise havoc with the ATC system, general aviation aircraft are required to use only transponders that conform to TSO specifications. Furthermore, transponder equipment is subject to biennial operational testing to assure proper operation.

DELEGATION OPTION AUTHORIZATION

Fundamentally, the FAA is ultimately responsible for the preservation of flight safety. This obligation is carried out (in part) by the issuance of type certificates, production certificates, airworthiness certificates, TSO's, etc. However, Title III of the Act which established the FAA contains a unique clause that permits the FAA to delegate to a private person the authority to grant type certificates, airworthiness certificates, and experimental certificates. This authority, known as Delegation Option Authorization (DOA), permits a manufacturer to act in behalf of FAA in the inspection, test, and certification of products for which he has received DOA. A manufacturer that is granted delegation option authorization shares with the FAA the responsibility for the airworthiness and safety of his products.

AIRCRAFT OWNERSHIP

The ownership of an aircraft as an individual, a partnership, a club, or as a corporation is becoming an increasingly common occurrence. Aircraft ownership is in many respects similar to ownership of an automobile. Both are subject to registrations, licensing, "rules of the road," and accident reporting procedures. Furthermore, in many states automobiles are required to have an annual inspection for safety purposes just as aircraft are required to have an annual inspection. The primary differences between the two are the requirements for controlled maintenance of aircraft and registration by the federal rather than the state government. As an owner of an aircraft, you are responsible for the following:

1. Registering and recording your aircraft title in accordance with Parts 47 and 49 of the Federal Aviation Regulations.

2. Keeping your airworthiness certificate current by having your aircraft inspected annually and complying with applicable AD notices.

3. Maintaining your aircraft and avionics equipment in an airworthy condition.

4. Assuring that all maintenance is properly recorded (in both the engine and airframe log books).

5. Keeping abreast of current regulations concerning the operation and maintenance of your aircraft.

6. Notifying the FAA Aircraft Registry immediately of any change of permanent mailing address or of the sale or export of your aircraft.

Aircraft Purchase And Operation

In buying an aircraft there is no substitute for examining the aircraft records to obtain a history of the ownership of the aircraft and to determine if there are any outstanding liens or mortgages—*before* purchase! This procedure will help avoid delay in registering an aircraft, and the headaches many aircraft purchasers have suffered because they failed to take this one important step. Naturally, the purchase of an aircraft from a reputable distributor or dealer is desirable. However, even these people may fail to determine the true title status of an aircraft.

When buying a used aircraft, it is desirable to have the aircraft inspected by a qualified person or facility *before* signing a sales agreement. The condition of the aircraft and the state of its maintenance records can be determined by persons knowledgeable on that particular make and model. These include a certificated Airframe and Power plant (A & P) mechanic or an approved repair station. A check of public records may likewise be of value. All aircraft public records are maintained by the FAA and are on file at the following address:

> Department of Transportation
> FAA Aeronautical Center
> Aircraft Registration Branch
> AC-250, P. O. Box 25082
> Oklahoma City, OK 73125

As a minimum, the purchaser of either a new or a used aircraft may expect to receive at time of purchase the following legal documents:

1. Bill of Sale (Aircraft Title).
2. Airworthiness certificate.
3. All aircraft and engine log books.
4. A list of equipment installed.
5. Weight and balance data.
6. Appropriate maintenance manual, service letters, bulletins, etc.
7. Airplane flight manual or operating limitations.

The following documents must be carried aboard the aircraft at all times:

1. Airworthiness certificate (must be displayed within the aircraft).
2. Certificate of registration.
3. Operations limitations and/or airplane flight manual. For aircraft over 6,000 pounds certificated gross weight, an FAA approved airplane flight manual is required; under 6,000 pounds it is the manufacturer's option to provide one.
4. FCC radio station license if the aircraft is equipped with a radio transmitter.

These items are *required* whether you *own, lease,* or *rent* an aircraft!

An aircraft title lists legal ownership of property and specifies any obligations that may be placed against the property. Ideally a "clear title" implies that there are no encumbrances such as liens, chattel mortgages, or other unsatisfied claims against the aircraft. A title search must be conducted to establish the exact title conditions of an aircraft. This may be accomplished a number of different ways. First, you may search the aircraft records yourself. The FAA GADO will be able to assist you in organizing and conducting your search. You may also write to the FAA Aircraft Registry; a list of title search companies AC Form 8050-55 will be furnished upon request. Second, an aircraft title search may be conducted by an attorney or a qualified aircraft title search company. Third, your local banker may be of assistance. Most lending institutions that finance aircraft require a title search before granting a loan. The purchaser of a home wouldn't consider such a transaction without a title search; an aircraft buyer should do no less when purchasing an aircraft.

In general, a bill of sale which meets the requirements of FAA Form 8050-2 is adequate to describe aircraft ownership. Other conveyances are a contract of conditional sale, mortgage, assignment of mortgage, or other instruments affecting title to or interest in the property. Part 49 of the FAR's requires that all aircraft titles be recorded. If a person submits a title (a conveyance) for recording and wants the original returned to him, he must submit a true copy with the original. After recording, the copy is kept by the FAA and the original is returned to the applicant stamped with the time and date of recording. The copy must be imprinted on paper permanent in nature, including dates, and signatures to which is attached a certificate by the person submitting the conveyance stating that the copy has been compared with the original and that it is a true copy. If the seller of an aircraft is not shown on the FAA record as the owner of the aircraft, a conveyance, including a contract of conditional sale, submitted for recording must be accompanied by a bill of sale or similar document showing consecutive transfers from the last registered owner, through each intervening owner, to the seller. Aircraft titles apply to aircraft, engines developing more than 750 horsepower, and propellers able to absorb greater than 750 rated takeoff shaft horsepower. A fee for recording an aircraft title is required.

If you purchase an aircraft, before you fly it you must apply for a certificate of registration in accordance with FAR 47. An aircraft is eligible for registration only if it is owned by a citizen of the

United States, and is not registered under the laws of any foreign country. When applying for a registration certificate, an aircraft bill of sale or other evidence of ownership must be submitted. The certificate of registration does not indicate aircraft ownership as such. The FAA issues a certificate of aircraft registration to the person who appears to be the owner on the basis of the evidence of ownership submitted with the application for aircraft registration, or recorded at the FAA aircraft registry.

To apply for a certificate of aircraft registration it is only necessary to fill out FAA Form 8050-1. The "pink" copy of the application is placed in the aircraft until the permanent certificate of registration is received from the FAA. This certificate, FAA Form 8050-3, replaces the "pink" copy in the aircraft. The certificate of registration expires when:

1. The aircraft is registered under the laws of a foreign country;

2. The registration is cancelled at the written request of the holder of the certificate;

3. The aircraft is totally destroyed or scrapped;

4. Ownership of the aircraft is transferred;

5. The holder of the certificate loses his United States citizenship; or

6. Thirty days have elapsed since the death of the holder of the certificate.

When an aircraft is sold the previous owner (seller) must notify the FAA by filling in the back of his certificate of registration and mailing it to the FAA Aircraft Registry.

The holder of a certificate of aircraft registration is required to submit an Aircraft Registration Eligibility, Identification, and Activity Report. This form must be submitted to the FAA Aircraft Registry by April 1 each year. Part I is mandatory, Part II is voluntary. Refusal or failure to submit the required information may be cause for suspension or revocation of the holder's certificate of aircraft registration.

Within thirty days after any change in his permanent mailing address, the holder of a certificate of aircraft registration must notify the FAA Aircraft Registry of his new address. A revised certificate of aircraft registration will then be issued without charge.

An aircraft radio station license, FCC Form 556, is a required item of aircraft legal paperwork if the aircraft contains any radio transmitter. Application for an aircraft radio station license (FCC

Form 404) may be obtained from your local FCC office or from the Federal Communications Commission, Gettysburg, Pennsylvania, 17325. A new station license is required when the aircraft is first licensed, aircraft ownership is changed, new transmitter equipment is added, or a period of five years has elapsed from original issue (use FCC Form 405-B for renewal, if no change from previous license). When an aircraft is sold the seller's radio station license (if valid at the time of sale) remains valid for thirty days for the benefit of the buyer, provided the buyer applies for a new aircraft radio station license at the time of sale.

The FAR's require any owner or operator who sells a U.S. registered aircraft to transfer to the purchaser, at the time of sale the maintenance records of the aircraft. The records must contain at least the following information:

1. The total time in service of the airframe;

2. The current status of life-limited parts of each airframe, engine, propeller, rotor, and appliance;

3. The time since last overhaul of all items installed on the aircraft which are required to be overhauled on a specified time basis;

4. The identification of the current inspection status of the aircraft, including the times since the last inspections required by the inspection program under which the aircraft and its appliances are maintained;

5. The current status of applicable airworthiness directives, including the method of compliance;

6. A list of current major alterations to each airplane, engine, propeller, and appliance.

Maintenance records must contain a description of the work performed, the dates of completion of the work performed, and the signature and certificate number of the person approving the aircraft for return to service. An aircraft owner or operator shall make all maintenance records available to the NTSB if so requested.

In order to fly your new aircraft the FAR's require that a current FAA approved aircraft flight manual*, placards, listings, and instrument markings which describe operating limitations for that aircraft, be available to the pilot. As a minimum, power plant limitations, airspeed operating ranges, aircraft weight and balance limits, minimum flight crew, kinds of operation permitted, and maximum operating altitude must be described.

*If required for that model of aircraft.

AIRCRAFT MAINTENANCE

The term "maintenance" applied to an airplane can be subdivided into three categories: inspection, preventive maintenance, repairs and alterations. FAR 91.163 states, "The owner or operator of an aircraft is primarily responsible for maintaining that aircraft in an airworthy condition . . ." Furthermore, FAR 91.165 goes on to state, "each owner or operator of an aircraft shall have the aircraft inspected, etc. . . . In addition, he shall insure that maintenance personnel make appropriate entries in the aircraft and maintenance records indicating the aircraft has been released to service." Thus it becomes very clear that the *aircraft owner or operator* is the individual responsible to have the aircraft inspected, preventive maintenance performed, and repairs or alterations made as necessary to maintain the aircraft in an airworthy condition in accordance with its type certificate!

Inspections Required

An airplane must have an annual inspection every twelve calendar months. This inspection may be performed by a certificated airframe and power plane mechanic holding an FAA Inspection Authorization (IA), an FAA certificated repair station, or the manufacturer of the aircraft if he meets the requirements of the regulations. In the event more than twelve calendar months have elapsed from the last annual, the aircraft is no longer airworthy and cannot be flown unless a special FAA ferry permit is obtained.

An aircraft used to carry passengers for hire, or for flight instruction for hire, must be inspected within each 100 hours of time in service be a certificated A & P or IA mechanic, an appropriately rated FAA repair station, or the manufacturer of the aircraft. The annual inspection is acceptable as a 100-hour inspection, but the reverse is not true. The 100-hour limitation may be exceeded by not more than 10 hours if necessary to reach a place at which the inspection can be done. The excess time, however, is included in computing the next 100 hours of time in service. When renting an aircraft for a cross-country, take care to check the maintenance records. Assure yourself that adequate time remains to complete your cross-country before the next 100-hour inspection is due. In the event IFR operation is anticipated, additionally check to assure that the altimeter and static system has been inspected within the preceding 24 calendar months. A transponder check is also required every 24 calendar months.

Following an inspection it is the responsibility of the IA mechanic to certify that the aircraft is again airworthy. This is accomplished by entering in the aircraft maintenance records a description of the type of inspection, the date of the inspection, and the signature and certificate number of the person approving the aircraft for a return to service. In the event an aircraft fails an inspection the IA is obligated to provide a list of discrepancies and unairworthy items to the aircraft owner or lessee. The local FAA district office may similarly be provided with a list of discrepancies.

Surprise! The letter from FAA containing form 8320 states that you have been selected to have your aircraft examined for compliance with applicable maintenance standards. Would you kindly grant permission to FAA to conduct an inspection of your aircraft? The use of a "spot check" is a valuable tool in any quality control program and its utilization by FAA is in keeping with a basic responsibility for maintaining standards of aircraft maintenance. Should you be solicited for a spot check of your aircraft, the inspection cost will be borne by FAA and the work will be accomplished at a time convenient to you.

Preventive Maintenance

A significant factor in reducing flying costs is the fact that certain maintenance is not classified as major and thus may be performed by the pilot himself. Federal Aviation Regulations state that "the holder of a pilot certificate issued under FAR 61 may perform preventive maintenance on any aircraft owned or operated by him that is not used in air carrier service." Preventive maintenance means simple or minor preservation operations and replacement of small standard parts not involving complex assembly operations. Work of the following type is classified as preventive maintenance:

1. Removal, installation, and repair of landing gear tires.
2. Replacing elastic shock absorber cords on landing gear.
3. Servicing landing gear shock struts by adding oil, air, or both.
4. Servicing landing gear wheel bearings, such as cleaning and greasing.
5. Replacing defective safety wiring or cotter keys.
6. Lubrication not requiring disassembly other than removal of nonstructural items such as cover plates, cowlings, and fairings.

7. Making simple fabric patches not requiring rib stitching or the removal of structural parts or control surfaces.

8. Replenishing hydraulic fluid in the hydraulic reservoir.

9. Refinishing decorative coating of fuselage, wings, tail group surfaces (excluding balanced control surfaces), fairings, cowling, landing gear, cabin, or cockpit interior when removal or disassembly of any primary structure or operating system is not required.

10. Applying preservative material to components where no disassembly of any primary structure or operating system is involved and where such coating is not prohibited or is not contrary to good practices.

11. Repairing upholstery and decorative furnishings of the cabin or cockpit interior when the repairing does not require disassembly of any primary structure or operating system or interfere with an operating system or affect primary structure of the aircraft.

12. Making small simple repairs to fairings, nonstructural cover plates, cowlings, and small patches and reinforcements not changing the contour so as to interfere with proper airflow.

13. Replacing side windows where that work does not interfere with the structure or any operating system such as controls, electrical equipment, etc.

14. Replacing safety belts.

15. Replacing seats or seat parts with replacement parts approved for the aircraft, not involving disassembly of any primary structure or operating system.

16. Trouble shooting and repairing broken circuits in landing light wiring circuits.

17. Replacing bulbs, reflectors, and lenses of position and landing lights.

18. Replacing wheels and skis where no weight and balance computation is involved.

19. Replacing any cowling not requiring removal of the propeller or disconnection of flight controls.

20. Replacing or cleaning spark plugs and setting of spark plug gap clearance.

21. Replacing any hose connection except hydraulic connections.

22. Replacing prefabricated fuel lines.

23. Cleaning fuel and oil strainers.

EMERGENCY AIRWORTHINESS DIRECTIVE
DEPARTMENT OF TRANSPORTATION
FEDERAL AVIATION ADMINISTRATION

FLIGHT STANDARDS SERVICE
FLIGHT STANDARDS NATIONAL FIELD OFFICE
P.O. BOX 25082
OKLAHOMA CITY. OKLAHOMA 73125

June 7, 1979

Pursuant to the authority of the Federal Aviation Act of 1958, delegated to me by the Administrator, the following Airworthiness Directive is issued and applicable to all aircraft equipped with Airborne Aviation Products Corporation dry air pumps. This directive is effective immediately upon receipt of this letter. Compliance required prior to next flight.

Action requires review of serial numbers of dry air pumps and removal of affected serial numbered pumps.

AIRBORNE AVIATION PRODUCTS CORPORATION: Applies to the below listed part number dry air pumps installed on piston engine aircraft certificated in all categories.

AIRBORNE PART NUMBER	SERIAL NUMBERS
211CC	5E9318 thru 5E9347
	5E9407 thru 5E11419
211CC-9	5E616 thru 5E715
211CC TR	5E1264 thru 5E1406
212 CW	5E3403 thru 5E4197
	5E9129E thru 5E9131E
212 CW-6	5E9 thru 5E25
242 CW-4	5E8 thru 5E11
441 CC	5E332 thru 5E401
	5E450 thru 5E483
441 CC-7	5E911 thru 5E981
441 CC-9	5E75 thru 5E80
441 CC-11	5E4
441 CC-13	5E7 thru 5E12
441 CC-17	5E106 thru 5E116
442 CW	5E926 thru 5E1023
442 CW-4	5E137 thru 5E149
442 CW-6	5E765 thru 5E785
442 CW-8	5E114
442 CW-12	5E431 thru 5E435

These pumps were not available for installation before May 15, 1979, therefore dry air pumps installed previous to that date are exempt from this AD. Compliance is required prior to next flight. To preclude loss of vacuum source due to possible pump bearing seizure, remove above listed dry air pumps from service and replace with an airworthy pump of the same part number. An airworthy pump is one which has a serial number not listed above or if listed above also has an "A" or "2" ink stamped by the manufacturer with black ink on the periphery of the body near the mounting flange. The aircraft may be flown, under day VFR conditions, in accordance with FAR 21.197 to a base where the corrective action can be performed.

Airborne Aviation Products Corporation Service Letter, Number 22A, dated June 5, 1979, applies to the subject matter of this AD.

FOR FURTHER INFORMATION CONTACT:

C.L. Smalley, Engineering and Manufacturing Branch, Flight Standards Division, AGL-213, Federal Aviation Administration, 2300 East Devon Avenue, Des Plaines, IL 60018, Telephone (312) 694-4500, Extension 379.

W.J. BARLOW
Acting Director, FAA Great Lakes Region

Fig. 3-12. Example of FAA Emergency Airworthiness Directive.

24. Replacing batteries and checking fluid level and specific gravity.

25. Removing and installing glider wings and tail surfaces that are specifically designed for quick removal and installation and when such removal and installation can be accomplished by the pilot.

Major Repairs or Alterations

When a major repair or alteration to an airplane is accomplished, it must be inspected and the airplane returned to service by an A & P mechanic with an FAA Inspection Authorization, or a properly certificated repair station, manufacturer, air carrier or commercial operator. Repair and alteration work may be accomplished by a mechanic, a repair man, or a person working *under the supervision of a mechanic or repair man* providing the supervisory mechanic *personally observes* the work being done to the extent necessary to insure that it is accomplished properly. Furthermore, the supervisor must be readily available, *in person*, for consultation during the process. However, the authority to participate in a major repair or alteration does not extend to inspections. Major repairs include such items as the strengthening, reinforcing, splicing, and manufacturing of primary structural members of an aircraft, the overhaul of a power plant, the overhaul of a propeller, and the calibration and repair of instruments and radio equipment. Persons performing a major repair or major alteration are required to describe the service on FAA Form 337; one copy is presented to the aircraft owner and a second copy is forwarded to the local FAA district office.

AIRWORTHINESS DIRECTIVES

Airworthiness directives (commonly referred to as AD's or AD notes), are intended to alert aircraft owners of unsafe aircraft conditions that exist or are likely to exist. The AD note identifies the condition and specifies the required corrective action, and the conditions or limitations, if any, under which the aircraft may continue to be used. FAR's require a chronological record be maintained of all AD's on which action has been taken. This record must include the date, AD number, a brief description of the method of compliance, and the signature and certificate number of the repair station or mechanic who compiled the AD. It is the aircraft owner's responsibility to assure compliance with all pertinent AD's. This includes those AD's that require recurrent or continuing action. For example, an AD may require a certain inspection every 50 flight hours, which means that the particular inspection must be accomplished *and recorded* for every 50 hours of flight.

Airworthiness directives are published in the *Federal Register* as a formal means of distributing information. In addition, AD's are distributed to certified repair stations and published in summary form in two volumes. Volume I includes directives applicable to small aircraft (12,500 pounds or less); Volume II includes directives applicable to large aircraft. These summaries may be obtained from the Superintendent of Documents, Washington, D.C., by asking for the *Summary of Airworthiness Directives* Vol. I, Small Aircraft, or Vol. II, Large Aircraft. Cost of the summaries is currently $21.00 and $18.00 respectively.

The Airworthiness Directive is one of the finest systems of preventive maintenance yet devised. The rapid distribution of maintenance information throughout the aviation community is in keeping with the basic FAA charter to uphold public safety. AD's are sometimes judgmental in nature and, as such, may be subject to question. However, when equated against the exorbitant cost of funerals even the most questionable AD is still a bargain. AD's are issued by the FAA based upon recommendations of the NTSB, recommendations of manufacturers, or as a result of accident statistics. An example AD is shown in Fig. 3-12.

Chapter 4
Preflight Planning

Aviation regulations are sometimes short in their narrative form and at the same time extensive in coverage. The regulations which define the responsibility and authority of the Pilot-in-Command (PIC) are of this exact nature; brief, concise, and far-reaching in meaning and intent.

FAR 91.3 states: *"The pilot-in-command of an aircraft is directly responsible for, and is the final authority as to, the operation of that aircraft."* This brief but important regulation defines without doubt that the pilot-in-command is the individual responsible for the operation and safety of the aircraft. Just who is the pilot-in-command?

1. Student Pilot. A student pilot is the pilot-in-command during all authorized solo flights of his training and during the flight test as an applicant for a pilot certificate.

2. Private and Commercial Pilots. A private, commercial, or air transport pilot is considered pilot-in-command during such time he (or she) is the sole manipulator of the controls of an aircraft for which he is rated, or when the sole occupant of an aircraft, or when acting as pilot-in-command of an aircraft on which more than one pilot is required under the type certification for the aircraft.

3. Flight Instructors. A certificated flight instructor is the pilot-in-command during all flight time in which he acts in the capacity of a flight instructor.

4. Flight Examiners. FAA inspectors or other authorized flight examiners are *not* pilot-in-command of an aircraft during a

Fig. 4-1. A beautiful pair, the Beechcraft Bonanza F33A (foreground) and A36.

flight test unless required to act in that capacity for the flight, or a portion of the flight. Basically, the inspector observes the applicant's ability to perform procedures and maneuvers of the flight test.

5. All Pilots. To act as pilot-in-command one must have a valid pilot certificate for the type of aircraft involved, possess a current medical certificate, and meet recency of flight experience requirements.

When a pilot elects to act as pilot-in-command, the authority granted by this position carries with it the undeniable responsibilities of the post.

PREFLIGHT RESPONSIBILITY OF THE PILOT-IN-COMMAND

FAR 91.5 states: *"Each pilot-in-command, shall before beginning a flight, familiarize himself with all available information concerning that flight."* As an absolute minimum this information must include weather reports and forecasts, fuel requirements, and alternate landing fields for flights extending a distance greater than 25 statute miles from the departure airport. In addition, the pilot-in-command must familiarize himself with runway lengths at airports of intended use as well as takeoff and landing distances taking into account airport elevation, runway slope, aircraft gross weight, wind conditions, temperature and density altitude. As the person responsible for, and final authority as to the operation of an aircraft, the pilot-in-command is also obligated to determine that the aircraft is airworthy prior to flight. FAR's 91.3 and 91.5 leave no doubt as to the responsibility of the PIC to accomplish a thorough job of preflight planning.

The degree of preflight planning required to fulfill the obligations of a pilot-in-command varies with the type of flight involved (business, recreational, training, etc.), type of aircraft (Figs. 4-1 through 4-6) and flight route. In reality, each flight is unique in one way or another and requires individual preflight planning. However, all flights require attention to at least the following areas of planning:

1. Route planning
2. Weather briefing
3. Alternate destinations
4. Aircraft operation limitations
5. Aircraft airworthiness

The next few paragraphs constitute a suggested pilot-in-command preflight action list for cross-country flight.

Fig. 4-2. The Cessna Skyhawk II combines 122 knot cruise with standard nonstop range of 558 miles. The Skyhawk/172 is the world's best selling airplane.

Fig. 4-3. Big brother of the 172 series, the Cessna Hawk XP II boasts a 6-cylinder 195 hp fuel-injected engine and a constant speed prop.

Fig. 4-4. A new entry into the retractables, the Cessna Skylane RG is available in normally aspirated and turbocharged versions.

Fig. 4-5. The Cessna Skylane II instrument panel is typical of a panel which meets FAA requirements for VFR and IFR flight (compass not shown).

Route Planning

The requirement that a pilot-in-command "familiarize himself with *all available information* concerning that flight . . . ", demands careful attention to planning the flight route and possible alternates. As a minimum the following items are suggested:

1. Obtain current Sectional or World Aeronautical Charts which cover the intended route of flight and alternates or route deviations that may be required. Review the intended route of flight for obstacles; don't be surprised to see hazards such as 1000 feet high TV antennas directly on airways or in the very near vicinity of airports! Using the distances involved make a preliminary (no wind) estimate of the fuel required to accomplish the flight. Plan at least a 45 minute fuel reserve. Prior to departure upgrade the estimate using forecast winds aloft information. Compute required compass headings. Although modern navigation is conducted basically by the use of radio aids, the process of dead reckoning is a valuable back-up.

2. Review thoroughly all radio aids available on the intended route of flight including alternate routes or destinations. This information is available in brief on sectional charts. However, a more thorough source of information is the *Airport/Facility Directory*. This reference contains a listing of all major airports with control towers, including a tabulation of radio navigation aids and frequencies. The *Airport/Facility Directory* as well as all other parts of the AIM are available for

84

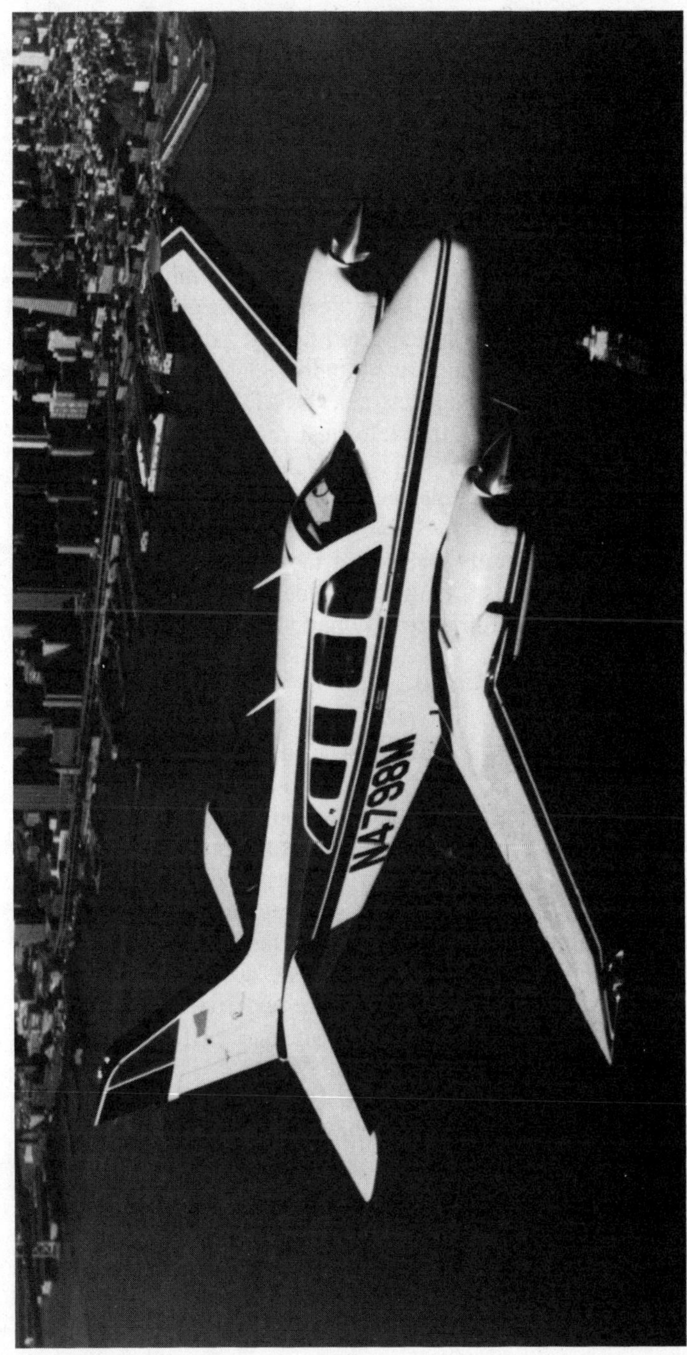

Fig. 4-6. Another high-performance aircraft of the Beechcraft line, the Baron 58P sports two 325 hp Continental engines.

use free of charge by either telephoning or visiting your local Flight Service Station.

3. To obtain information on VOR receiver checkpoints or VOR restrictions (such as shadowing due to mountains) it is necessary to consult the Graphic Notices and Supplemental Data. This part of the AIM contains a tabulation of parachute jump areas, military routes, radar service areas, and other pertinent navigational data.

4. It is required by FAR that each pilot-in-command determine the runway length(s) at airports of intended use as well as calculate the expected takeoff and landing distance based upon the density altitude and aircraft performance characteristics. Unfortunately, the AIM and sectional charts list only the length of the longest runway. Complete airport information, however, is available in several commercial publications. The *Jeppesen Airway Manual* as used for IFR flight is an extensive source of airport information. This manual is generally available at most fixed base operators. A second source is the Aircraft Owners and Pilot's Association (AOPA) Airport directory. For information write

AOPA,
Air Rights Building,
7315 Wisconsin Ave.
Washington D.C. 20014

Computing takeoff distance and rate of climb? Naturally your aircraft flight manual is a desired source. However, a handy aid for this purpose is the Denalt Performance Computer illustrated in Fig. 4-7. These computers are for sale by the Superintendent of Documents, U.S. Government Printing Office, Washington, D.C., 20402, at a current price of approximately 75c each. Be sure and specify whether you want a computer for a fixed pitch propeller aircraft or a constant speed propeller aircraft.

5. If your flight route contains segments which are off established airways or into wilderness areas, consider carrying an appropriate survival kit. For night flight at least one flashlight is an absolute must.

6. Make sure your airplane isn't due an annual inspection (or a 100 hour inspection in the case of a rented aircraft). Remember, it's up to the pilot-in-command to determine the airworthiness of the aircraft.

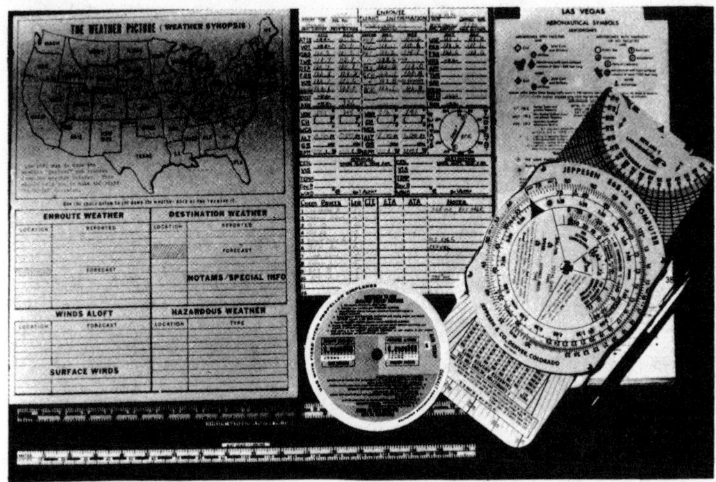

Fig. 4-7. Flying doesn't require an aviation library to be carried aloft. A weather briefing record, flight log, map (current issue), computer, and plotter will do for VFR flight. A Denalt Computer is convenient for calculating takeoff and rate of climb performance.

7. Now check the weight and balance of your aircraft. Both items *must* be within the manufacturer's allowable limits.

WEATHER BRIEFING

The weather briefing is perhaps the single most important item to be accomplished as a part of preflight planning. Aviation weather information is available 24 hours-a-day, seven-days-a-week, with distribution via radio, telephone, or by person-to-person briefing at your local FSS. Of these means for obtaining a weather briefing the person-to-person review with a flight service specialist is by far the preferred method. To obtain a briefing suited to your needs, advise the weather briefer of your qualifications as a pilot (student, private, instrument-rated or not), time and place of departure, type aircraft; N number of aircraft, proposed route of flight, destination, and estimated time of arrival. This will advise the briefer of the type and extent of weather information required. As a minimum determine the following:

1. Expected weather at departure airport at time of departure.
2. Forecast weather on route.
3. Forecast weather at destination airport.
4. Locations and movement of major weather systems such as fronts, areas of precipitation, thunderstorms, icing, fog, turbulence, etc.

5. In the event weather along your intended route is marginal, plan an alternate route or destination and obtain a weather briefing accordingly. Uncertain VFR weather at a destination airport may necessitate an extra fuel stop as a precautionary move. Whatever the case plan "a way out" before flight departure.

6. Write down the pertinent facts of your weather briefing.

A Flight Log

Cross-country flight planning requires reference to a number of charts, books, manuals, the weather briefer, and a flight computer. The pay-off resulting from good flight planning is that the library of information consulted to perform the planning process need not be taken in the aircraft. The results of proper flight planning can be entered on a single sheet of paper not much larger than the size of a page of this book. Figure 4-7 illustrates a typical flight plan and enroute flight log. Good flight planning results in the following minimal material being carried during flight:

1. Sectional or Wac Charts of the flight route.
2. A flight computer and plotter.
3. A flight planning log (sheet of paper No. 1), and
4. A weather briefing log (sheet of paper No. 2).

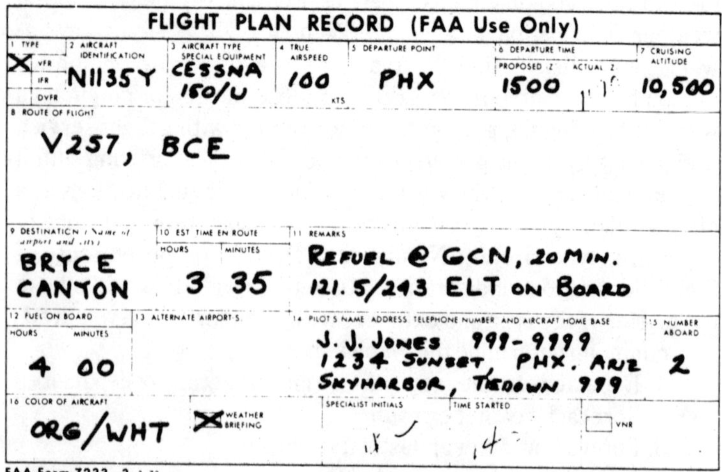

Fig. 4-8. This simple flight plan record form is all that needs to be filled out to insure the services of FSS and Air Rescue should an emergency arise.

Filing A Flight Plan

Every pilot is urged to file a flight plan prior to departure. Although not required by regulation for VFR flight operation, the flight plan is the best form of term insurance any pilot could take and the cost is absolutely zero. For those VFR pilots who elect to use this service, FAR 91 lists the information required on the flight plan. For the most part the information is self-evident. However, certain of the blocks (Fig. 4-8) deserve additional explanation. Block 1 is checked for VFR flight with Block 2 the N number of the aircraft. Block 3 is for specifying the type of aircraft including any special equipment carried. Although listing of transponder/DME equipment is only required for IFR flight plans, it is an excellent VFR practice. See Table 4-1.

Block 4 is the true air speed of the aircraft in knots and Block 5 is the point of departure. Block 6 is used to denote the proposed Zulu departure time. Upon opening your flight plan (by radio), flight service will fill in your actual Zulu departure time. Block 7 provides for a record of your proposed MSL Altitude on the initial leg of your flight. Particular attention should be given to Block 8 to describe an accurate record of your flight route. Block 9 is self-evident. Block 10 is a *total* time estimate beginning at the estimated time of departure and ending at the estimated time of arrival including ground time spent enroute for purposes of

Table 4-1. Special Equipment Codes.

/X	no transponder.
/T	transponder with no altitude encoding capability.
/U	transponder with altitude encoding capability.
/D	DME, no transponder.
/B	DME, transponder with no altitude encoding capability.
/A	DME, transponder with altitude encoding capability.
/M	TACAN-only, no transponder.
/N	TACAN-only, transponder with no altitude encoding capability.
/P	TACAN-only, transponder with altitude encoding capability.
/C	RNAV, transponder with no altitude encoding capability.
/F	RNAV, transponder with altitude encoding capability.
/W	RNAV, no transponder.

refueling, etc. Block 11 provides the pilot an opportunity to list special conditions pertaining to his flight. Typical examples are:

1. Refueling at________airport.
2. Touch and go practice landings will be made at________airport enroute.
3. 121.5/243 MHz emergency locator beacon on board.
4. VIP on board.
5. Student cross-country, solo.
6. Radio inoperative.
7. 90 channel radio.
8. Ambulatory case aboard.
9. Will avoid all restricted areas.
10. Advise customs of arrival.

Block 12 is self-evident. Block 13 is for listing alternate airports and is used primarily for IFR flights. Block 14 is to list the full name of the pilot-in-command, home address, telephone number, and aircraft home base (including tie-down number). Should no one be present at the listed address (as in the case of a family vacation) it is desirable that Block 14 contain the name, address, and telephone number of a friend or a relative who would be familiar with the pilot. Blocks 15 and 16 are self-evident. The remaining blocks are for use by flight service personnel. However, special attention should be directed to the letters VNR in the lower right hand block. These stand for *Visual Flight Not Recommended!* This formally records the fact that a flight service specialist advised against a VFR flight!

After a flight plan has been filed and opened, the pilot-in-command is responsible to notify the nearest FAA Flight Service Station to close the flight plan upon completion of his flight. VFR flight plans are closed only by FSS, not by tower or radar as for an IFR flight plan. In the event a pilot fails to close his flight plan the following actions are initiated by FSS:

1. Thirty minutes after ETA: FSS initiates a radio telephone communication search of all airports along the proposed flight route and in the local area of the destination airport listed on the flight plan. (In some instances a local area includes as many as 50 airports). The radio telephone communication search also includes contact with local law enforcement agencies.

2. One and one-half hours after ETA: An Alert Notice is issued. The result is an extended communications search for information throughout the area within flight range of the

aircraft. Search assistance is requested from flights traversing the search area.

3. Two and one-half hours after ETA: Search and rescue is notified of a missing aircraft; appropriate civilian air patrol and military search action is initiated.

The procedure for closing a flight plan is not difficult since a Flight Service Station is as close as the aircraft radio or a telephone. Attention to this simple matter at the destination of a flight enables search and rescue procedures to be utilized for the purpose for which they were developed—to help the pilot in distress.

Aircraft Instruments and Equipment Required For VFR Flight

As a part of the preflight process it is necessary that the pilot-in-command examine his aircraft to assure conformance with FAR's which describe instruments and equipment required. For VFR flights during the day in a standard category aircraft the following instruments and equipment are required:

1. Airspeed indicator
2. Altimeter
3. Compass
4. Tachometer
5. Oil pressure gauge (one for each engine).
6. Temperature gauge (one for each liquid-cooled engine).
7. Oil Temperature gauge (one for each air-cooled engine).
8. Manifold pressure gauge (one for each engine). This instrument is required for engines which are supercharged or aircraft which have been certified with a manifold pressure gauge.
9. Fuel gauge(s) (as required to indicate the quantity of fuel in each tank).
10. Landing gear position indicator (retractable aircraft only).
11. Approved flotation gear for each occupant and at least one pyrotechnic signaling device if the aircraft is to be operated for hire over water and beyond power-off gliding distance from shore. This applies to any aircraft regardless of the number of engines.
12. Safety belts for all occupants who have reached their second birthday, and shoulder harnesses for all crew members. The latter applies principally to the pilot and copilot seats of a light aircraft.

If VFR flight is to be conducted at night the following are additionally required:

1. A red left and green right wing light and a white tail light.

2. A red or white anti-collision light (rotating beacon or strobe).

3. Electric landing light (if the aircraft is operated for hire).

4. A generator whose capacity is adequate to power all electrical and radio equipment required to conduct the flight.

5. One spare set of fuses, or three spare fuses of each kind required.

For flight involving aerobatics or unusual attitudes each occupant of the aircraft must wear an approved parachute. (Unusual attitudes are defined as bank angles exceeding 60 degrees relative to the horizon and pitch angles exceeding 30 degrees.) If a chair type parachute is employed, it must have been packed by an appropriately rated parachute rigger within the preceding 120 days. If other types of parachutes are employed, such must have been packed by an appropriately rated rigger within the preceding 60 days.

Flights to be made at high altitudes require additional planning as to oxygen for the pilot and passengers. Federal Aviation Regulations require that oxygen be provided the flight crew for all aircraft operating at cabin pressure altitudes of 12,500 feet MSL up to and including 14,000 feet MSL for any part of the flight of more than 30 minutes duration at these altitudes. Above 14,000 feet MSL, the flight crew is required to use supplimental oxygen during the entire time at these altitudes; and, above 15,000 feet MSL, each occupant of the aircraft must be provided supplemental oxygen.

In reality, the altitude limits for supplemental oxygen as prescibed by FAR are quite liberal. Studies indicate that all persons begin to deteriorate in alertness and mental efficiency to some degree above 12,000 feet without supplemental oxygen. Above 14,000 feet distinct impairment of mental facilities occurs—especially with respect to mathematical reasoning capabilities. Night vision is sharply impaired at higher altitudes, even though other symptoms of hypoxia may not be apparent. It is said that a pilot flying at night without supplemental oxygen is 24% blind at 8,000 feet and 50% blind at 12,000 feet. Individual physical fitness and other factors may change a person's tolerance to hypoxia. Smoking at 10,000 feet produces effects equivalent to those experienced at 14,000 feet without smoking.

Congressional legislation has established the requirement that all aircraft be equipped with an Emergency Locater Transmitter (ELT). For civil usage, emergency locater transmitters may be of either the fixed or deployable types; units must be attached to the airplane in a manner that the probability of damage to the transmitter is minimized in the event of a crash impact. Batteries used in the emergency locater transmitters must be replaced when (a) the transmitter has been in use for more than one cumulative hour or, (b) when 50% of the useful life of the batteries has expired (per mfgr. ratings). Exceptions to the requirements for an emergency locater transmitter apply in cases of ferrying an aircraft for the installation or repair of an emergency locater transmitter, training flights conducted within a 50-mile radius of the airport from which the flight began, or in the case of agricultural aircraft operations. Turbojet and air carrier aircraft are not required to have ELT's; similarly for aircraft not equipped to carry more than one person.

Preflight preparations are not complete until a thorough check of the adequacy and operability of all necessary radio equipment has been made by the pilot-in-command. A communications transmitter-receiver is required for flight into fields where traffic is controlled by a tower. Similarly a 4,096 code transponder is required for operation in Group I and Group II Terminal Control Areas. A check on the operability of your communications transmitter-receiver can be accomplished by calling ground control, FSS, or Unicom prior to taxi. A request for a "radio check" will bring a reply of the signal strength and modulation quality of your transmitter. Navigational receiver equipment may be checked in many cases prior to flight by tuning to the Very High Frequency Omni Test signal (VOT) provided at many major airports. The AIM lists airports so equipped and the frequency for use. As an alternate technique, VOR receiver check points may be used at airports not equipped with a VOT. Although formal VOR receiver accuracy limits are not specified for VFR flight, the FAR requirements for IFR receiver accuracy provide an excellent guide. These requirements are ±4 degrees for a ground VOR receiver check and ±6 degrees for an airborne VOR receiver check. Transponder equipment may be checked prior to flight by utilizing the self-test feature or, far better, by requesting a "transponder check" from ATC (line of sight conditions permitting).

The Final Step

The final step in the process of preflight planning is an aircraft examination and engine run-up. Assuming that the engine(s) and propeller(s) operate in accordance with the manufacturer's run-up specifications and the associated instruments operate properly, the aircraft preflight is complete. Now let's set the altimeter to the field altitude and check the barometric pressure shown in the Kollsman window with the altimeter setting. An error greater than ±1″ Hg (100 feet) indicates a need for altimeter servicing.

Federal aircraft regulations require that the pilot keep his seat belt fastened during all phases of flight. In addition his shoulder harness must be fastened during takeoff and landing. Each person on board an aircraft must occupy a seat with a safety belt properly secured about him. However, a child who has not reached his second birthday may be held by an adult. In the event the purpose of the flight is for sport parachuting, parachutists on board may use the floor of the aircraft as a seat. As pilot-in-command it is your responsibility to advise passengers to fasten their seat belts prior to takeoff or landing and in the event of anticipated and actual turbulence conditions.

The pilot-in-command is the individual responsible for *all* phases of safety during both aircraft ground and flight operations. He accomplishes the necessary preflight planning, inspects the aircraft as required to determine its airworthiness, and is considerate of passenger safety from boarding to exit at the destination airport.

LACK OF PREFLIGHT PLANNING

The following case histories relate to preflight planning. These histories illustrate the depth to which the courts consider preflight operations a vital aviation function.

The Unread NOTAM

At the airport in question a 7½ foot high mound of sand was located at the approach end of a runway. As soon as the sand was placed in this position, the airport proprietor issued a NOTAM giving the dimensions of the sand pile and the position of a new threshold located some 500 feet further down the runway to allow clearance over the sand pile. An aircraft attempting to land on the runway in question struck the sand pile and was substantially damaged. As a result, the owners and insurers sued both the proprietor of the airport and the U.S. government. The owners claimed the tower operator should have warned the pilot of the sand pile and the displaced threshold. An investigation disclosed the fact that the sand pile in question had been in place nearly *two years*. Further, the NOTAM describing the sand pile and relocated threshold appeared regularly in the Airmen's Information Manual up to and including the issue current at the time of the incident. The court held that the true cause of the accident was failure of the crew to familiarize themselves with the NOTAM. The court found: "It was the custom and practice of controllers to advise of new, temporary, variable or short-term conditions about which a pilot could not know without an advisory, but not about longterm conditions which had been duplicated by NOTAM, such as the pile of sand and the displaced runway which had existed and was published 1½ years before the incident."

The moral to the story is very clear: A study of NOTAMS prior to flight is a required preflight function on the part of the pilot-in-command. NOTAM information is available in the AIM or by simply calling Flight Service and requesting NOTAM information for your intended route of flight.

A Problem of Airworthiness

A pilot and his companion engaged in a flight in a club airplane. The pilot failed to notice, prior to take off, that the rudder was missing. As a result, the airplane crashed and the companion was injured. Suit was brought against the flying service claiming that they were at fault in not warning the pilot of the missing rudder. However, the court ruled that the pilot's failure to preflight the aircraft was the sole cause of the accident, the flying service was not responsible for injuries sustained. So let it be emphasized that the pilot-in-command is indeed the party responsible for *determining the airworthiness of an aircraft!*

Chapter 5
The Airspace

Airspace needs of the pilot, the air traveller, and the community are many and varied. Airspace restrictions result for reasons of safety and equality in sharing a national resource. In certain instances the density of aircraft movements necessitates the airspace be subdivided into carefully defined regions with special operating procedures. In other instances special use airspace is reserved for training of military personnel, testing of missiles, etc. Due to the many and varied airspace needs a number of different airspace subdivisions result. The first part of this chapter is devoted to definitions of the various airspace subdivisions and the second part to general flight operations within our national airspace.

DISTRIBUTION OF THE NATIONAL AIRSPACE

The national airspace system can be diagrammed as shown in Fig. 5-1. As illustrated the national airspace is divided into the two major categories, controlled and uncontrolled airspace. Since uncontrolled airspace represents a minimal regulation condition, there are no sub-categories in the classification. Controlled airspace, on the other hand, is subdivided into a number of basic and special use types of airspace. East of the Mississippi River about 70 % of the airspace is controlled; in the West, about 25%, and along the West Coast, about 70%. A third subdivision is voluntary flight procedure airspace. Typically volunteer flight

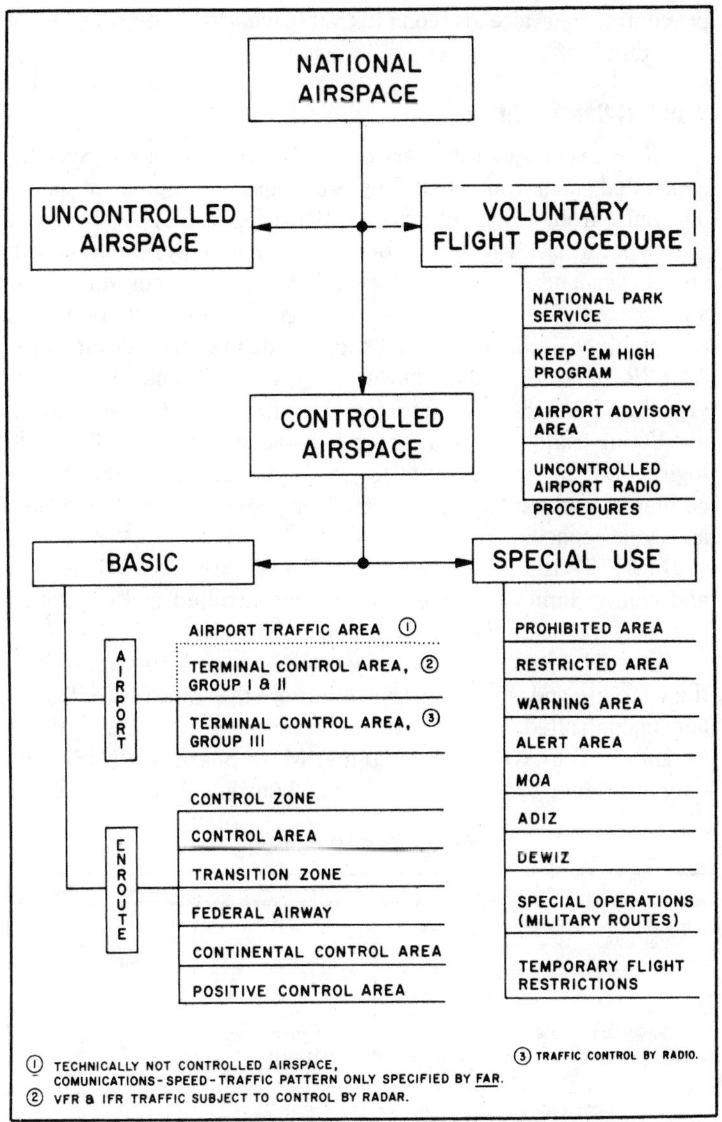

Fig. 5-1. Our National Airspace is divided into various major and minor categories depending upon the use and/or aerial activity involved.

procedure airspace is located over national parks or major cities. Pilots are requested to fly at altitudes greater than the minimum FAR limit for ecology and noise abatement reasons. Adherence to voluntary flight procedures is highly desirable as a means of preventing our already complex airspace from becoming even more difficult.

Visibility Considerations

The two major divisions of the National Airspace System, controlled and uncontrolled (Fig. 5-2) segregate regions of sparse air traffic from blocks of airspace that support large volumes of flight operations. The word "control" *does not* imply operation with an ATC ground facility (except for IFR flight and certain special VFR cases). In reality, the word *control* denotes minimum acceptable visibility criteria for flight while in controlled airspace. As VFR flight is based on the "see and be seen" principle, cockpit visibility must be adequate for the pilot to detect and avoid conflicting traffic as well as ground obstacles (Fig. 5-3). Thus it is logical that minimum limits for visibility, cloud clearance, and ceiling be established; limits which are consistent with the "see and avoid" concept.

FAR 91.105 provides basic VFR visibility, cloud clearance, and ceiling limits for flight in both uncontrolled and controlled airspace. These limits are illustrated in Table 5-1.

It will be noticed that when operating above 10,000 feet MSL, the visibility and cloud clearance limits are the same for controlled and uncontrolled airspace. In the zone from 1,200 feet Above Ground Level (AGL) to 10,000 feet Mean Sea Level (MSL) the only difference between controlled and uncontrolled airspace is a

Table 5-1. Visual Flight Rules.

Altitude	Flight visibility	Distance from clouds
1,200 feet or less above the surface (regardless of MSL altitude)—		
Within controlled airspace	3 statute miles	500 feet below 1,000 feet above. 2,000 feet horizontal
Outside controlled airspace	1 statute mile except for helicopters	Clear of clouds
More than 1,200 feet above the surface but less than 10,000 feet MSL—		
Within controlled airspace	3 statute miles	500 feet below 1,000 feet above 2,000 feet horizontal
Outside controlled airspace	1 statute mile	500 feet below 1,000 feet above 2,000 feet horizontal
More than 1,200 feet above the surface and at or above 10,000 feet MSL.	5 statute miles	1,000 feet below 1,000 feet above 1 mile horizontal

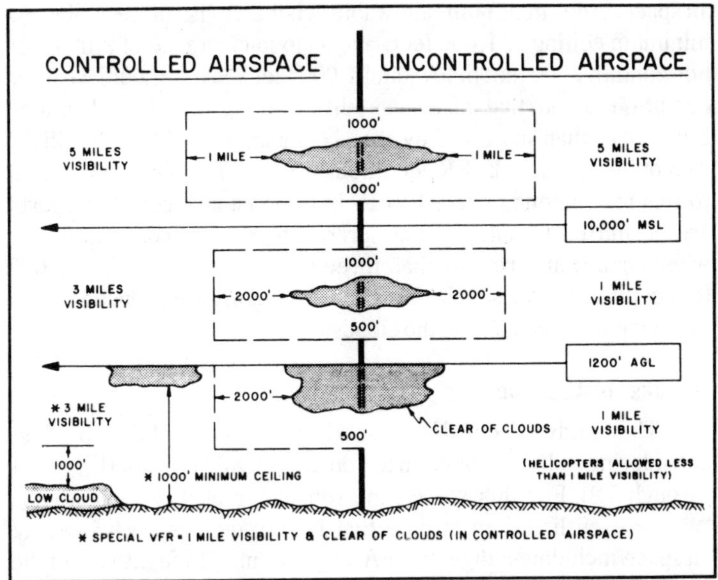

Fig. 5-2. The primary difference between controlled and uncontrolled airspace is a requirement for greater visibility and cloud clearance limits in controlled airspace (below 10,000' MSL).

visibility requirement of three miles when in controlled airspace and one mile in uncontrolled airspace. For operations from the surface level to 1,200 feet AGL in uncontrolled airspace flight requirements are simply one mile visibility and clear of the clouds. However, for operating in the same altitude region in controlled

Fig. 5-3. For true VFR flying, how can one beat the Evans VP-1 Volksplane! Powered by a Volkswagon engine, the VP-1 is all wood construction, weighs 440 lbs., and will cruise at 75 mph.

airspace, the minimum allowable visibility is three miles, a minimum ceiling of 1,000 feet, and a cloud clearance of 2,000 feet horizontally, 500 feet below, and 1,000 feet above. Should weather conditions exist that make visibility, ceiling or cloud clearance limits less than indicated by the accompanying table, VFR flight cannot be conducted; IFR conditions prevail. The single exception to this regulation is special VFR. In this instance certain airports are permitted to authorize an aircraft to enter a controlled area when conditions are less than three miles visibility and/or 1,000 feet ceiling. Under special VFR the minimums are reduced to one mile visibility and clear of the clouds.

"Blocks" of Airspace

The study of our National Airspace is simplified by first considering "blocks" of airspace on an individual basis (Figs. 5-4 through 5-8). For this reason the remainder of this section is an "airspace outline"; an outline that briefly defines each block of airspace including a digest of FAR's pertaining to flight within the particular block.

    I. Uncontrolled Airspace. Airspace in which minimal VFR visibility and cloud restrictions apply. Airspace of this nature is generally located in remote areas and extends in altitude from the surface nominally to 14,500 feet MSL (floor of the Continental Control Area). Often the ceiling of uncontrolled airspace is limited to 700 or 1,200 feet AGL by an overlaying transition area or control area. Aerobatics are permitted in uncontrolled airspace.

    II. "Controlled Procedure" Airspace. The following two common airspace allocations are characterized by various FAR procedures but are technically uncontrolled airspace from the standpoint of ceiling and visibility limits.

        1. Airport Traffic Area.

        The airspace within a horizontal radius of 5 statute miles from the geographical center of any airport at which a control tower is operating, extending from the surface of the airport up to an altitude of 3,000 feet above the elevation of the airport. Within an airport traffic area the following regulations apply:

            a. Flight operations are intended for purposes of takeoff and landing only. Should it be desirable to penetrate an airport traffic area (perhaps due to a

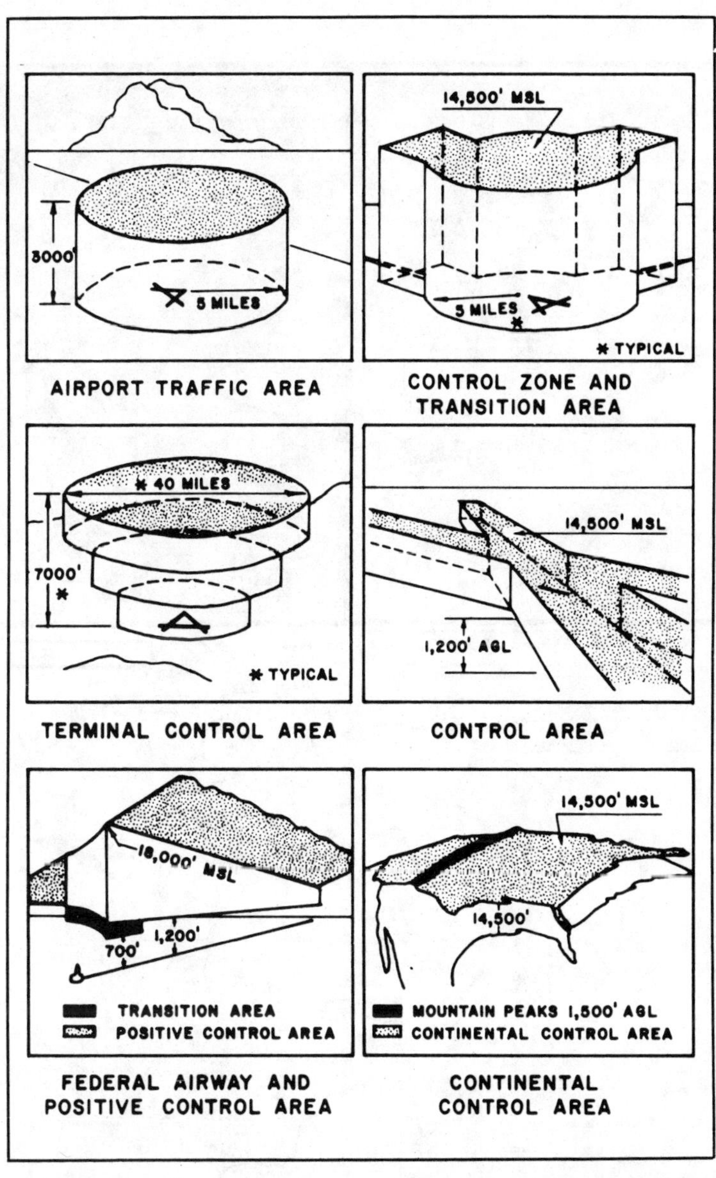

Fig. 5-4. Controlled airspace comes in a variety of sizes and shapes. Except for the airport traffic area, the above figures illustrate basic controlled airspace configurations. The airport traffic area is technically not controlled airspace; FAR's apply only to communications, speed limits, and landing/takeoff procedures. Visibility and cloud clearance limits in an airport traffic area are the same as for uncontrolled airspace (unless the airport traffic area is located in a control zone).

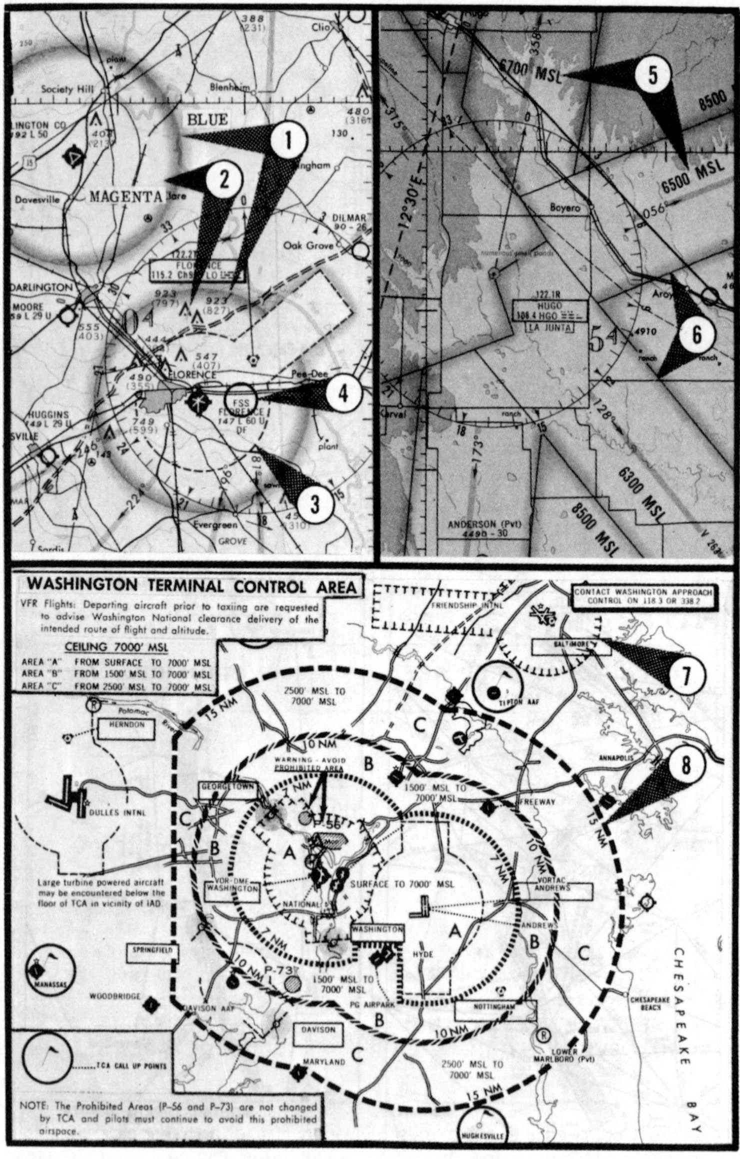

Fig. 5-5. 1. Control Areas (blue shading); 2. Transition Areas (magenta shading); 3. Control Zone; 4. FSS Advisory Area; 5. Nonstandard Federal Airway Floors; 6. Federal Airway (lateral limits); 7. Special VFR not permitted; 8. Terminal Control Area.

low ceiling), it is necessary to obtain an ATC clearance from the tower before entering the area (initial callup should be made 15 miles out).

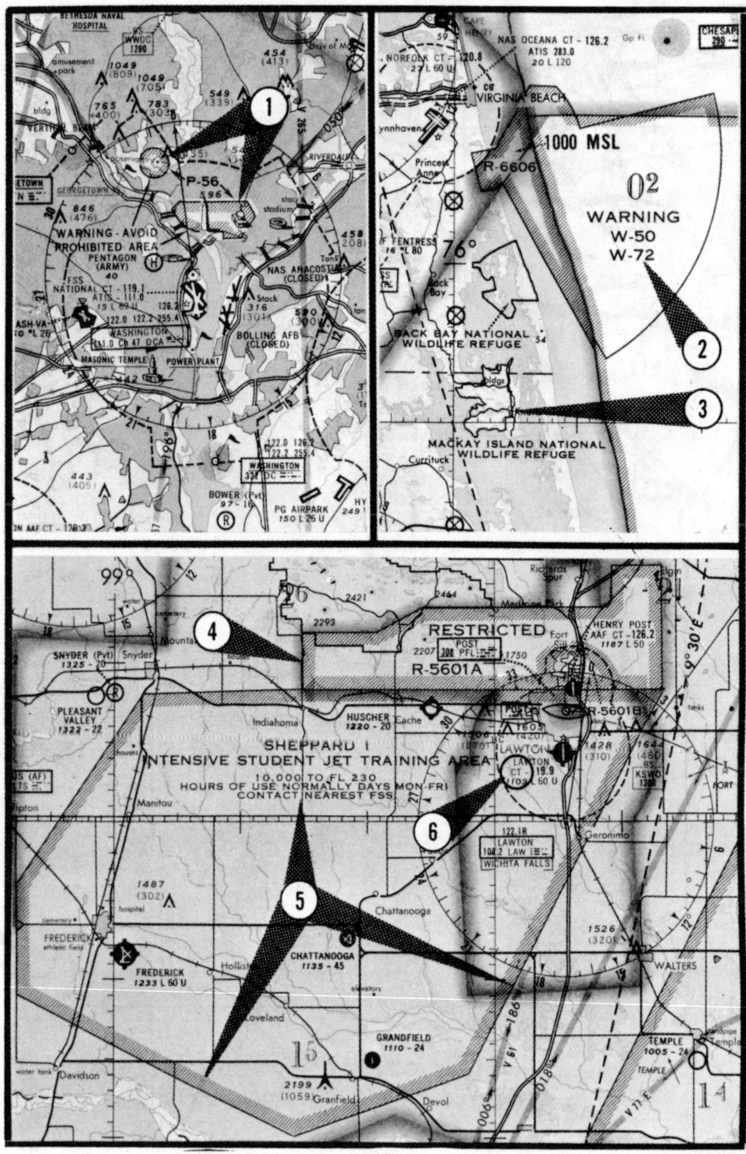

Fig. 5-6. 1. Prohibited Area; 2. Warning Area; 3. Wildlife Refuge; 4. Restricted Area; 5. Military Operational Area (MOA); 6. Control Zone and Airport Traffic Area.

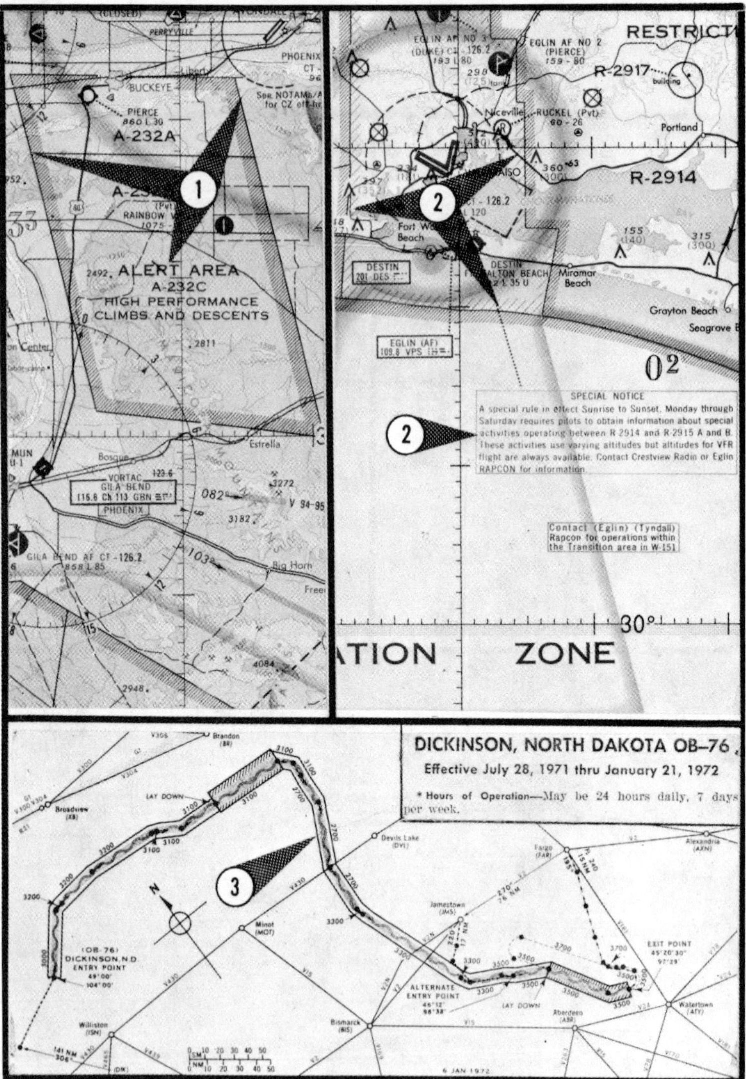

Fig. 5-7. 1. Alert Area, 2. Special Air Traffic Rules Area, 3. Special Operations (Olive Branch Route).

Two-way communications with the tower are required while in an airport traffic area (unless landing at a non-tower airport within an airport traffic area of another airport).

b. Turbine powered aircraft shall maintain a traffic pattern altitude of at least 1,500 feet above the

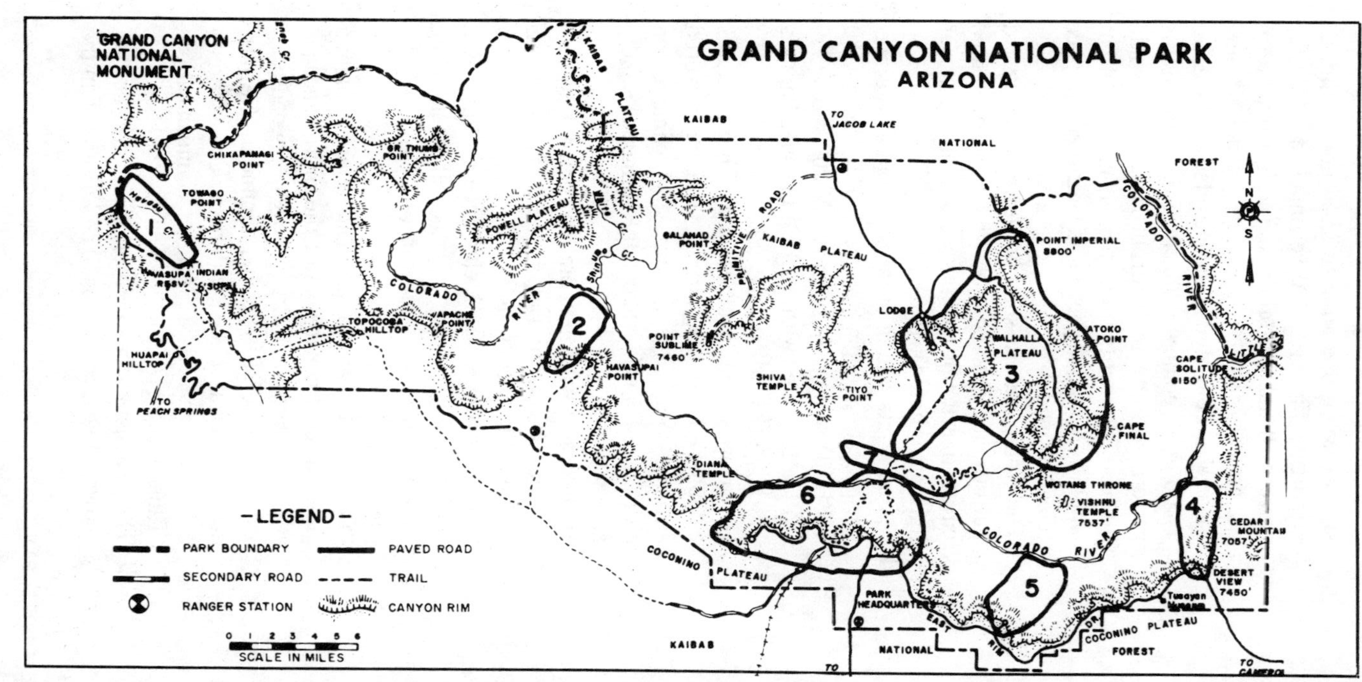

Fig. 5-8. Voluntary flight procedure airspace. Numbered areas list MSL altitude minimums of 5000' (1 & 5); 6000' (7); 6500' (2); 8500' (4, 6, and 2); and 10,000' (3):

surface when operating to an airport within an airport traffic area. A recommended* traffic pattern altitude for other aircraft is 1,000 feet if the airport is located in a control zone; 800 feet for an airport located in uncontrolled airspace. If a 1,000 foot ceiling were to exist, the traffic pattern altitude in controlled airspace would become 500 feet (minimum cloud clearance) but would remain 800 feet in uncontrolled airspace (clear of clouds).

c. The traffic pattern shall be flown as a lefthand pattern unless noted otherwise by a segmented circle, instructions from tower, a flashing amber light, or as listed in the AIM.

d. Speed limits within an airport traffic area are: reciprocating engine aircraft: 156 knots (180 MPH) Turbine powered aircraft: 200 knots (230 MPH). If located within a TCA the speed limit is 250 knots (288 MPH).

e. Departure routing shall comply with procedures established for that airport; turbine powered airplanes and large aircraft shall climb to an altitude of 1,500 feet above the surface as rapidly as practicable.

f. At an airport with an operating control tower no pilot may taxi an aircraft on a runway, or takeoff, or land an aircraft unless he has received a clearance from ATC. A clearance to "taxi *to*" a runway is a clearance to cross all intersecting runways but is not a clearance to "taxi *on*" the assigned runway.

g. In the event of an aircraft radio failure in flight a pilot may continue to operate the aircraft and land at a tower controlled airport if weather conditions are at or above VFR weather minimums. He must maintain visual contact with tower and adhere to the light signals as a means of communications (Table 5-2).

h. The dimensions of an airport traffic area are not shown on sectional maps. The letters CT in the

---

*Except for special cases, FAR's do not specify traffic pattern altitudes; the values noted are conventions in use.

Table 5-2. Light Signals.

Color and Type of Signal	On the Ground	In flight
STEADY GREEN	Cleared for take-off	Cleared to land
FLASHING GREEN	Cleared to taxi	Return for landing (to be followed by steady green at proper time)
STEADY RED	Stop	GIVE WAY TO OTHER aircraft and continue circling
FLASHING RED	Taxi clear of landing area (runway) in use	Airport unsafe—do not land
FLASHING WHITE	Return to starting point on airport	
ALTERNATING RED & GREEN	General Warning Signal—Exercise Extreme Caution	

airport data description block indicate that an airport traffic area is located at that airport.

2. *Airport Advisory Area.*

The area within five statute miles of an airport where a control tower is not operating but where a Flight Service Station is located (a ceiling limit is not specified). At such locations the FSS provides advisory service to arriving and departing aircraft. The following regulations apply:

  a. Arriving aircraft shall make all traffic pattern turns to the left unless the airport displays light signals or visual markings indicating that turns should be made to the right.

  b. Departing aircraft shall comply with FAA traffic patterns for that airport.

Although it is not mandatory that pilots participate in the airport advisory service program it is strongly recommended that they do so. The following procedure is suggested:

  c. Arriving aircraft operating VFR contact FSS 15 miles from the airport; advise position, altitude, and intentions. Continue to advise FSS of posi-

tion entering downwind, base, final, and clear of
the active runway.
d. Departing aircraft contact FSS for pre-taxi ad-
visories. Advise FSS when clear of the active
runway and leaving the airport advisory area.

The same procedure is useful at airports where there is neither a
control tower or FSS. In such instances the Unicom frequency
should be used to advise the local airport operator as well as other
possible air traffic.

III. Controlled Airspace. Airspace in which increased
visibility, ceiling, and cloud clearance limits generally
apply is termed controlled airspace; additional restrictions may also apply depending upon the reason for
control.

1. *Control Zone*

   Airspace which extends upward from the surface
   nominally to 14,500 feet (the base of the Continental
   Control Area). In general, control zones are located
   about airports and are normally circular in shape with a
   radius of five statute miles plus extensions as necessary for instrument departure and arrival paths. A
   control zone may encompass more than one airport.
   Control zones are depicted on charts by a dotted blue
   line.

2. *Control Areas and Federal Airways*

   Control areas consist of all federal airways plus
   airspace areas needed to interconnect the federal
   airway system. A federal airway is eight nautical miles
   in width and typically extends from 1,200 feet AGL to
   18,000 feet MSL. Sectional maps use a shaded blue
   outline to denote control area floors of 1,200 feet
   above the surface. Control area floors other than 1,200
   feet above the surface are noted with an appropriate
   numerical value.

3. *Transition Area*

   Controlled airspace which extends upward generally
   from 700 feet or more above the surface to the floor of
   the overlying controlled airspace. The basic function
   of the transition area is to provide a zone of controlled
   airspace for IFR operations in process of descent or
   climb while in the near vicinity of an airport. Transi-

tion zones are noted by a magenta colored outline on sectional maps.

4. *Continental Control Area*

The continental control area consists of the airspace above the United States and Alaska (excluding the Alaska Peninsula) at and above 14,500 feet MSL but does not include the airspace less than 1,500 feet above the surface of the earth or certain prohibited or restricted areas. For all practical purposes, the continental control area may be thought of as a large blanket covering the United States and Alaska with provisions for a 1,500-foot clearance where it passes over mountainous regions. The ceiling of the continental control area is infinity.

5. *Positive Control Area*

Airspace so designated that flight must be conducted only under instrument flight rules. For operations within positive control areas, aircraft must be:

    a. Equipped with instruments and equipment for IFR operation and flown by a pilot rated and current for instrument flight.

    b. Equipped with a coded radar beacon transponder.

    c. Equipped with communications radio transmitter- receiver equipment necessary to the flight.

    d. Equipped with Distance Measuring Equipment (DME) if the operation is above flight level 240. Positive control area exists from 18,000 feet MSL to flight level 600 throughout the United States.

6. *Terminal Control Area (TCA)*

Terminal Control areas are in reality "super air traffic control areas" where *all aircraft* are subject to direction by ATC. Terminal control areas vary in shape in that they are designed to accommodate the particular airport they service. In general, they extend from the surface level to an altitude of 7,000 feet MSL. Group I terminal control areas are located at the busiest airports in the United States with Group II terminal control areas at less congested locations. The following regulations apply to Group I terminal control areas:

    a. Traffic clearance is required prior to flight in a TCA.

  b. Student flights are prohibited.

  c. An operable two-way radio capable of communicating with ATC is required.

  d. An operable VOR (or TACAN) receiver is required.

  e. An operable radar beacon transponder having at least a mode A/3, 409B-code capability, replying to A/3 interrogations is required.

Group II terminal control areas are similar in their requirements to Group I with the exception that student flight is permitted. Group III terminal control areas simply require that radio communications be established between the pilot and the ATC facility in order to enter the TCA.

  The terminal control area is basically an anti-collision flight procedure. Fundamentally this is accomplished by radar control of *all aircraft* within the boundaries of a TCA. As a VFR pilot you must obtain a clearance from ATC (approach control) before entering the TCA. Be prepared to hold outside the TCA in the event traffic conditions do not permit an immediate entry. Remember that the VFR pilot still has the obligation of remaining in VFR weather conditions. Should instructions from approach control direct a pilot too close to clouds or other non-VFR weather, advise approach control immediately and obtain an amended clearance. While so doing, remain clear of all clouds and other non-VFR weather. Radar cannot see weather as you can.

  IV. Special Use Airspace. Special use airspace consists of airspace domains where airborne activity must be confined because of the specialized nature of the activity. The following describes various classes of specialized use airspace:

  1. Prohibited Area

  Airspace within which the flight of aircraft is not allowed for security or other reasons associated with the national welfare unless prior permission has been granted by the cognizant government authority. An example of a prohibited area is the area that encompasses the White House and the Capitol buildings in Washington, D.C. Prohibited areas are also established to safeguard the forest and wildlife in the few remaining wilderness areas of the United States. Avoid all prohibited areas!

2. *Restricted Area*

Airspace within which flight, while not wholly prohibited, is subject to restrictions. The function of a restricted area is to confine or segregate activities considered to be hazardous to non-participating aircraft. Restricted areas denote the existence of unusual, often invisible, hazards to aircraft such as artillery firing, aerial gunnery, or guided missiles. Penetration of restricted areas without authorization from the using or controlling agency may be extremely hazardous to the aircraft and its occupants. Restricted areas vary in their hours of use. Flight information to describe a restricted area can be obtained from; (a) sectional charts, (b) by contacting Flight Service, and (c) by contacting the cognizant agency which is given authority over the restricted area in question.

3. *Warning Area*

A warning area is airspace, within international airspace, established to contain hazardous operations conducted by U.S. military forces. The activities conducted within warning areas may be hazardous to nonparticipating aircraft. However, no restriction to flight is imposed because flight within international airspace cannot legally be restricted. To alert nonparticipants to the existence of possible hazardous conditions, warning areas are depicted on aeronautical charts. Most warning areas lie within three statute miles of the coast line and are located over ocean areas.

4. *Alert Area*

Airspace which may contain a high volume of pilot training activity or an unusual type of aeronautical activity—neither of which is uncommonly hazardous to aircraft. All flight activity in an alert area shall be conducted in accordance with FAR's, without waiver, and pilots of participating aircraft as well as pilots of aircraft transiting the area shall be equally responsible for collision avoidance. The establishment of alert areas does not impose any unique flight restrictions or communications requirements.

5. *Military Operations Area (MOA)*

Airspace which contains intensive flight training activities of military student pilots and in which restrictions are imposed on IFR flights only. All VFR flights within a MOA shall be conducted in accordance with FAR's without waiver, and all participating military pilots as well as pilots of aircraft transiting the area shall be equally responsible for collision avoidance. Information on MOA's may be obtained from any FSS within 200 miles of the area.

6. *Special Operations*
Airspace which is devoted to military training routes termed Olive Branch routes. Refer to the *Airmen's Information Manual* for route descriptions. Treat these regions as alert areas.

7. *Terminal Area Graphic Notice*
High density terminal area suggested routes for transiting aircraft. Refer to the *Airmen's Information Manual* for route descriptions.

8. *Special Air Traffic Rules and Air Traffic Patterns*
Airspace in which special rules or procedures apply. Pilots must obtain an ATC advisory concerning operations being conducted therein before entering the area.

9. *Temporary Flight Restrictions*
Temporary airspace flight restrictions may be put into effect in the vicinity of any incident or event which by its nature may generate such a high degree of public interest that the likelihood of a hazardous congestion of air traffic exists. Examples are major sporting events, parades, air shows, and similar functions. Forest fires, floods, and disaster situations may likewise result in temporary flight restrictions. Information concerning temporary flight restrictions is distributed by NOTAMS and by Flight Service.

V. **Voluntary Flight Procedure Airspace**. Under terms of federal legislation that establishes wilderness areas, the controlling agency, be it the U.S. Forest Service, the National Park Service, or the Fish and Wildlife Service, is in a position to close or severely limit use of existing airport facilities within these areas. Rather than create new regulations to control aircraft operations in wilderness areas, a program of voluntary cooperation between the aviation community and the various agencies involved in the wilderness area program is being enacted. The purpose of the program is to minimize noise over wilderness areas; pilots are requested to "keep 'em high" while over such regions. As shown in the accompanying map of Grand Canyon National Park, areas 1 through 7 represent airspace designations where the Grand Canyon National Park Service and the FAA have instituted a program of voluntary restricted minimum altitude flight. Remember that no person is permitted to land an aircraft on land or water within an area administered by the National Park Service, except (a) at officially designated landing sites, (b) in the event of an emergency, or (c) on official business for the federal government. Should your route of flight involve a wilderness area check with your local FSS to determine voluntary procedures that may be in use. As a rule of thumb consider at least a 2,000 feet minimum AGL altitude when over-flying congested and recreational areas and national parks.

VI. **Air Defense Identification Zones.** Air Defense Identification Zones (ADIZ's) are areas of airspace over land or water in which the ready identification, location, and control of civil aircraft is required in the interest of national security. As shown by the accompanying map there are four zones which surround the continental United States. These are the Pacific Coastal ADIZ, Southern Border Domestic ADIZ, Gulf of Mexico Coastal ADIZ and Atlantic Coastal ADIZ (Fig. 5-9). In addition to the ADIZ, Distant Early Warning Identification Zones (DEWIZ's) are established to provide a similar aircraft identification at long ranges. A zone of this nature is located in Alaska. If your route of flight requires the penetration of an ADIZ or DEWIZ the following regulations in general apply:

1. A flight plan is required for both VFR and IFR flight. A VFR flight plan shall be designated DVFR.
2. Two-way radio is required for penetrating an ADIZ or DEWIZ. In the event an aircraft is not equipped with a two-way radio on a flight between Mexico and the United States, the aircraft must land at a designated airport of entry nearest the point of entry into the United States and file an arrival or a completion notice.
3. Position reports are required. DVFR flights must give their estimated time of penetration of an ADIZ at least

15 minutes before penetration takes place. DVFR aircraft entering the United States through a DEWIZ shall report before penetration takes place. Reports must contain the time, position, and altitude at which the aircraft passed the last reporting point before penetration and the estimated time of arrival at the next reporting point or penetration point. Time estimates should be within five minutes and course position estimates within ten miles of the center line for a domestic ADIZ and within twenty miles of course center line or a coastal ADIZ or DEWIZ.

4. With the exception of flight between Mexico and the United States, coastal or domestic ADIZ requirements do not apply if operating North of 25 degrees latitude or West of 85 degrees West longitude at a true air speed of less than 180 knots. DEWIZ procedures do not apply in Alaska when operating at a true speed of less than 180 knots while the pilot maintains a listening watch on an appropriate frequency. ADIZ procedures do not apply within the continental United States, over or within three miles of any island in Hawaii, and for flights that remain within ten miles of a departure point.

5. In an emergency situation that requires immediate action for the safety of flight, a pilot-in-command of an aircraft may deviate from ADIZ and DEWIZ procedures to the extent required by the emergency. He must report the reasons for the deviation to the appropriate communications facility as soon as possible. In the event of radio failure an aircraft operating under a DVFR flight plan may proceed in accordance with the original flight plan or land as soon as practicable. The pilot is required to report the radio failure to the appropriate facility as soon as possible.

AIRWAY FLIGHT ROUTES

The National Airspace is divided into two major systems of airways (Fig. 5-10). The airspace from surface level to 18,000 feet MSL is referred to as the low altitude structure and contains the "Victor" system of airways. The airspace from 18,000 feet MSL to flight level 600 (60,000 feet MSL on a standard day) is the high altitude structure of "Jet" airways. Fundamentally, the high

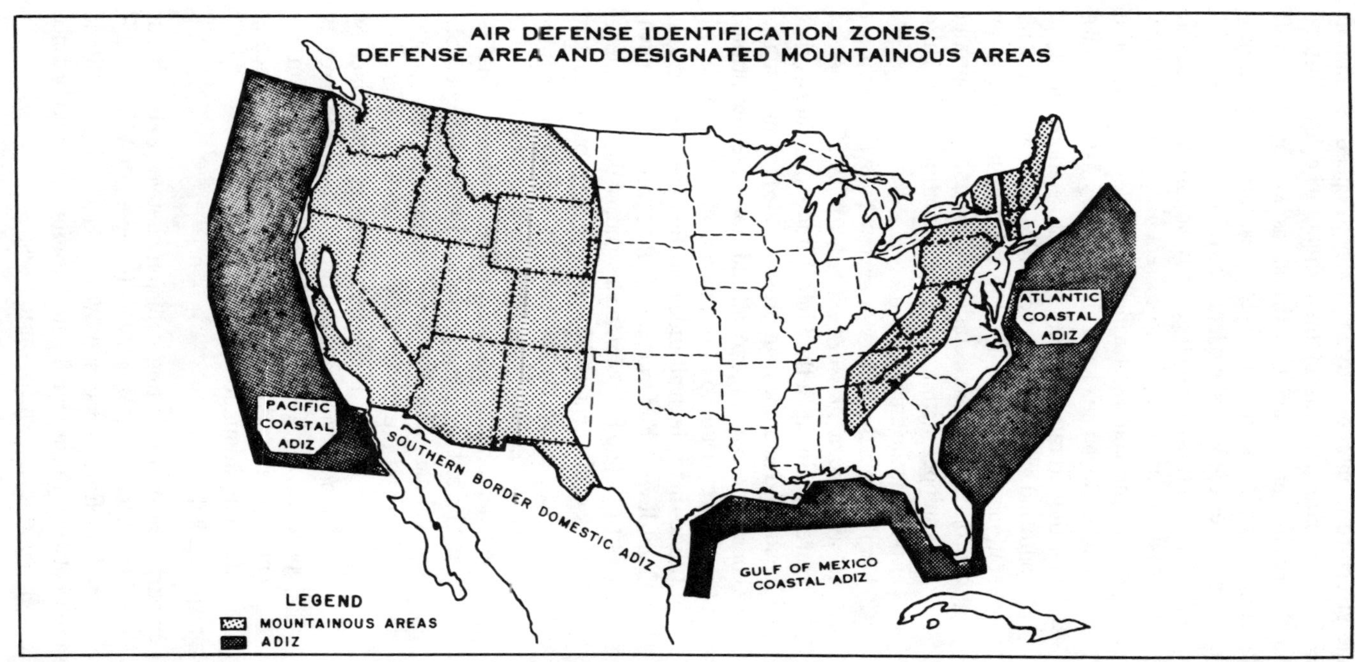

Fig. 5-9. The United States is surrounded by four Air Defense Identification Zones. This map also shows designated Mountainous and Flat areas. A good altitude rule to follow is 1000 ft. minimum over flat areas and 2000 ft. over mountainous areas.

altitude structure is intended to service high speed jet aircraft during the en route portions of their flight. The low altitude structure services general aviation aircraft as well as all aircraft during arrivals and descents in a terminal area. Operational differences between the Victor and Jet airways are as follows:

1. Altitudes flown on Victor airways are MSL altitudes. It is necessary for a pilot to correct his altimeter setting during the course of a flight in order to maintain a constant mean sea level altitude. Regulations require that the aircraft altimeter be set to stations along the route and within 100 nautical miles of the aircraft. If no stations are within the 100 nautical mile range an appropriate available station may be used. For aircraft not equipped with a radio, the elevation of departure airport may be employed.

2. For flights in the Jet airway system the altimeter is set to a constant *pressure altitude* of 29.92 inches of mercury. Flight in the high altitude structure is thus conducted at a constant pressure altitude instead of a constant mean sea level altitude as in the low altitude structure. Due to this fundamental difference, altitudes in the high altitude structure are referred to as flight levels. For example, flight level 190 is equivalent to 19,000 feet MSL on a standard day, flight level 200 is equivalent to 20,000 feet MSL on a standard day, etc. On non-standard days (when the altimeter is other than 29.92) flight levels will not be located at their numerically equivalent MSL altitudes. That is, flight level 190 will not be located at 19,000 feet MSL. To maintain the division between the Jet airway system and the Victor airway system at 18,000 feet MSL the lowest usable flight level is determined by the atmospheric pressure in the intended area of operation per Table 5-3. The Jet airway system lies entirely within the region of positive control, hence all flight operations are IFR at all times.

The Hemispheric Rule

For purposes of preventing collisions between aircraft our airspace is divided into a series of layers to segregate VFR and IFR traffic as well as East bound and West bound traffic. Altitude separation between these four flight operations is in accordance with the hemispheric rule which is as follows:

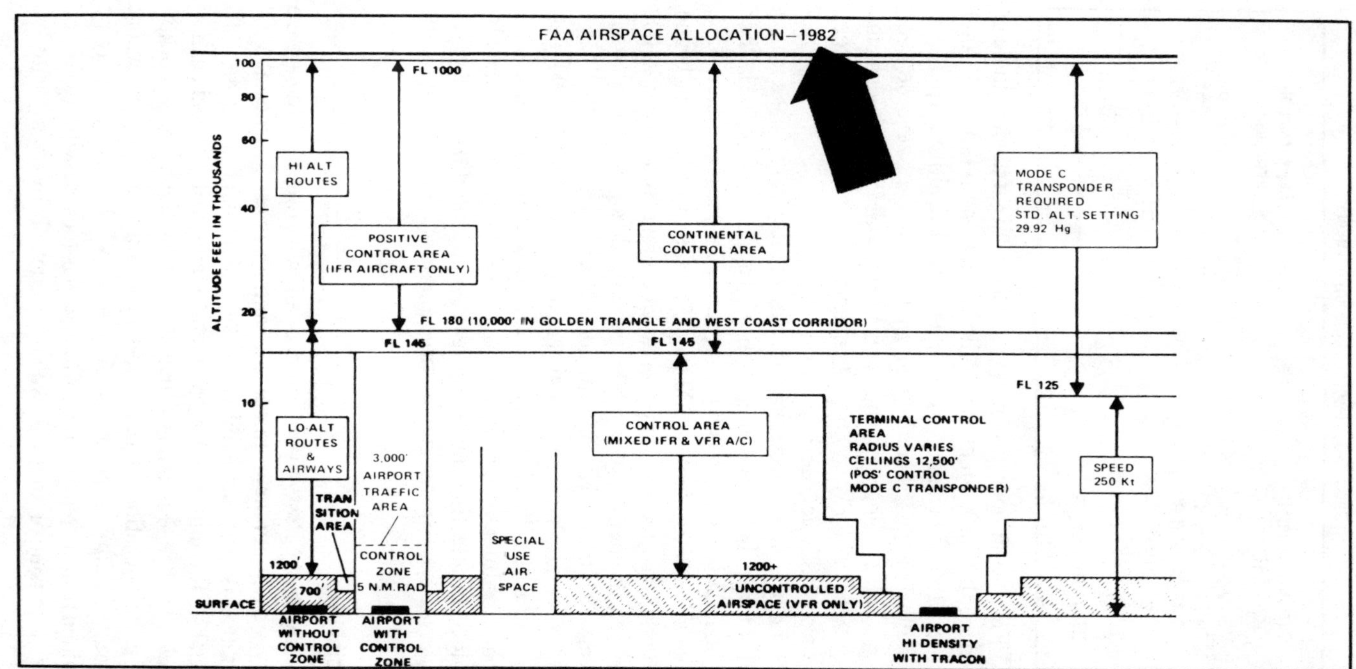

Fig. 5-10. 1982 airspace; increased areas of positive control; more TCA's and a lower positive floor in the Golden Triangle and West Coast corridor.

Table 5-3. Lowest Usable Flight Levels.

Current Altimeter Setting	Lowest Usable Flight Level
29.92 (or higher)	180
29.91 thru 29.42	185
29.41 thru 28.92	190
28.91 thru 28.42	195
28.41 thru 27.92	200
27.91 thru 27.42	205
27.41 thru 26.92	210

VFR aircraft operating in level cruising flight at an altitude of more than 3,000 feet above the surface and less than 18,000 feet MSL shall maintain odd 1,000 foot MSL altitudes plus 500 feet when East bound (a magnetic course of zero degrees through 179 degrees) and even 1,000 foot MSL altitudes plus 500 feet when West bound (a magnetic course of 180 degrees through 359 degrees).

Who flies at the even and odd level of altitudes? That's where the IFR traffic is located in uncontrolled airspace (Fig. 5-11). For East bound IFR level cruising flight below 18,000 feet typical MSL altitudes are 3,000 feet, 5,000 feet, 7,000 feet, etc. West bound IFR flights are located typically at 4,000 feet, 6,000 feet, 8,000 feet, etc. The net result is a 500 feet altitude separation between VFR and IFR traffic. Holding altitude while VFR is important!

The hemispheric rule applies at altitudes above 18,000 feet, however, since the region above 18,000 feet is positive control airspace, altitudes are assigned by ATC and flown accordingly by all pilots. In selecting your en route cruising altitude remember it's the *magnetic course* that counts (true course plus or minus magnetic variation).

Special VFR

Special VFR weather minimums allow a lower cloud clearance and visibility minimum in certain control zones if a special VFR clearance is first obtained from ATC. For fixed wing aircraft, special VFR minimums are one statute mile visibility and clear of the clouds. These minimums may be determined from the cockpit except for operations to or from an airport in a control zone where ground visibility is reported. A special VFR clearance may be obtained from a control tower when such is located within a control zone. For control zones in which no control tower is located, a

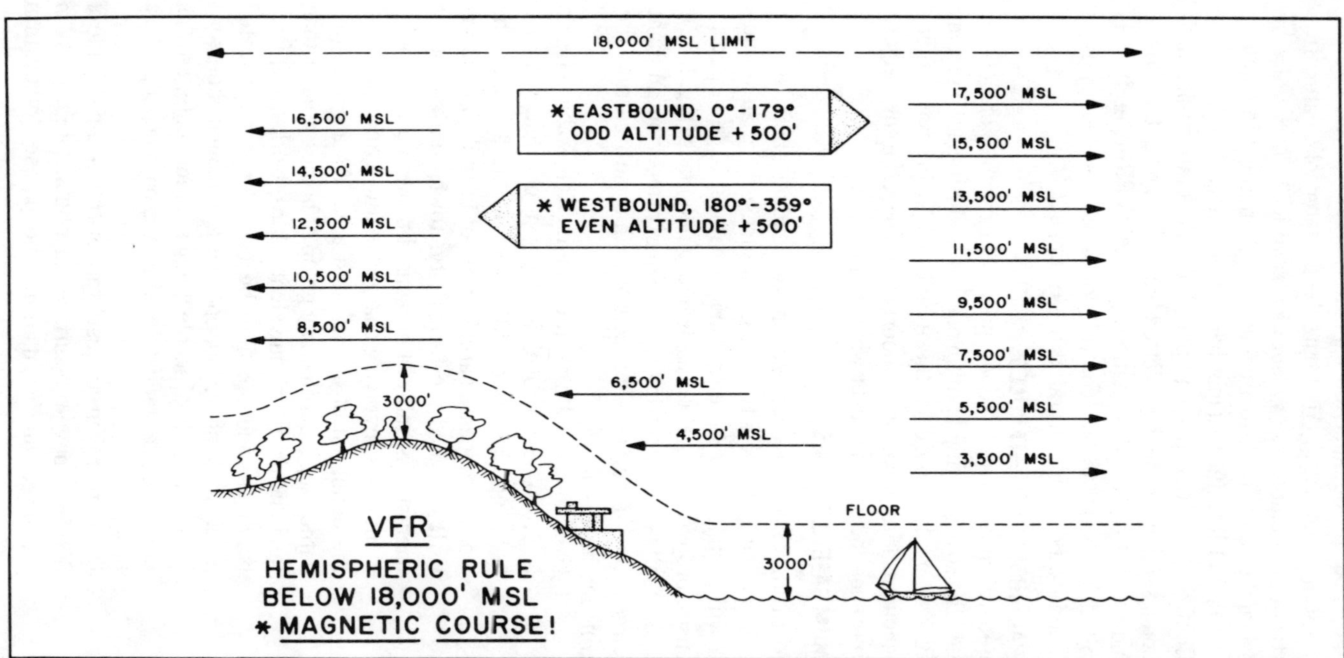

Fig. 5-11. When flying at an altitude greater than 300 ft. above ground level, VFR traffic is segregated into Eastbound and Westbound layers in accordance with the Hemispheric Rule. This segregation is one means of collision avoidance.

clearance may be obtained through the nearest tower, nearest Flight Service Station, or Air Route Traffic Control Center. It is important to remember that *special VFR is an ATC clearance*. The purpose of the clearance is to provide separation from IFR traffic or other special VFR traffic that may be in the vicinity.

Control zones that permit special VFR operations are indicated on a sectional chart by a dotted blue line. Those that do not are indicated by "T" designation. Special VFR operation is prohibited at night in all control zones unless the pilot is instrument-rated and his aircraft is IFR equipped.

GENERAL AIRSPACE FLIGHT RULES

Federal aviation regulations provide a number of general "rules of the road" for flight within our National Airspace System. The fundamental purpose of these rules is to promote safety throughout aviation and, just as important, respect the sovereignty of the persons and property over which we fly.

MINIMUM SAFE ALTITUDES

With the exception of takeoff or landing there is one overriding minimum safe altitude rule; we must fly at an altitude which will permit an emergency landing without undo hazard to persons or property on the surface if an engine(s) fails (Fig. 5-12). This rule applies anywhere but is particularly important in flight over large metropolitan areas. In addition to this one basic rule, the following minimums also apply:

1. Over congested areas the minimum allowable altitude is 1,000 feet above the highest obstacle within a horizontal radius of 2,000 feet of the aircraft. The words "congested area" have not been defined in the FAR's beyond the terms city, town, settlement, or open air assembly of persons. Furthermore, enforcement cases involving charges of low-flying vary as to the definition of a congested area. As a guide, this author suggests that any group of homes greater than ten in number (or equivalent cabins, workshops, farm buildings, and the like) be considered a congested area. Remember, too, open air assemblies of persons include crowds of people lining a parade route, crowds of people in a stadium, or simply groups of people on a beach.

2. Over other than congested areas a minimum altitude of 500 feet above the surface must be maintained. As in the preceding regulation, the phrase "other than

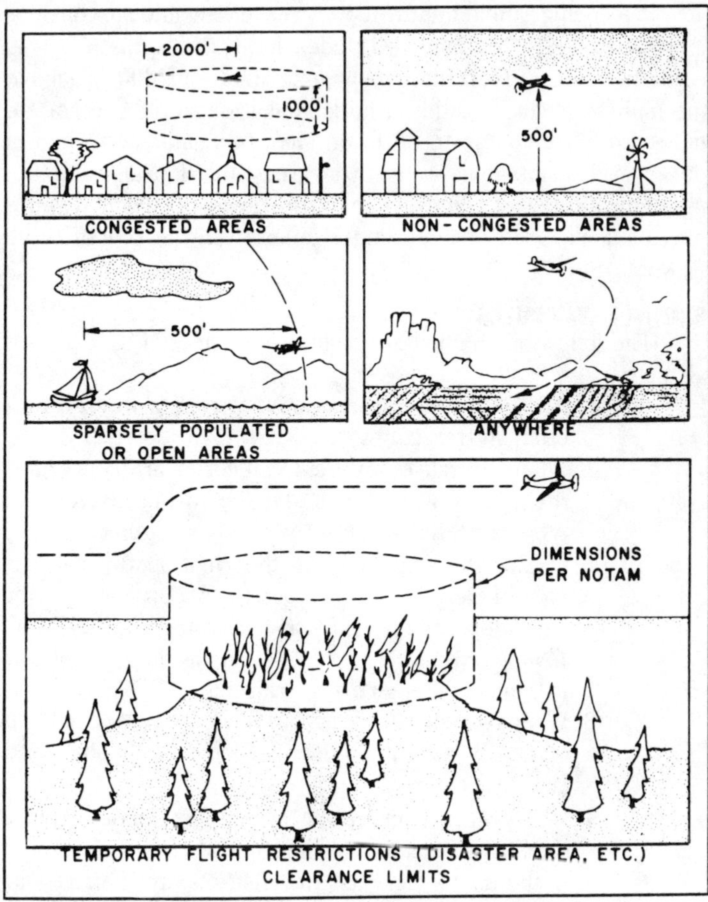

Fig. 5-12. Minimum altitude clearance limits for VFR aircraft.

congested areas" is not defined by FAR. The author suggests that groupings of one or two houses located a significant distance from like groupings (as in the case of open farm country) be considered "other than congested areas."

3. Over open water or sparsely populated areas aircraft may not be operated closer than 500 feet to any person, vessel, vehicle, or structure. In this instance FAR's do not imply a minimum altitude but rather a minimum *distance* from a structure or person. The term "sparsely populated" is interpreted herein as meaning a single isolated dwelling in the middle of a desert or a forest.

In planning a minimum altitude, an excellent guide to consider is FAR 91.119, "Minimum Altitudes for IFR operation." This regulation states in part that a minimum altitude of 2,000 feet above the highest obstacle within a horizontal distance of five statute miles from the course to be flown shall be maintained when in designated mountainous areas. In flat areas aircraft shall maintain a minimum altitude of 1,000 feet above the highest obstacle within a horizontal distance of five statute miles from the course to be flown.

RIGHT-OF-WAY RULES

The following summarizes right-of-way rules (Fig. 5-13) for aircraft:

1. An aircraft in distress has the right-of-way over *all* other air traffic.
2. When aircraft of different categories are converging the right-of-way is granted to the *least* maneuverable. A balloon has the right-of-way over any other category of aircraft. A glider has the right-of-way over an airship, airplane, or rotocraft. An airship has the right-of-way over an airplane or a rotocraft. An aircraft towing or refueling other aircraft has the right-of-way over all other engine driven aircraft.
3. When two aircraft are approaching each other head-on, or nearly so, each pilot of each aircraft shall alter course to the *right*.
4. When two aircraft are converging at approximately the same altitude at a right angle (or nearly so) the aircraft to the pilot's right has the right-of-way. This regulation is similar to the unmarked intersection in driving, that is, the auto to the right has the right-of-way. Technically speaking, a Cessna 150 would have the right-of-way over a Boeing 747 should the two be on a converging course with the 150 located to the right of the Boeing 747. While the FAR's do not differentiate in right-of-way rules applicable to light and heavy aircraft, we might consider the advice of Shakespeare who wrote, "The better part of valor is discretion...."
5. Each aircraft that is being overtaken has the right-of-way and each pilot of an overtaking aircraft shall alter course to the right and pass well clear.
6. Aircraft, while on final approach to land, or while landing, have the right-of-way over other aircraft in

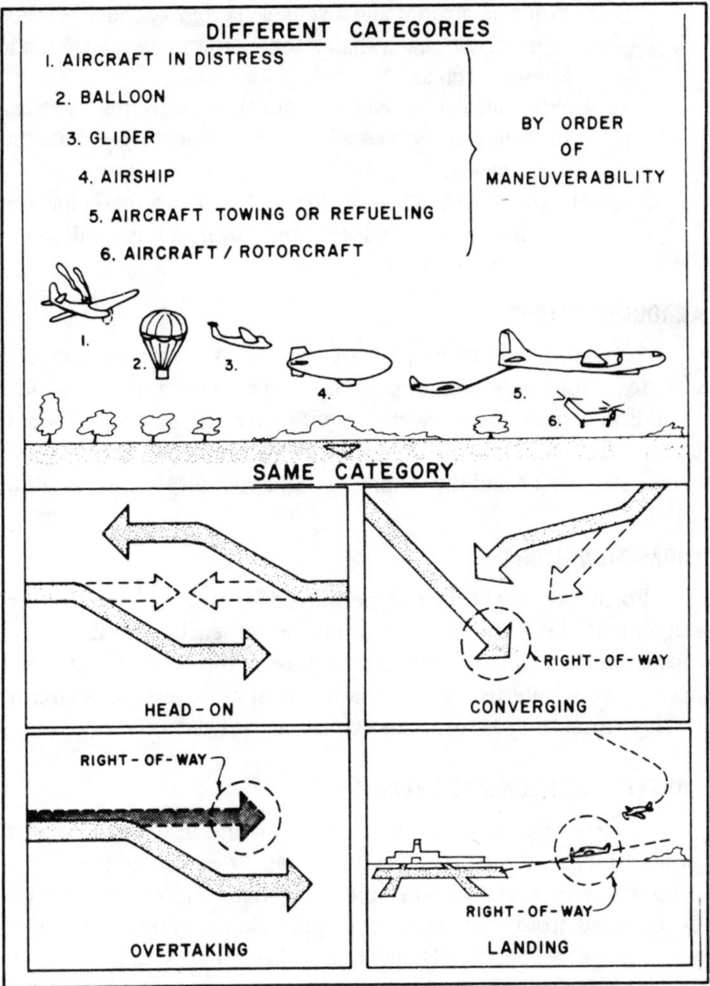

Fig. 5-13. Aviation "rules of the road" for airborne vehicles of the same and different categories.

flight or operating on the surface. When two or more aircraft are approaching an airport for the purpose of landing, the aircraft at the lower altitude has the right-of-way, but it shall not take advantage of this rule to cut in front of another which is on final approach to land, or to overtake that aircraft.

7. Aircraft to operating over water shall, insofar as possible, keep clear of all vessels and avoid impeding their navigation.

8. When an aircraft and a vessel are approaching head-on or nearly so, each shall alter its course to the right and keep well clear.
9. When an aircraft and a vessel are on crossing courses, the aircraft or vessel to the other's right has the right-of-way.
10. An aircraft or vessel that is being overtaken has the right-of-way, and the one overtaking shall alter course to keep well clear.

ACROBATIC FLIGHT

Acrobatic flight is not permitted over any congested area of a city, town, or settlement or over an open air assembly of persons. In addition, such is not permitted within a control zone or federal airway. The minimum altitude for acrobatic flight is 1,500 feet above the surface and the minimum flight visibility is three statute miles.

FORMATION FLIGHT

Formation flight is not permitted except by previous arrangement with the pilot-in-command of each aircraft in the formation. Carrying passengers for hire in formation flight is not permitted. A pilot flying in formation must not operate his aircraft so close to another as to create a collision hazard.

CARELESS OR RECKLESS OPERATIONS

A person may not operate an aircraft in the air or on the ground in a careless or reckless manner so as to endanger the life or property of another. No pilot-in-command may allow any object to be dropped from that aircraft in flight that creates a hazard to persons or property. (Objects may be dropped if reasonable precautions are taken to avoid injury or damage to persons or property).

LIQUOR AND DRUGS

No person may act as a crew member of a civil aircraft within eight hours after the consumption of any alcoholic beverage, while under the influence of alcohol, or while using any drugs that affect his faculties in any way contrary to safety. Furthermore, except in an emergency, no pilot may allow a person who is obviously under the influence of intoxicating liquors or drugs (except a medical patient under proper care) to be carried in that aircraft.

FLIGHTS BETWEEN MEXICO AND THE UNITED STATES

No person may operate a civil aircraft between Mexico and the United States, with knowledge that narcotic drugs, marijuana, or depressant or stimulant drugs or substances as defined in federal statutes are carried in the aircraft. Furthermore, persons operating a civil aircraft between Mexico and the United States must comply with requirements for penetrating the ADIZ as noted in Part I of this chapter. If the aircraft does not have a two-way radio, that person shall, in addition to complying to ADIZ requirements, land at the designated airport of entry nearest the point of entry into the United States, and file an arrival or completion notice.

The FAA is granted the authority to suspend or revoke, as well as deny application for, pilot certificates and/or operating certificates of all persons convicted of violating any federal or state statute, "relating to the growing, processing, manufacturing, sale, disposition, possession, transportation or importation of narcotic drugs, marijuana, and depressant or stimulant drugs or substances." Flight plans are required for persons operating a civil aircraft on a flight between Mexico or Canada and the United States. See Apppendix A, Mexican Federal Aviation Regulations, for additional detail.

Many countries, including Canada and Mexico, require advance notice of the intent of pilots to arrive in those countries. Under agreements between the United States, Canada, and Mexico, operators of private planes may, in most cases, include this advance notice in a flight plan to be filed prior to departure from the United States with the nearest FAA communications station. The station will then relay the message to the proper authorities in the country of destination without further action on the part of the pilot. In a similar manner, operators of private planes upon reentering the United States from a foreign country must provide an advance notice of the estimated time of arrival to U.S. Customs *for each flight*. In general, one hour advance notice is sufficient. Advance notice may be provided by a direct phone call to Customs, by filing a flight plan with a request to "advise customs," or by radio contact with the nearest Flight Service Station after takeoff. For Customs services Sundays, holidays, or outside regular duty hours, expect to be charged an overtime fee, payable in cash. For detailed information on customs regulations contact your local FSS or write for the following booklet:

Customs Guide for Private Flyers
Stock No. 4802-0029

Superintendent of Documents
U.S. Government Printing Office
Washington, D.C. 20402

AIRCRAFT LIGHTING

Regulations require that aircraft position lights be turned on from official sunset to official sunrise. This applies regardless of whether the aircraft is in flight or taxiing on the ground. The aircraft anti-collision rotating beacon or strobe light is likewise required to be turned on. Hours for sunset and sunrise can be determined by calling FSS, tower, or other such ATC facility. Tables of sunrise and sunset are available for almost all cities of over 50,000 population. To obtain information on your particular city request a table of sunrise and sunset times from the following agency:

Nautical Almanac Office
United States Naval Observatory
Washington, D.C. 20390

For those in Alaska, aircraft position lights must additionally be turned on during periods when a prominent unlighted object cannot be seen from a distance of three statute miles or when the sun is more than six degrees below the horizon.

COMPLIANCE WITH ATC CLEARANCES

When a pilot requests and is granted an air traffic control clearance there exists, in effect, a verbal contract between the pilot and the air traffic controller. The contract is "signed" when the pilot acknowledges and accepts the ATC clearance. Each contract (clearance and acknowledgment) so formed is unique; it exists between the *individual* pilot and his air traffic controller (Fig. 5-14). A clearance given to one pilot *does not* apply to another even though their situations may be similar. Furthermore, when an ATC clearance has been obtained, no pilot-in-command may deviate from that clearance, except in an emergency, unless he obtains an amended clearance.

EMERGENCY AUTHORITY OF THE PILOT-IN-COMMAND

FAR 91.3 grants a pilot-in-command emergency authority; that is, authority to deviate from flight rules in an emergency to the extent required to meet that emergency. What constitutes an emergency? The situations are many and varied. A few examples follow:

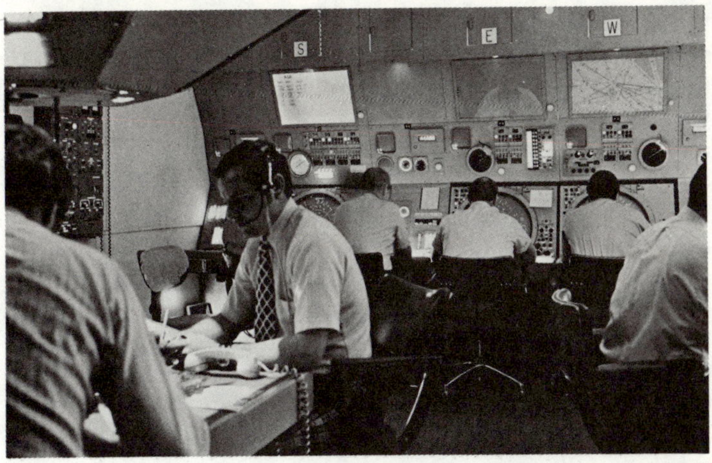

Fig. 5-14. As aviation grows, additional dependence will be placed on ARTS II and III facilities (Automated Terminal Radar Service) for both IFR and VFR flight. Radar traffic advisory, vectoring, and sequencing service in the vicinity of sizeable airports is a highly desirable VFR procedure. A call to radar 15-25 miles out is all that is required to obtain service.

1. Getting caught in a weather situation that puts a VFR pilot in an IFR situation.
2. Night flight can provide unexpected experiences for the low-time VFR pilot. On a dark night clouds are not visible, distant auto headlights and stars appear similar, and the horizon is often indistinct if visible at all. IFR situations may exist even though VFR conditions prevail.
3. Unexpected illness on the part of the pilot or a passenger.
4. A failure of the aircraft power plant or instruments that would affect the safety of the flight.
5. Becoming lost or disoriented.
6. Becoming low on fuel.

Whatever the situation—uncertainty, alert, or distress— three basic actions by the pilot-in-command will bring the ATC network to his assistance. First, if transponder equipped, squawk Code 7700. This rings bells and lights at all ATC radar facilities within range. In addition, your radar return signal is automatically identified thus giving the radar controller your location. Second, contact the air traffic control agency nearest you (FSS, tower, center, etc.). If unable to determine the appropriate frequency, use the emergency frequency 121.5 MHz. Identify who you are, where you are, the nature of the distress, and the type of assistance desired. Advise ATC that you are "declaring an emergency." This will grant you priority over other aircraft under ATC control in your local area. Third, comply with advice and instructions received.

When a pilot is in doubt of his position, or feels apprehensive for his safety, he should not hesitate to request assistance. Search and rescue facilities, including radar, radio, and Direction Finding stations are ready and willing to help. There is no penalty for using this service. Delay in so doing has caused accidents and cost lives. A pilot-in-command who is given priority by ATC in an emergency, *if requested*, shall submit a detailed report of that emergency within 48 hours to the chief of the particular ATC facility involved.

SPEED LIMITS

The following aircraft speed limits apply:

1. Surface to 10,000 feet MSL: 250 knots maximum.
2. Below layer of a Terminal Control Area: 200 knots maximum.
3. Within an aircraft traffic area: 156 knots—reciprocating engine aircraft, 200 knots—jet aircraft. If located within a TCA, 250 knots—all aircraft.

Military aircraft which cannot slow to the specified speeds are permitted to operate at safe minimum speeds for the type of aircraft involved.

LANDING AREAS

With over 12,500 public airports, there is little doubt that such is the intended terminal area for aircraft. Landing at private or restricted airports is subject to conditions that may be imposed by the owner. Furthermore, airports of this nature often do not meet national standards for over-runs, obstacles, and other conditions which may make landings and takeoffs hazardous. The FAA does not have jurisdiction over such airports.

Except in an emergency, a landing in open country is open to question. In a national park or wildlife preserve the regulations previously noted apply. On municipal or state-owned property conditions imposed by the cognizant government body apply. Some states prohibit landing in open country except in the case of an emergency. Consult your local FAA GADO and state aviation authority regarding landing in areas other than at public airports.

FLIGHT INSTRUCTION

An aircraft used for VFR flight instruction must have fully functioning dual controls. However, instrument flight instruction may be given in a single engine airplane equipped with a throwover control wheel. If used in simulated instrument flight, an appropriately rated pilot must occupy the co-pilot seat as a safety pilot. Furthermore, the safety pilot must have adequate vision forward and to each side of the aircraft. In the event his vision is not adequate a competent observer is additionally required in the aircraft to supplement the vision of the safety pilot.

FLIGHT RESTRICTIONS IN THE
PROXIMITY OF THE PRESIDENTIAL PARTY

Regulations prohibit operating an aircraft over or in the vicinity of any area to be visited or travelled by the President, the Vice-President, or other public figures contrary to the restrictions established by the FAA and published in a Notice to Airmen. Civil aircraft flight is similarly prohibited in areas being utilized for space flight recovery operations; designation thereof is by NOTAM.

Chapter 6
Accident Reports
and Investigations

The National Transportation and Safety Board was established as a part of the FAA Act of 1958 to make rules and regulations governing the notification and reporting of accidents involving civil aircraft (Fig. 6-1) and to investigate such accidents and report the facts and probable cause. The Board is responsible to recommend to the FAA actions designed to prevent similar accidents in the future, issue reports to the public where such will enhance safety, and determine techniques or procedures that will eliminate or reduce the possibility of accidents. The primary function of the NTSB is to promote safety in transportation. A second function of the NTSB is to review airmen appeals relating to suspension, amendment, revocation, or denial of certificates (such as pilot's certificate).

The findings of the NTSB are reported directly to the Secretary of the Department of Transportation and the administrator of the FAA where applicable. The NTSB investigates rail, highway, and pipeline accidents in addition to those associated with the civil aviation.

The NTSB delegates authority to the FAA to investigate certain aircraft accidents and to submit reports to the Board from which the NTSB may determine the probable cause of the accident. Authority is granted the FAA to investigate an furnish reports to the NTSB on (1) all non-fatal general aviation aircraft accidents; (2) all aerial application accidents; (3) all amateur-built aircraft

Fig. 6-1. The Cessna Titan is available in three versions. The Ambassador is an executive aircraft; the courier serves general aviation; the Freighter is outfitted for cargo transportation. Shown here is the Ambassador.

accidents; and (4) all restricted category aircraft accidents. In addition, the FAA is granted authority to investigate and report on fixed-wing aircraft accidents for vehicles which are not engaged in air carrier or air taxi operations and weigh less than 12,500 pounds.

A major service performed by the NTSB is the compilation of accident statistics with subsequent publication of results and recommendations for corrective action. Reports of this nature are available to the public. In addition, the NTSB publishes safety notices and recommendations on a day by day basis. These, too, are available to the public. Recommendations of the Board result in AD notices, airman training criteria and operational procedures.

The NTSB establishes rules pertaining to aircraft accidents, incidents, over-due aircraft, and safety investigations. The operator of an aircraft is required to immediately notify the National Transportation Safety Board, Bureau of Aviation Safety Field Office in case of any of the following aircraft accidents or incidents:

1. Aircraft accidents;
2. Flight control system malfunction, or failure;
3. Inability of any required flight duties as a result of injury or illness;
4. Turbine engine rotor failures excluding compresser blades and turbine buckets;
5. In-flight fire;
6. Aircraft collide in flight;

7. An aircraft is overdue and is believed to have been involved in an accident.

The notification shall include the following information if available:

    a. Type, nationality, and registration marks of the aircraft;

    b. Name of owner, and operator of aircraft;

    c. Name of the Pilot-in-command;

    d. Date and time of accident;

    e. Last point of departure and point of intended landing of the aircraft;

    f. Position of the aircraft with reference to some easily defined geographical point;

    g. Number of persons aboard, number killed and number seriously injured;

    h. Nature of the accident including weather and the extent of damage to the aircraft so far as it is known;

    i. A description of any explosives, radioactive materials, or other dangerous articles carried.

The NTSB may be notified through the local FAA office or, in major cities, by direct telephone contact with the NTSB field office.

A formal report is required within ten days after an accident or reportable incident has occurred; or when after seven days an overdue aircraft is still missing. An "aircraft accident" means an occurrence associated with the operation of an aircraft which takes place between the time any person boards the aircraft with the intention of flight until such time as all such persons have disembarked. An accident is defined as a situation in which any persons suffered death or serious injury as a result of being in an aircraft or by direct contact with the aircraft or anything attached thereto, or the aircraft received substantial damage. "Fatal injuries" imply any injury which results in death within seven days. A "serious injury" means any injury which (1) requires hospitalization for more than 48 hours, commencing within seven days from the date the injury was received; (2) results in a fracture of any bone (except simple fractures of fingers, toes, or nose); (3) involves lacerations which cause severe hemorrhages, nerve, muscle, or tendon damage; (4) involves injury to any internal organ; or (5) involves second- or third-degree burns, or any burns affecting more than 5% of the body surface. The term "substantial damage" implies aircraft damage which adversely affects structural strength, performance, or flight characteristics of the aircraft resulting in a major repair.

The preceding paragraph, although long and involved to read, has been included to reinforce the responsibility associated with being a pilot-in-command.

A report on an incident for which notification is required shall be filed only if requested by a representative of the NTSB. "Incidents" are many and varied. Typically incidents relate to simple engine failure, bent fairings or cowlings, dented skin, ground damage to propeller blades, damage to landing gear, wheels, tires, flaps, engine accessories, brakes, wing tips, etc.

In summary, NTSB regulations require immediate notification in cases of major accidents or overdue aircraft with formal reports following ten days after an accident and seven days after an overdue aircraft is still missing. Reports on incidents are only filed if requested. NTSB Form 6120.1 is the proper document for filing an aircraft accident report for small aircraft (under 12,500 pounds).

Chapter 7
Enforcement Procedures

Under Title IX of the Federal Aviation Act of 1958, FAA is granted authority to enforce regulations and levy penalties for both civil and criminal actions. Safety, economic, and postal offenses are subject to civil penalties; perjury of certificates, false marking of aircraft, interference with air navigation, falsification of records, transportation of dangerous articles, aircraft piracy, carrying weapons aboard an aircraft, and the like are subject to criminal penalties. In enacting the Federal Aviation Act of 1958, Congress granted the FAA and the NTSB authority to investigate violations, hold hearings, subpoena witnesses, collect evidence, and levy penalties commensurate with the violation. Thus the FAA and NTSB become successively the first two levels in a system of "courts" applicable to aviation. Cases that are appealed beyond the NTSB level enter the federal district court as the third level in the hierarchy of justice. At this point and beyond legal proceedings are conducted in a traditional fashion.

Any person who knows of a violation of the Federal Aviation Act of 1958, or of any regulation or order issued under it may report in to an FAA regional or district office. Each report is investigated by FAA personnel. The results of that investigation are the basis for determining enforcement actions that the FAA will take.

The purpose of this chapter is to outline the processes involved in handling minor violations of FAR's. Criminal actions and procedures are beyond the scope of this text. Let us explore situations involving the simple fallibility of being human. What happens when a pilot enters a Terminal Control Area—unannounced? What happens when a pilot lands on a taxiway instead of the runway at a controlled airport? What happens when a VFR pilot becomes involved in IFR conditions and must declare an emergency? Let us also explore some of the more serious violations such as a student pilot carrying a passenger, or reckless flying.

MINOR VIOLATIONS

If it is found that a violation does not require legal enforcement action, a flight standards inspector or other appropriate FAA official may issue a safety compliance notice including a letter of reprimand to the violator, or a letter of correction that confirms decisions and states the corrective action agreed to as acceptable to the FAA. If the agreed-upon corrective action is successfully completed, the case is terminated. If, however, the agreed upon corrective action is not successfully completed, legal enforcement action may be initiated.

Let us take an example to examine the process involved. Assume that a non-instrument rated private pilot carrying passengers gets "caught" in IFR conditions and requests help by declaring an emergency. Further assume that the pilot is current in the aircraft, possesses a valid medical, and has done an average job of checking weather prior to flight. Although he was either unfortunate or possibly wrong in proceeding into IFR conditions he was absolutely *right* in declaring an emergency. By so doing he correctly considered the welfare of his passengers and also other aircraft that might legally be IFR in the local vicinity. After landing most likely the following procedural actions would occur.

First the FAA gathers the relevent facts of the matter. Who is the pilot-in-command? What is the status of his medical? Who owns the aircraft? What was the history of the flight (and flight planning)? What weather information was available en route? Were all charts current? The facts are assembled to answer the basic question, "was the emergency of the pilot's *own making* or did it occur as a result of an unforecast weather situation?"

At this point, the FAA has discretionary authority. Assuming that the emergency was not of the pilot's own making, an informal discussion of the matter with an FAA accident prevention specialist may terminate the issue. Let us assume that the pilot was found to be a trifle careless. He may then be requested to brush up by taking some ground school work or he may be required to take a written and/or flight test reexamination if there is evidence of incompetence.

Should it be determined that the pilot was truly negligent—the emergency *was* of his own making—he may be subject to suspension of his flying privileges for a period of time or a fine (but not both).

In general it may be stated that the investigative-enforcement action resulting from a violation of regulations is conducted from the standpoint of positive motivation. A serious and detailed effort is made to determine the underlying cause of a regulatory violation. Counseling and/or a review of aeronautical requirements and practices is first accomplished. Should reprimand be required, as in the case of repeated FAR violations by a pilot, the fine or certificate suspension is tailored to the individual and the violation involved. For persons who make their living by flying, a fine may be levied so as not to jeopardize their occupation through suspension of their certificate. Conversely, for persons who are of financial means or fly for sport, the removal of their pilot privileges for a period of time is often the more meaningful reprimand.

SERIOUS VIOLATIONS

Serious violations of Federal Aviation Regulations are subject to legal enforcement actions and penalties. Violations of civil matters such as the nationality and ownership of aircraft, safety regulations, or aircraft access investigations are subject to penalties of $1,000 for each violation. If such violation is a continuing one, *each day* of the violation may constitute a separate offense. If a civil penalty is considered advisable by the FAA, a formal letter is sent to the person charged with the violation, advising him of the charges against him and the law, regulation, or order that he is charged with violating and, if appropriate, an offer to compromise the penalty. The person charged with the violation may present, to the official who signed the letter, any oral or written material or information in answer to the charges, explaining, mitigating, or denying the violation or showing extenuating circumstances. Material or information so presented is considered in making final determination as to probable liability for a civil penalty, or the amount for which it will be compromised. If the person charged with the violation offers to compromise for a specific amount, he may send a certified check or money order for that amount payable to the Federal Aviation Administration. Assuming the compromise

Fig. 7-1. A true world's favorite aircraft, the Cessna Skylane/182 carries four persons at a comfortable 144 knots cruise.

amount is acceptable the person charged with the violation is notified, by letter, that the acceptance is full settlement of the civil penalty for the violation. If the compromise settlement is not acceptable, the FAA may instigate proceedings to collect the penalty.

As another form of enforcement procedure the FAA may reinspect any civil aircraft (Figs. 7-1 through 7-3) or reexamine any civil airman at any time. In cases where the FAA is considering the suspension or removal of a pilot's certificate, the pilot will be advised of the charges and the proposed action by letter. The letter, a notice of proposed certificate action, allows the holder to answer the charges by checking the appropriate box on the form. The holder may thus:

☐ Admit the charges and surrender his certificate;

☐ Answer the charges in writing;

☐ Request an order be issued in accordance with the notice of proposed certificate action so that he may appeal to the National Transportation Safety Board;

☐ Request an opportunity to be heard in an informal conference with the FAA council; or

☐ Request a formal hearing if the charges concern a matter under Title V of the Act (Nationality and Ownership of aircraft).

The holder must return the form with his answer not later than 15 days after the date it is received, otherwise the FAA will issue the certificate action as proposed. If the holder has requested an informal conference with the FAA council and the charges concern nationality and ownership of aircraft, he may after that conference also request a formal hearing in writing by letter with postmark not later than ten days after the close of the conference. A formal hearing is conducted in the manner of a trial court with the hearing officer acting as the judge. Witnesses are called and testify under oath, motions are allowed, and parties are given an opportunity to present arguments usually through an attorney. The hearing officer listens to the evidence and makes the final judgment as to whether the airman's certificate should be amended, suspended, or revoked.

In the event a person is not satisfied with the results of a hearing, he may file an appeal with the National Transportation and Safety Board for a retrial. Where possible, hearings with the FAA or NTSB will be held at a federal facility near the home location of the accused so as to minimize travel on the part of persons involved. Should a certificate holder believe the action of the FAA

Fig. 7-2. The six seats of the Cessna Stationair 6 may be arranged to face forward or in a club configuration with the center row facing aft.

to be unreasonable, arbitrary, or contrary to law after he has exhausted his right of appeal to the NTSB, he may then appeal to a federal district court for further retrial.

Penalties for serious crimes range from seizure of the aircraft to a fine of $1,000 and/or three years imprisonment for forgery of certificates and false marking of aircraft; a $5,000 fine and/or five years imprisonment for interference with air navigation; $10,000 fine and/or ten years imprisonment for illegal transportation of explosives and other dangerous articles, and twenty years imprisonment or death* for aircraft piracy.

Serious violations require the FAA initiate a thorough investigation of *all* facts involved. Not only are the aeronautical data of the accident investigated but related civil facts are assembled. Has the airman ever been cited for a criminal violation? Does he have any record of misdemeanors? What is his driving record? What is his exact medical history? Is a possible psychological problem involved? The set of data so assembled is accomplished with exceeding thoroughness to provide a case that will withstand a test of the courts. Facts relating to both the guilt *and the innocence* of the accused are assembled. A report is filed with the FAA regional office and the legal procedural machinery is put into action.

Of course, results depend upon the details of a situation. To illustrate, let us assume a pilot-in-command is involved in an aircraft accident for which alcoholic impairment of efficiency and judgement was the cause.** The nature of the reprimand? Probably revocation of the pilot's certificate for at least a year. At the end of the revocation period the pilot may apply to have his certificate reinstated. However, he may be required to retake all of the medical, written, and flight test examinations applicable to his certificate in order to gain reinstatement! Should there be indications that problems leading to the accident continue to persist, it is most likely that the FAA will not grant a reinstatement; the pilot may be denied his certificate for an additional period of time.

Fundamentally, it is not the policy of the FAA to discipline through revocation of a pilot's certificate. Rather, the matter is to ground the violator until the fundamental cause factor can be overcome; perhaps by additional training; perhaps by medical aid;

---

*Per the FAA Act of 1958; subject to current court decisions relating to capital punishment.
**Typically 44 fatal accidents per year.

Fig. 7-3. A four to six place, twin-engine, high performance, turbocharged, pressurized monoplane, the Beech Duke B60 is an aircraft built for operation in positive control airspace.

or by requiring a "breather" to reevaluate one's outlook on flight safety. With the problem area overcome, reinstatement of a pilot's certificate is likely provided the airman can demonstrate that he meets the aeronautical standards required. Is the procedure effective? It must be, for the FAA has a record of fewer repeat violators than any other comparable regulatory agency.

In conclusion, we have dealt in this chapter with enforcement procedures from the standpoint of the federal government. However, this is but a part of the story. Violations which result in property damage or bodily injury may be compounded by additional legal action. Personal liability law suits may well extract an enforcement penalty far *greater* than that levied by the federal government.

Let us remember the law represents the minimum limit of acceptability that society will condone. Enforcement results when that limit is exceeded.

Chapter 8
Commercial Operations

Commercial air carriers are regulated in virtually all phases of their operations. Federal standards apply to the type of aircraft used (Figs. 8-1 through 8-5), maintenance, air crew, routes flown, and rates that may be charged. Air taxi operators and charter services are subject to similar regulations. Aviation training schools, maintenance facilities, and aviation equipment manufacturers are governed by federal minimum standards for services and products. And last, but not least, a large number of people within the aviation community are subject to minimum federal standards as a condition of employment. The pilot, aviation mechanic, flight engineer, repairman, flight instructor, air traffic controller, ground school instructor, parachute rigger . . . all must meet federally specified minimum proficiency standards to render services on a commercial basis.

This chapter is a brief review of Federal Aviation Regulations applicable to certificated personnel performing services on a commercial basis in the field of aviation. The purpose of this chapter is to answer the question, "Where to from here? What qualifications must I have to enter aviation on a commercial basis?" Typical industry minimum requirements, current at the time of preparing this text, have been included as an additional factor for consideration.

FLIGHT TRAINING

The aviation flight training school is often the first step in the career of a commercial pilot. Flight training schools are certified

Fig. 8-1. The Beechcraft Super King Air features pressurization, air conditioning, reversible props, and twin P & W 850 shp turboprops.

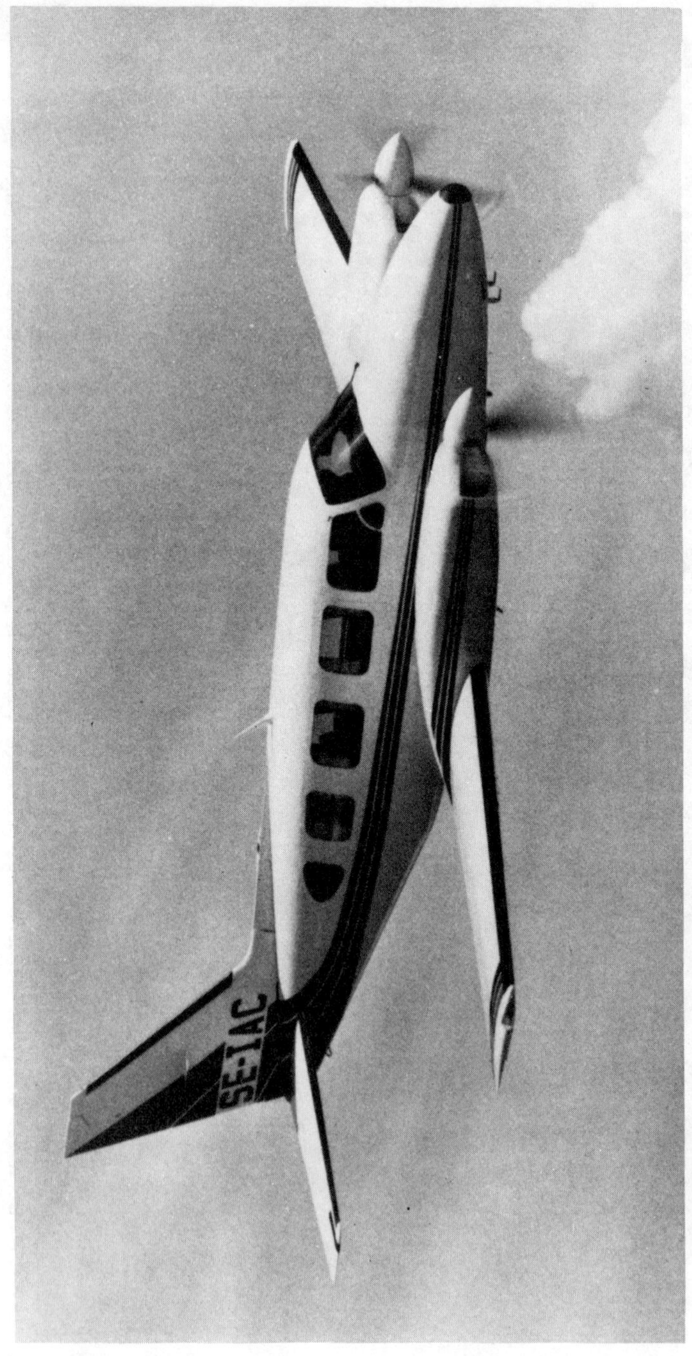

Fig. 8-2. The Piper Navajo carries six at 215 knots with a pair of 310 hp Lycoming engines.

Fig. 8-3. Tailwheel aircraft are still being manufactured. The Cessna 180 Skywagon has been serving for years on unimproved landing strips and also as a floatplane.

Fig. 8-4. The pressurized cabin of the Cessna 340 holds six passengers.

under FAR Part 141. This FAR establishes minimum limits for the classroom, the training equipment used, and the quality of instruction offered. Curriculum requirements as well as instructor requirements are specified for FAA approved ground schools and flight schools. Aircraft maintenance facility requirements are additionally specified for flight schools. Fundamentally, the flight school offers aviation opportunities at three basic levels; namely, the *ground school instructor*, the *flight instructor*, and the *chief flight instructor*. The following is a brief synopsis of minimum federal aviation requirements for each of these levels.

Ground School Instructors

FAR Part 143 describes the basic requirements for a ground school instructor. To be eligible for a certificate a person must be at least 18 years of age, of good moral character, and able to show his practical and theoretical knowledge of the subject for which he seeks a rating by passing a written test on that subject. Ground school instructor ratings are as follows:

Ground Instructor-Basic: This rating qualifies the holder to instruct in a basic pilot ground school (private pilot).

Ground Instructor-Advanced: This rating qualifies the holder to instruct in a basic or advanced pilot ground school (private and commercial pilots).

Ground Instructor-Instrument: This rating qualifies the holder to give ground instruction in an instrument flying school. In addition, it qualifies the holder to operate instrument procedures training devices.

146

Applicants for a ground instructor certificate must take a written test on fundamentals of instruction and a written test for the rating desired. Holders of a ground instructor certificate with one or more ratings are not required to retake the fundamentals of instruction test to obtain additional ratings. An applicant who holds a currently effective teacher certificate issued by a state, county, or city authorizing him to instruct in a junior or senior high school, or who is regularly employed as an instructor in an accredited college or university, is not required to take the written test on the fundamentals of instruction.

The ground instructor rating is perhaps the simplest and most certainly the lowest cost aviation certificate that is offered. The only real requirement is a bit of serious home study relating to the theory and practice of flight. The following reference material is highly recommended for anyone considering the ground instructor certificate:

1. *Aviation Weather*, AC 00-6A
2. *Aviation Weather Services*, AC 00-45A
3. *Aviation Instructor's Handbook*, AC 60-14
4. *Flight Instructor Airplane, Answer Book*, 61-72A
5. *Flight Training Handbook*, AC 61-21A
6. *Flight Instructor Airplane, Written Test Guide*, AC 61-72A
7. *Fundamentals of Instructing, Flight and Ground Instructor Test Guide*, AC 61-90
8. *Ground Instructor Instrument Written Test Guide*, AC 143-2A

Fig. 8-5. One of the smaller turboprops, the Beechcraft King Air E90 seats six to ten people and cruises at 245 knots at 21,000 feet on only 65% power.

9. *Ground Instructor Written Test Guide*, AC 143-1E

10. *Instrument Flying Handbook*, AC 61-27B

11. *Instrument Rating, Written Test Guide*, AC 61-8D

All of these are available from:

> Superintendent of Documents
> U.S. Government Printing Office
> Washington, DC 20402

Ground school instruction opportunities exist at flight schools, junior colleges, community colleges, and, in some cases, at the university level. Earnings vary, depending upon the course and level, but are typically in the vicinity of $5.00 to $10.00 per hour. To retain currency, the holder of a ground instructor certificate must show that he has served for at least three months as a ground instructor out of the previous 12 months or demonstrate to the FAA that he meets the standards prescribed for the certificate and rating. The ground instructor rating is an excellent certificate to have; it is one which many flight instructors acquire as a normal part of their activities. To be a truly effective ground instructor, flight experience is a must.

Flight Instructor

The flight instructor is a foundation stone of aviation. On his shoulders rests the quality and knowledgability of today's pilots. Chapter 2 lists minimum federal requirements for the ratings of flight instructor-airplane and flight instructor-instrument. Typical wages range from $10.00 to $15.00 per flight hour. Considering the fact that one hour in the air usually requires 1½ to 2 hours of time, and waits between students bring little earnings, the flight instructor is at best poorly paid overall. Only large flight schools place their flight instructors on a salary plan.

Chief Flight Instructor

FAR Part 141 specifies that a chief flight instructor must be at least 21 years of age and have a good record as pilot and flight instructor. For a primary flying school, a chief flight instructor must have at least a commercial pilot's certificate with a flight instructor certificate and ratings for the category of aircraft to be used. In addition, at least 1,000 hours pilot-in-command time is required. Primary flight instruction experience acquired as either a certificated flight instructor or an instructor in a military pilot primary flight training program is required to the extent of 1,000 flight instruction hours to two years experience and a total of 500 flight instruction hours. Furthermore, within the year preceding designation as a chief flight instructor, and applicant must have given at least 100 hours of primary flight instruction as a certificated flight instructor in the category of aircraft used in the course.

For a commercial flying school or flight instructor school, a chief flight instructor must have at least a commercial pilot's certificate and a flight instructor's certificate, each with the rating for the category of aircraft to be used in the course. In addition, an instrument rating and 2,000 hours as pilot-in-command are required. Flight instruction experience must consist of at least three years experience and a total of 1,000 flight hours (or 1,500 flight hours in lieu of the three year time requirement). Within the year preceding designation as a chief flight instructor, the applicant must have given at least 100 hours of pilot instruction as a certificated flight instructor in the category of aircraft to be used in the course. In addition, one year of active service as chief flight instructor of an approved primary flight course or one year of active service as an FAA designated pilot examiner is required.

For an instrument flying school, a chief flight instructor must have at least a commercial pilot's certificate and a flight instructor certificate each with an instrument rating. In addition, 100 hours of flight time under actual or simulated instrument conditions are required and 1,000 hours as pilot-in-command. Instrument flight instruction experience must consist of at least two years experience and a total of 250 flight hours (or 400 flight hours in lieu of the two year time requirement). Within the year preceding designation as a chief flight instructor, the applicant must have given at least 100 hours of instrument flight instruction as a certificated instrument flight instructor or have acted in the capacity of an FAA designated instrument rating examiner during the course of the preceding year.

AIR TAXI AND COMMERCIAL OPERATORS

Airborne operations relating to air taxis and commercial operation of aircraft have reached a high degree of sophistication during the past few years. The holder of an Air Taxi/Commercial Operator (ATCO) certificate is required to adhere to rigorous standards of operations, maintenance, and reporting. To qualify as pilot-in-command for VFR flights, and applicant must have at least 500 hours flight time as a pilot including at least 100 hours of cross-country time, at least 25 hours of which were at night. In addition, a commercial certificate and an instrument rating or ATP is required.

To qualify as pilot-in-command for IFR flight an applicant must have at least 1,200 hours flight time as a pilot including 500 hours of cross-country flight time, 100 hours of night flight time, and 75 hours of actual or simulated instrument flight time, at least 50 of which were in actual flight.

The pilot's proficiency requirements noted represent the minimum as specified by FAR 135. In practice it is common that air taxi operators require an Air Transport Pilot Rating (ATP) to qualify for a pilot position: An applicant for the position of co-pilot (second-in-command) is required to have at least a current commercial pilot certificate with appropriate airplane category and class ratings. In the case of flights under IFR, an instrument rating is additionally required.

FAR Part 135 requires that all ATCO pilots retain currency by virtue of recent flight experience. A pilot-in-command must pass a flight evaluation performance check at 12 month intervals. Additionally, to carry passengers, a pilot must have made at least three takeoffs and three landings to a full stop within the preceding 90 days in an aircraft of the same category, class, and type (if a type rating is required). IFR currency requires that the pilot pass an instrument proficiency check every six months.

AGRICULTURAL AIRCRAFT OPERATION

Specialized aviation activities such as crop dusting are governed by specific federal aviation regulations designed to describe the special activity to be accomplished. FAR Part 137, Agricultural Aircraft Operation, is the regulation which governs the dispensing of either poisons or nutrients intended for plant nourishment, soil treatment, propagation of plant life, or pest control. A person who wishes to use an aircraft for any of these operations must apply for an agricultural aircraft operator certificate to the FAA district office having jurisdiction over the area in which the applicant's home base of operations is located. To qualify as pilot-in-command, an applicant must hold a current private, commercial, or airline transport pilot certificate and be properly rated for the aircraft used. Permission to operate with a private pilot certificate is intended for the individual who wishes to dispense agricultural chemicals *only* over *his own* property and *not* for compensation or hire. A private operator is not permitted to fly over congested areas.

Unless an applicant has a record of proven safety in his flight operation and competence in dispensing agricultural materials or chemicals, he may be required to pass a test demonstrating his knowledge and skill to the chief supervisor of agricultural aircraft operation. The test of knowledge may consist of the following:

1. Steps to be taken before starting operations, including a survey of the area to be worked.
2. Safe handling of economic poisons and the proper disposal of used containers for such poisons.
3. The general effects of economic poisons and agricultural chemicals on plants, animals, and persons, with emphasis on those normally used in the areas of intended operations; and the precautions to be observed in using poisons and chemicals.
4. Preliminary symptoms of poisoning of persons from economic poisons, the appropriate emergency measures to be taken, and the locatin of poison control centers.
5. Performance capabilities and operating limitations of the aircraft to be used.
6. Safe flight and application procedures.

In addition, a test of flying skill to demonstrate short field and soft field takeoffs, approaches to the working area, flare-outs, swath runs, pull-ups and turn-arounds may be additionally required.

The application of agricultural chemicals from aircraft has become a specialized art. Converted World War II airplanes are being replaced by aircraft specially designed for the function. A modern agricultural aircraft typically costs in the vicinity of $50,000 to $100,000.

FLIGHT ENGINEERS

To be eligible for a flight engineer certificate, a person must be at least 21 years of age, able to read, speak and understand the English language, and hold at least a second-class medical certificate. Flight engineers certificates are offered for reciprocating engine aircraft, turboprop powered aircraft, and turbojet aircraft. An applicant for a flight engineer certificate must pass a written test on regulations concerning the duties of a flight engineer, the theory of flight, aero-dynamics, meteorology, and the various aspects of airplane equipment, systems, operating procedures, etc. Before taking the written test an applicant must demonstrate that he has experience which will meet one of the following criteria:

1. At least three years of diversified practical experience in aircraft and aircraft engine maintenance (of which at least one year was in maintaining multiengine aircraft with engines rated at least 800 horsepower each, or the equivalent in turbine engine powered aircraft), and at least five hours of flight training in the duties of a flight engineer.
2. Graduation from at least a two-year specialized aeronautical training course in maintaining aircraft and aircraft engines (of which at least six calendar months were in maintaining multiengine aircraft with engines rated at least 800 horsepower each, or the equivalent in turbine engine powered aircraft), and at least five hours of flight training in the duties of a flight engineer.
3. A degree in aeronautical, electrical, or mechanical engineering from a recognized college, university, or engineering school; at least six calendar months of practical experience in maintaining multiengine aircraft with engines rated at least 800 horsepower each, or the equivalent in turbine engine powered aircraft; and at least five hours of flight training in the duties of a flight engineer.
4. At least a commercial pilot certificate with an instrument rating and at least five hours of flight training in the duties of a flight engineer.
5. At least 200 hours of flight time in a transport category airplane (or in a military airplane with at least two engines and at least equivalent weight and horsepower) as pilot in command or second in command performing the function of a pilot in command under the supervision of a pilot in command.
6. At least 100 hours of flight time as a flight engineer.
7. Within the 90 day period before application, successful completion of an approved flight engineer ground and flight course of instruction must be accomplished.

All references to flight experience refer to an aircraft that has at least two engines that are rated at least 800-horsepower each or the equivalent in turbine powered engines.

In addition to the written examination, an applicant must pass a practical test on the duties of a flight engineer to demonstrate that he can satisfactorily perform pre-flight and in-flight functions required for the position.

Typical airline standards for the position of flight engineer require men between 21 and 35 years of age, 5 feet 6 inches to 6 feet 4 inches in height, with at least two years of college training. A commercial pilot's certificate plus 500 to 1,000 hours of flight experience may also be required. For additional information on certification requirements relating to flight engineers and also flight navigators refer to FAR Part 63.

AIR TRAFFIC CONTROL

Our system of skyways is greatly dependent upon the "flying skill" of a vast armada of tower operators, radar operators, and flight service personnel. The term "flying skill" is not without basis for today more than 50 percent of our air traffic control operators have received pilot training of one form or another in the course of their work. The following is a brief description of federal requirements for basic air traffic control positions.

Tower Operator

To be eligible for an air traffic control tower operator certificate, a person must be at least 18 years of age, of good moral character, able to read, write, and understand the English language and hold at least a second-class medical certificate. An applicant must pass a written examination on Federal Aviation regulations, airport traffic control procedures, communications procedures, flight assistance service, air navigation, and aviation weather. In addition, a practical test is required relating to the control tower equipment and its use. This test includes procedures relating to the airport, runways, taxiways, control zones, traffic patterns, operations with ATC center, emergency procedures, and associated duties. Having passed both knowledge and skill requirements, an applicant is generally subjected to apprenticeship training of approximately two years. To remain current it is necessary that he serve at least three of the preceding six months as an air traffic control operator at the control tower to which his facility rating applies, or at the operating position for which he has qualified. A beginning tower operator is typically classified as a GS-5 (Government Service Grade 5); experienced tower operators generally earn a GS-9 classification.

Air Traffic Control Specialist (Radar)

The air traffic control specialist who operates the terminal radar or center air route traffic control radar represents basically the "next step up" in the FAA career ladder of traffic control specialists. Radar controllers issue instructions, clearances, advice, and information to aircraft; maintain a progressive check of aircraft movement; and initiate search and rescue action if needed. In addition to a comprehensive knowledge of the laws, rules, and regulations governing air traffic control, positions in this specialization require:

1. The ability to control aircraft operating at very high speeds over great distances;
2. The skill to arrange air traffic in patterns that assure maximum safety and minimum delay at points where such aircraft are "handed-off" or transferred to other facilities or other sectors within the center;
3. The judgement to estimate when and where traffic congestion will build to a point that necessitates changing patterns and to plan accordingly.

Experience as a tower operator provides much of the basic training required to qualify for the position of a radar air traffic control specialist.

Flight Service Station Specialists

Flight service station specialists provide information about weather, air navigation, and airport conditions to pilots before and during flight. They relay traffic control instructions and other information between pilots and air traffic control facilities; provide information to pilots in distress; orient lost pilots; and initiate search and rescue action if needed. Applicants at the GS-5 and GS-7 levels must pass a written test unless they meet the requirements of GS-9. A minimum of three years of general experience in administrative, technical, or other work which demonstrates the potential for learning and performing air traffic control work is required. (A four year recognized college degree will satisfy this requirement.) In addition, specialized experience, one to two years, is required for the GS-7 and GS-9 levels. Specialized experience of the required type may be gained in a military or civilian air traffic control facility.

GS-5 and GS-7 Qualifications Requirements

GS-5 - Pass a written test plus three years general experience *or* four years of college *or* combination of both.

GS-7 - Pass a written test with a very high score plus three years general experience *or* four years of college *or* combination of both.

OR

Pass a written test plus *one of the following:* Civilian or military facility rating in ATC involving active control or air traffic in center or terminal; past or present FAA air carrier dispatcher certificate; past or present instrument flight rating; past or present FAA navigator/Bombardier; past or present co-pilot or pilot rating or equivalent military rating with 350 hours of flight time; past or present rating as an Aerospace Defense Command Intercept Director.

General Experience is progressively responsible work which *demonstrates the candidate's potential for learning and performing Air Traffic Control work.* The work could be administrative, technical or other types of employment. A rigid physical examination will be made by a medical examiner designated by the Regional Flight Surgeon. Applicants must meet the established medical standards before appointment. Costs of this examination together with consultations as ordered by the Federal Aviation Administration will be borne by the agency.

AIRLINE TRANSPORT PILOT

Top of the line—the airline transport pilot.

This lofty goal is reached only with a great deal of endeavor. FAR Part 61 describes the minimum requirements necessary to obtain an Airline Transport Pilot certificate, the first major step to an airline pilot career. As indicated by the FAR, a candidate must be at least 23 years of age, of good moral character, able to read, write, and understand the English language, be a high school graduate (or equivalent), and have a first-class medical certificate. In addition, an applicant must pass a comprehensive written examination on the fundamentals of air navigation, weather, regulations, radio communications, and other items pertaining to the airline transport field. In terms of experience, an applicant must have a commercial pilot certificate without limitations or must be a pilot in an armed force of the United States whose military experience qualified him for a commercial pilot certificate. The applicant must have had at least 250 hours of flight time as pilot-in-command, 100 hours of which were cross-country and 25 hours which were night flights. In addition, 1,500 hours of total flight time is required. Of this quantity at least 500 hours must be cross-country flight time, 100 hours night flight time, and 75 hours of actual or simulated instrument time, at least 50 hours of which were actual. In certain instances second-in-command time or flight engineer time may be credited against the requirement for 1,500 hours of flight time (see FAR Part 61 for details). A comprehensive oral examination and flight test is also required to obtain an Airline Transport Pilot certificate.

To reach the goal of airline transport pilot requires a great deal of skill and determination as well as the gift of good health and sufficient financial backing to acquire the minimum flight time experience. Employment as a flight instructor is one technique often used to obtain the necessary flight hours and skills.

Typical industry standards for airline transport pilot candidates are 20 to 35 years of age (21-28 preferred), 5 feet 6 inches to 6 feet 4 inches in height, 140 to 170 lbs., and a four year college degree.

accelerate-stop distance: the distance required to accelerate an airplane to a specified speed and, assuming failure of the critical engine at the instant that speed (V_1) is attained, to bring the airplane to a stop.

air carrier: a person who undertakes directly by lease, or other arrangement, to engage in air transportation.

aircraft: a device that is used or intended to be used for flight in the air.

aircraft engine: an engine that is used or intended to be used for propelling aircraft. It includes turbosuperchargers, appurtenances, and accessories necessary for its functioning, but does not include propellers.

airframe: the fuselage, booms, nacelles, cowlings, fairings, airfoil surfaces (including rotors but excluding propellers and rotating airfoils of engines), and landing gear of an aircraft and their accessories and controls.

airplane: an engine-driven fixed-wing aircraft heavier than air, that is supported in flight by the dynamic reaction of the air against its wings.

airport: an area of land or water that is used or intended to be used for the landing and takeoff of aircraft, and includes its buildings and facilities, if any.

airport traffic area: unless otherwise specifically designated in Part 93, that airspace within a horizontal radius of 5 statute miles from the geographical center of any airport at which a control tower is operating, extending from the surface up to, but not including, an altitude of 3,000 feet above the elevation of the airport.

airship: an engine-driven lighter-than-air aircraft that can be steered.

air traffic: aircraft operating in the air or on an airport surface, exclusive of loading ramps and parking areas.

air traffic clearance: an authorization by air traffic control, for the purpose of preventing collision between known aircraft, for an aircraft to proceed under specified traffic conditions within controlled airspace.

air traffic control: a service operated by appropriate authority to promote the safe, orderly, and expeditious flow of air traffic.

air transportation: interstate, overseas, or foreign air transportation or the transportation of mail by aircraft.

alternate airport: an airport at which an aircraft may land if a landing at the intended airport becomes inadvisable.

area navigation (RNAV): a method of navigation that permits aircraft operations on any desired course within the coverage of station-referenced navigation signals or within the limits of self-contained system capability.

area navigation high route: an area navigation route within the airspace extending upward from, and including, 18,000 feet MSL to flight level 450.

area navigation low route: an area navigation route within the airspace extending upward from 1,200 feet above the surface of the earth to, but not including, 18,000 feet MSL.

calibrated airspeed: indicated airspeed of an aircraft, corrected for position and instrument error. Calibrated airspeed is equal to true airspeed in standard atmosphere at sea level.

category: (1) As used with respect to the certification, ratings, privileges, and limitations of airmen, means a broad classification of aircraft. Examples include: airplane; rotorcraft; glider; and lighter-than-air; and (2) As used with respect to the certification of aircraft, means a grouping of aircraft based upon intended use or operating limitations. Examples include: transport; normal; utility; acrobatic; limited; restricted; and provisional.

category II operation: with respect to the operation of aircraft, means a straight-in ILS approach to the runway of an airport under a Category II ILS instrument approach procedure issued by the Administrator or other appropriate authority.

ceiling: the height above the earth's surface of the lowest layer of clouds or obscuring phenomena that is reported as "broken," "overcast," or "obscuration," and not classified as "thin" or "partial."

class: (1) As used with respect to the certification, ratings, privileges, and limitations of airmen, means a classification of aircraft within a category having similar operating characteristics. Examples include: single engine; multiengine; land; water; gyroplane; helicopter; airship; and free balloon; and (2) As used with respect to the certification of aircraft, means a broad grouping of aircraft having similar characteristics of propulsion, flight, or landing. Examples include: airplane; rotorcraft; glider; balloon; landplane; and seaplane.

controlled airspace: airspace designated as a continental control area, control area, control zone, terminal control area, or transition area, within which some or all aircraft may be subject to air traffic control.

crewmember: a person assigned to perform duty in an aircraft during flight time.

critical altitude: the maximum altitude at which, in standard atmosphere, it is possible to maintain, at a specified rotational speed, a specified power or a specified manifold pressure. Unless otherwise stated, the critical altitude is the maximum altitude at which it is possible to maintain, at the maximum continuous rotational speed, one of the following: (1) The maximum continuous power, in the case of engines for which this power rating is the same at sea level and at the rated altitude. (2) The maximum continuous rated manifold pressure, in the case of engines, the maximum continuous power of which, is governed by a constant manifold pressure.

critical engine: the engine whose failure would most adversely affect the performance or handling qualities of an aircraft.

decision height: with respect to the operation of aircraft, means the height at which a decision must be made, during an ILS or PAR instrument approach, to either continue the approach or to execute a missed approach.

equivalent airspeed: the calibrated airspeed of an aircraft corrected for adiabatic compressible flow for the particular altitude. Equivalent airspeed is equal to calibrated airspeed in standard atmosphere at sea level.

extended over-water operation: (1) With respect to aircraft other than helicopters, and operation over water at a horizontal distance of more than 50 nautical miles from the nearest shoreline; and (2) With respect to helicopters, an operation over water at a horizontal distance of more than 50 nautical miles from the nearest shoreline and more than 50 nautical miles from an off-shore heliport structure.

fireproof: (1) With respect to materials and parts used to confine fire in a designated fire zone, means the capacity to withstand at least as well as steel in dimensions appropriate for the purpose for which they are used, the heat produced when there is a severe fire of extended duration in that zone; and (2) With respect to other materials and parts, means the capacity to withstand the heat associated with fire at least as well as steel in dimensions appropriate for the purpose for which they are used.

fire resistant: (1) With respect to sheet or structural members means the capacity to withstand the heat associated with fire at least as well as aluminum alloy in dimensions appropriate for the purpose for which they are used; and (2) With respect to fluid-carrying lines, fluid system parts, wiring, air ducts, fittings, and powerplant controls, means the capacity to perform the intended functions under the heat and other conditions likely to occur when there is a fire at the place concerned.

flame resistant: not susceptible to combustion to the point of propagating a flame, beyond safe limits, after the ignition source is removed.

flammable: with respect to a fluid or gas, means susceptible to igniting readily or to exploding.

flap extended speed: the highest speed permissible with wing flaps in a prescribed extended position.

flash resistant: not susceptible to burning violently when ignited.

flight crewmember: a pilot, flight engineer, or flight navigator assigned to duty in an aircraft during flight time.

flight level: a level of constant atmospheric pressure related to a reference datum of 29.92 inches of mercury. Each is stated in three digits that represent hundreds of feet. For example, flight level 250 represents a barometric altimeter indication of 25,000 feet; flight level 255, an indication of 25,500 feet.

flight plan: specified information, relating to the intended flight of an aircraft, that is filed orally or in writing with air traffic control.

flight time: the time from the moment the aircraft first moves under its own power for the purpose of flight until the moment it comes to rest at the next point of landing. ("Block-to-block" time.)

flight visibility: the average forward horizontal distance, from the cockpit of an aircraft in flight, at which prominent unlighted objects may be seen and identified by day and prominent lighted objects may be seen and identified by night.

glider: a heavier-than-air aircraft, that is supported in flight by the dynamic reaction of the air against its lifting surfaces and whose free flight does not depend principally on an engine.

ground visibility: prevailing horizontal visibility near the earth's surface as reported by the United States National Weather Service or an accredited observer.

gyrodyne: a rotorcraft whose rotors are normally engine-driven for takeoff, hovering, and landing, and for forward flight through part of its speed range, and whose means of propulsion, consisting usually of conventional propellers, is independent of the rotor system.

gyroplane: a rotorcraft whose rotors are not engine-driven except for initial starting, but are made to rotate by action of the air when the rotorcraft is moving; and whose means of propulsion, consisting usually of conventional propellers, is independent of the rotor system.

helicopter: a rotorcraft that, for its horizontal motion, depends principally on its engine-driven rotors.

heliport: an area of land, water, or structure used or intended to be used for the landing and takeoff of helicopters.

IFR conditions: weather conditions below the minimum for flight under visual flight rules.

IFR over-the-top: with respect to the operation of aircraft, means the operation of an aircraft over-the-top on an IFR flight plan when cleared by air traffic control to maintain "VFR conditions" or "VFR conditions on top."

indicated airspeed: the speed of an aircraft as shown on its pitot static airspeed indicator calibrated to reflect standard atmosphere adiabatic compressible flow at sea level uncorrected for airspeed system errors.

landing gear extended speed: the maximum speed at which an aircraft can be safely flown with the landing gear extended.

landing gear operating speed: the maximum speed at which the landing gear can be safely extended or retracted.

large aircraft: aircraft of more than 12,500 pounds, maximum certificated takeoff weight.

lighter-than-air aircraft: aircraft that can rise and remain suspended by using contained gas weighing less than the air that is displaced by the gas.

load factor: the ratio of specified load to the total weight of the aircraft. The specified load is expressed in terms of any of the following: aerodynamic forces, inertia forces, or ground or water reactions.

mach number: the ratio of true airspeed to the speed of sound.

maintenance: inspection, overhaul, repair, preservation, and the replacement of parts, but excludes preventive maintenance.

major alteration: an alteration not listed in the aircraft, aircraft engine, or propeller specifications— (1) That might appreciably affect weight, balance, structural strength, performance, powerplant operation, flight characteristics, or other qualities affecting airworthiness; or (2) That is not done according to accepted practices or cannot be done by elementary operations.

major repair: a repair— (1) That, if improperly done, might appreciably affect weight, balance, structural strength, performance, powerplant operation, flight characteristics, or other qualities affecting airworthiness; or (2) That is not done according to accepted practices or cannot be done by elementary operations.

manifold pressure: absolute pressure as measured at the appropriate point in the induction system and usually expressed in inches of mercury.

medical certificate: acceptable evidence of physical fitness on a form prescribed by the Administrator.

minimum descent altitude: the lowest altitude, expressed in feet above mean sea level, to which descent is authorized on final approach or during circle-to-land maneuvering in execution of a standard instrument approach procedure, where no electronic glide slope is provided.

minor alteration: an alteration other than a major alteration.

minor repair: a repair other than a major repair.

navigable airspace: airspace at and above the minimum flight altitudes prescribed by or under this chapter, including airspace needed for safe takeoff and landing.

night: the time between the end of evening civil twilight and the beginning of morning civil twilight, as published in the American Air Almanac, converted to local time.

non-precision approach procedure: a standard instrument approach precedure in which no electronic glide slope is provided.

operate: with respect to aircraft, means use, cause to use or authorize to use aircraft, for the purpose of air navigation including the piloting of aircraft, with or without the right of legal control (as owner, lessee, or otherwise).

over-the-top: above the layer of clouds or other obscuring phenomena forming the ceiling.

parachute: a device used or intended to be used to retard the fall of a body or object through the air.

person: an individual, firm, partnership, corporation, company, association, joint-stock association, or governmental entity. It includes a trustee, receiver, assignee, or similar representative of any of them.

pilotage: navigation by visual reference to landmarks.

pilot in command: the pilot responsible for the operation and safety of an aircraft during flight time.

pitch setting: the propeller blade setting as determined by the blade angle measured in a manner, and at a radius, specified by the instruction manual for the propeller.

positive control: control of all air traffic, within designated airspace, by air traffic control.

precision approach procedure: a standard instrument approach procedure in which an electronic glide slope is provided, such as ILS and PAR.

preventive maintenance: simple or minor preservation operations and the replacement of small standard parts not involving complex assembly operations.

prohibited area: designated airspace within which the flight of aircraft is prohibited.

propeller: a device for propelling an aircraft that has blades on an engine-driven shaft and that, when rotated, produces by its action on the air, a thrust approximately perpendicular to its plane of rotation. It includes control components normally supplied by its manufacturer, but does not include main and auxiliary rotors or rotating airfoils of engines.

rated maximum continuous power: with respect to reciprocating, turbopropeller, and turboshaft engines, means the approved brake horsepower that is developed statically or in flight, in standard atmosphere at a specified altitude, within the engine operating limitations established under Part 33, and approved for unrestricted periods of use.

rated maximum continuous thrust: with respect to turbojet engines, means the approved jet thrust that is developed statically or in flight, in standard atmosphere at a specified altitude, within the engine operating limitations established under Part 33, and approved for unrestricted periods of use.

rated takeoff power: with respect to reciprocating, turbopropeller, and turboshaft engine type certification, means the approved brake horsepower that is developed statically under standard sea level conditions, within the engine operating limitations established under Part 33, and limited in use to periods of not over 5 minutes for takeoff operation.

rated takeoff thrust: with respect to turbojet engine type certification, means the approved jet thrust that is developed statically under standard sea level conditions, within the engine operating limitations established under Part 33, and limited in use to periods of not over 5 minutes for takeoff operation.

rated 30-minute power: with respect to helicopter turbine engines, means the maximum brake horsepower, developed under static conditions at specified altitudes and atmospheric temperatures, under the maximum conditions of rotor shaft rotational speed and gas temperature, and limited in use to periods of not over 30 minutes as shown on the engine data sheet.

rated 2½-minute power: with respect to helicopter turbine engines, means the brake horsepower, developed statically in standard atmosphere at sea level, or at a specified altitude, for one-engine-out operation of multi-engine helicopters for 2½ minutes at rotor shaft rotation speed and gas temperature established for this rating.

rating: a statement that, as a part of a certificate, sets forth special conditions, privileges, or limitations.

reporting point: a geographical location in relation to which the position of an aircraft is reported.

restricted area: airspace within which the flight of aircraft, while not wholly prohibited, is subject to restriction.

RNAV way point (W/P): predetermined geographical position used for route or instrument approach definition or progress reporting purposes that is defined relative to a VORTAC station position.

route segment: a part of a route. Each end of that part is identified by— (1) a continental or insular geographical location; or (2) a point at which a definite radio fix can be established.

second in command: a pilot who is designated to be second in command of an aircraft during flight time.

small aircraft: aircraft of 12,500 pounds or less, maximum certificated takeoff weight.

standard atmosphere: the atmosphere defined in *U.S. Standard Atmosphere, 1962* (Geopotential altitude tables).

stopway: an area beyond the takeoff runway, no less wide than the runway and centered upon the extended centerline of the runway, able to support the airplane during an aborted takeoff, without causing structural damage to the airplane, and designated by the airport authorities for use in decelerating the airplane during an aborted takeoff.

takeoff power: (1) With respect to reciprocating engines, means the brake horsepower that is developed under standard sea level conditions, and under the maximum conditions of crankshaft rotational speed and engine manifold pressure approved for the normal takeoff, and limited in continuous use to the period of time shown in the approved engine specification; and (2) With respect to turbine engines, means the brake horsepower that is developed under static conditions at a specified altitude and atmospheric temperature, and under the maximum conditions of rotorshaft rotational speed and gas temperature approved for the normal takeoff, and limited in continuous use to the period of time shown in the approved engine specification.

takeoff thrust: with respect to turbine engines, means the jet thrust that is developed under static conditions at a specific altitude and atmospheric temperature under the maximum conditions of rotorshaft rotational speed and gas temperature approved for the normal takeoff, and limited in continuous use to the period of time shown in the approved engine specification.

time in service: with respect to maintenance time records, means the time from the moment an aircraft leaves the surface of the earth until it touches it at the next point of landing.

traffic pattern: the traffic flow that is prescribed for aircraft landing at, taxiing on, or taking off from, an airport.

true airspeed: the airspeed of an aircraft relative to undisturbed air. True airspeed is equal to equivalent airspeed multiplied by $(po/p)^{1/2}$.

type: (1) As used with respect to the certification, ratings, privileges, and limitations of airmen, means a specific made and basic model of aircraft, including modifications thereto that do not change its handling or flight characteristics.. Examples include: DC-7, 1049, and F-27; and (2) As used with respect to the certification of aircraft, means those aircraft which are similar in design. Examples include: DC-7 and DC-7C; 1049G and 1049H; and F-27 and F-27F.

VFR over-the-top: with respect to the operation of aircraft, means the operation of an aircraft over-the-top under VFR when it is not being operated on an IFR flight plan.

Abbreviations and symbols.

ALS: approach light system.
ASR: airport surveillance radar.
ATC: air traffic control.

CAS: calibrated airspeed.
CAT II: Category II.
CONSOL or CONSOLAN: a kind of low or medium frequency long range navigational aid.

DH: decision height.
DME: distance measuring equipment compatible with TACAN.

EAS: equivalent airspeed.

FAA: Federal Aviation Administration.
FM: fan marker.

GS: glide slope.

HIRL: high-intensity runway light system.

IAS: indicated airspeed.
ICAO: International Civil Aviation Organization.
IFR: instrument flight rules.
ILS: instrument landing system.
IM: ILS inner marker.
INT: intersection.

LDA: localizer-type directional aid.
LFR: low frequency radio range.
LMM: compass locator at middle marker.
LOC: ILS localizer.
LOM: compass locator at outer marker.

M: mach number.
MAA: maximum authorized IFR altitude.
MALS: medium intensity approach light system.
MALSR: medium intensity approach light system with runway alignment indicator lights.
MCA: minimum crossing altitude.
MDA: minimum descent altitude.
MEA: minimum en route IFR altitude.
MM: ILS middle marker.
MOCA: minimum obstruction clearance altitude.
MRA: minimum reception altitude.
MSL: mean sea level.

NDB(ADF): nondirectional beacon (automatic direction finder).
NOPT: no procedure turn required.

OM: ILS outer marker.

PAR: precision approach radar.

RAIL: runway alignment indicator light system.
RBN: radio beacon.
RR: low or medium frequency radio range station.
RCLM: runway centerline marking.
RCLS: runway centerline light system.
REIL: runway end identification lights.
RVR: runway visual range as measured in the touchdown zone area.

SALS: short approach light system.
SSALS: simplified short approach light system.
SSALSR: simplified short approach light system with runway alignment indicator lights.

TACAN: ultra-high frequency tactical air navigational aid.
TAS: true airspeed.
TDZL: touchdown zone lights.
TVOR: very high frequency terminal omnirange station.

V_A: design maneuvering speed.
V_B: design speed for maximum gust intensity.
V_C: design cruising speed.
V_D: design diving speed.
V_{DF}/M_{DF}: demonstrated flight diving speed.
V_F: design flap speed.
V_{FC}/M_{FC}: maximum speed for stability characteristics.
V_{FE}: maximum flap extended speed.
VFR: visual flight rules.
V_H: maximum speed in level flight with maximum continuous power.
VHF: very high frequency.
V_{LE}: maximum landing gear entended speed.
V_{LO}: maximum landing gear operating speed.
V_{LOF}: lift-off speed.
V_{MC}: minimum control speed with the critical engine inoperative.
V_{MO}/M_{MO}: maximum operating limit speed.
V_{MU}: minimum unstick speed.
V_{NE}: never-exceed speed.
VOR: very high frequency omnirange station.

VORTAC: collocated VOR and TACAN.

V_R: rotation speed.

V_S: the stalling speed or the minimum steady flight speed at which the airplane is controllable.

V_{SO}: the stalling speed or the minimum steady flight speed in the landing configuration.

V_{S1}: the stalling speed or the minimum steady flight speed obtained in a specified configuration.

V_X: speed for best angle of climb.

V_Y: speed for best rate of climb.

V_1: critical-engine-failure speed.

V_2: takeoff safety speed.

$V_{2\,min}$: minimum takeoff safety speed.

VSSE: minimum safe single engine speed.

Appendix A
Mexican Federal
Aviation Regulations

Effective December 15, 1979, the government of Mexico issued new federal regulations which have a significant impact on the flight procedures for:

1. General aviation aircraft.
2. Privately owned and operated aircraft (for private purposes, recreation, or transit flights).
3. Single and multiengine (piston type) aircraft (maximum capacity up to 16 seats).
4. Aircraft of greater than 16 seating capacity (piston type).

Following are the new federal regulations issued by the government of Mexico:

I. This circular is applicable to private aircraft of foreign nationality and foreign registration of any type, whose maximum capacity is 16 seats for passengers and for which there is no transportation charge. On these aircraft, they can only transport people who have been invited and under no circumstance any cargo, merchandise, or articles of any kind that are not for personal use of the crew or their guests.

When the capacity of the aircraft is greater than 16 seats, in order to be able to operate in Mexican territory, application for approval must be applied for at least five working days before the date of the scheduled trip and written permission must be obtained from Directorate of Civil Aeronautics.

II. In order to enter Mexican territory, all private-owned foreign civil aircraft destined exclusively for private purposes, recreation, or transiting Mexico, must comply with the following requirements:

II-A. Jet or turbine aircraft can and should use for their entry/departure from Mexico any one of the aerodromes listed below, except Monterrey/Del Norte:

Acapulco, Gro.; Cancun, Q.R.; Ciudad Acuna, Coah.; Ciudad Juarez, Chih.; Cozumel, Q.R.; Chetumal, Q.R.; Chihuahua, Chih.; Guadalajara, Jal.; Guaymas, Son.; Hermosillo, Son.; La Paz, B.C.; Manzanillo, Col.; Matamoros, Tamps.; Mazatlan, Sin.; Mexicali, B.C.; Mexico, D.F.; Merida, Yuc.; Monterrey, N.L.; Nogales, Son.; Nuevo Laredo, Tamps.; Puerto Vallarta, Jal.; San Jose, Del Cabo, B.C.; Tampico, Tamps.; Tapachula, Chis.; Tijuana, B.C.; Torreon, Coah.; Veracruz, Ver.; Zihuatanejo, Gro.; and Piedras Negras, Coah.

II-B. One- and two-piston engine aircraft will only be able to use for their entry/departure from Mexico the following international border airports.

Northern Zone;

For flights from or to the northern border of Mexico: Tijuana/General Abelardo L. Rodriquez, B.C.; Mexicali/General Rodolfo S. Taboada, B.C.; Nogales, Son.; CD. Juarez, Chih,; CD. Acuna, Coah.; Piedras Negras, Coah.; Nuevo Laredo, Tamps.; Reynosa, Tamps.; and Matamoros, Tamps.; Hermosillo, Son.

South and Southeastern Zones;

For flights from or to the south and southeastern borders of Mexico: Tapachula, Chis.; Chetumal, Q.R.; Cozumel, Q.R.; and Cancun, Q.R.

II-C. The crew and passengers who travel on these aircraft will comply with the requirements and formalities of Customs, Immigration, and Health, at the international airports through which they will enter and depart the Republic of Mexico.

II-D. They must comply with the security standards established by the General Communications Systems Law and its regulations, as well as the provisions of their own country's regulations, regarding nationality and registration, weight, security, and first aid instruments, and have certificates of registry and airworthiness, licenses for flight personnel, and other pertinent documentation.

II-E. They must follow the air routes previously established by the Secretariate of Communications and Transports and must observe the regulatory procedures contained in the Mexican Aeronautics Information Publication (PIA/AIP).

II-F. Aircraft pilots should advise the aeronautical authorities, in advance and utilizing the most adequate and expeditious communications systems, that they are at the nearest international airport closest to the point at which they intend to cross the Mexican frontier. For jet and turbine aircraft that overfly border airports, the pilot should communicate with the nearest Air Traffic Control Center.

II-G. In view of what is set forth in articles 351 and 352 of the General Communications System Law, the operator of the aircraft should have insurance which will guarantee the damages which could be caused to third parties on the ground, due to operations effected in Mexican territory, according to the following list:

    Aircraft up to 5,000 kgs. gross weight.
    The amount of ---$60,000.00
    Aircraft up to 20,000 kgs. gross weight,
    The amount of --$150,000.00
    Aircraft up to 40,000 kgs. gross weight,
    The amount of --$600,000.00

This insurance can be arranged with a Mexican insurance company which usually furnishes its services to the international airports.

III. Upon arrival at the airport of entry, the aircraft pilots should fill out and sign form G.H.C.-001, requesting the proper authorization from Mexican Customs and Immigration. Form G.H.C.-001 should then be presented to the Airport Commandant so it can be duly approved and the original returned, which should be kept on board the aircraft for presentation upon request. At the airport of departure from Mexico, the original of form G.H.C.-001 must be surrendered to the Commandant of the airport.

Once these requirements have been met at the airport of entry, the foreign aircraft can operate freely in Mexican territory, subject to the provisions set forth in the permit and in the General Communications Systems Law, and its regulations.

IV. In instances where a rented aircraft without services of a pilot is involved, it will be permitted to enter the country only if it is a one- or two-engine aircraft with a maximum capacity of eight seats. If the aircraft is multimotor or of greater capacity, it will be necessary to obtain prior written permission from the Office of Civil Aeronautics, Department of International Air Transportation.

V. At the airport of departure from Mexican territory.

VI. The aircraft should operate and depart the country carrying on board the same crew shown on form G.H.C.-001. In case this is not done, the Airport Commandant will permit the aircraft to be piloted by a different crew other than the one listed, upon prior provision of a substantial reason for this action, and as long as the General Communication Systems Law and its regulations are not infringed.

VII. That which is set forth in this circular if subject to the principle of reciprocity and, therefore, will not be in effect for those aircraft registered in those countries that do not grant the same privileges to private service Mexican aircraft.

This circular revokes all previous ones that refer to the requisites for entry into the country of foreign private service aircraft and will enter into force on this date.

With respect to any matter relative to this circular, interested parties should contact the Department of International Air transportation, General Directorate of Civil Aeronautics, Avenida Universidad and Xola, Mexico 12, D.F., telephone numbers— (905) 519-81-83 and (905) 591-76-25.

Signed Ing. Jorge Cenoejas, Director General

1. Prior to deviating from any of the above regulations, it is the pilot's responsibility to obtain written authorization from the office of the Director General of Civil Aviation.

2. Aircraft originating on the United States' side of the border will be obliged, upon entering Mexico, to make their first stop at one of the designated Mexican border airports. When a pilot presents the flight plan for his aircraft for Mexico, say in Brownsville or McAllen, he will have to use one of the airports at Matamoros, Reynosa, or Nuevo Laredo and make the short flight to comply with Mexican Immigration, Customs, and Health requirements before proceeding further into the country. Since Mexican authorities will anticipate the aircraft's arrival at an indicated airport, its failure to arrive within a specified period of time will result in the Mexican authorities reporting the violation of the flight plan to the American authorities. Should an aircraft land at a different airport other than those indicated or at one in the interior of Mexico without having first landed at one of the designated Mexican airports along the border, the Mexican aviation authorities will immediately suspend its operation for violation and notify Mexican customs authorities.

Appendix B
Bibliography

Chapter 1. Our Regulatory System
1. FAR, Part II, *General Rule Making Procedures.*
2. *U.S. Government Organizational Manual,* 1970/1971, Office of the Federal Register, National Archives Records and Service.
3. *The National Aviation System Plan, Ten Year Plan,* 1973-1982.
4. *Briefs of Accidents Involving Weather as a Cause.* U.S. General Aviation, 1967, NTSB.
5. Fixel, Rowland. *The Law of Aviation,* Fourth Edition, The Michie Company, Charlottesville, Virginia.
6. Jackson, William. *The Federal Airways System,* Electrical and Electronic Engineers, Inc., Washington, D.C.
7. Robertson, Cliff. *Legal Guide for Pilots and Owners,* TAB Book 2219, TAB Books, Inc. Blue Ridge Summit, Pennsylvania.
8. Redford, Emmette. *The Regulatory Process,* Univ. of Texas Press.

Chapter 2, The Pilot
1. FAR, Part 61, Certification, Pilots and Flight Instructors
2. FAR, Part 67, Medical Standards and Certification

Chapter 3. The Aircraft
1. Federal Aviation Act of 1958
2. FAR, Part 21, Certification Procedures for Products and Parts.

3.	FAR, Part 23, Airworthiness Standards. Normal, Utility, and Acrobatic Category Airplanes.
4.	FAR, Part 36, Noise Standards: Aircraft Type Certification.
5.	FAR, Part 37, Technical Standard-Order Authorizations.
6.	FAR, Part 39, Airworthiness Directives.
7.	FAR, Part 43, Maintenance, Preventive Maintenance, Rebuilding and Alteration.
8.	FAR, Part 47, Aircraft Registration.
9.	FAR, Part 49, Recording of Aircraft Titles and Security Documents.

Chapter 4. Preflight Planning

1.	FAR, Part 91, General Operating and Flight Rules.
2.	Blum Bro. vs State of Louisiana, U.S. Dist. Court, E. Dist. of Louisiana, Oct. 4, 1971.
3.	Lock vs Packard Flying Service. Nebraska Supreme Court, Jan. 16, 1970.
4.	*Airman's Information Manual*, Parts 1, 2, 3, & 4.

Chapter 5. The Airspace

1.	FAR, Part 91, General Operating and Flight Rules.
2.	FAR, Part 93, Special Airport Traffic Rules and Airport Traffic Patterns.
3.	FAR, Part 99, Security Control of Aircraft.

Chapter 6. Accident Reports and Investigations

1.	NTSB, Part 430, Rules Pertaining to the Notification and Reporting of Aircraft Accidents, Incidents, and Overdue Aircraft, and Preservation of Aircraft Wreckage, Mail, Cargo, and Records.

Chapter 7. Enforcement Procedures

1.	FAR, Part 13, Enforcement Procedures.

Chapter 8. Commercial Operations

1.	FAR, Part 61, Certification: Pilots and Flight Instructors.
2.	FAR, Part 141, Pilot Schools.
3.	FAR, Part 143, Ground Instructors.
4.	FAR, Part 63, Certification: Flight Crew Members Other than Pilots.
5.	FAR, Part 135, Air Taxi Operators and Commercial Operators.
6.	FAR, Part 137, Agricultural Aircraft Operations.

Index